DATE DUE

PRINTED IN U.S.A.

VOLUME FIVE HUNDRED AND SEVENTY FIVE

METHODS IN ENZYMOLOGY

Synthetic Biology and Metabolic Engineering in Plants and Microbes
Part A: Metabolism in Microbes

METHODS IN ENZYMOLOGY

Editors-in-Chief

ANNA MARIE PYLE
*Departments of Molecular, Cellular and Developmental
Biology and Department of Chemistry
Investigator, Howard Hughes Medical Institute
Yale University*

DAVID W. CHRISTIANSON
*Roy and Diana Vagelos Laboratories
Department of Chemistry
University of Pennsylvania
Philadelphia, PA*

Founding Editors

SIDNEY P. COLOWICK and NATHAN O. KAPLAN

VOLUME FIVE HUNDRED AND SEVENTY FIVE

METHODS IN ENZYMOLOGY

Synthetic Biology and Metabolic Engineering in Plants and Microbes
Part A: Metabolism in Microbes

Edited by

SARAH E. O'CONNOR

The John Innes Centre
Department of Biological Chemistry
Norwich, NR4 7UH, UK

AMSTERDAM • BOSTON • HEIDELBERG • LONDON
NEW YORK • OXFORD • PARIS • SAN DIEGO
SAN FRANCISCO • SINGAPORE • SYDNEY • TOKYO
Academic Press is an imprint of Elsevier

Academic Press is an imprint of Elsevier
50 Hampshire Street, 5th Floor, Cambridge, MA 02139, United States
525 B Street, Suite 1800, San Diego, CA 92101-4495, United States
The Boulevard, Langford Lane, Kidlington, Oxford OX5 1GB, United Kingdom
125 London Wall, London, EC2Y 5AS, United Kingdom

First edition 2016

Copyright © 2016 Elsevier Inc. All rights reserved.

No part of this publication may be reproduced or transmitted in any form or by any means, electronic or mechanical, including photocopying, recording, or any information storage and retrieval system, without permission in writing from the publisher. Details on how to seek permission, further information about the Publisher's permissions policies and our arrangements with organizations such as the Copyright Clearance Center and the Copyright Licensing Agency, can be found at our website: www.elsevier.com/permissions.

This book and the individual contributions contained in it are protected under copyright by the Publisher (other than as may be noted herein).

Notices
Knowledge and best practice in this field are constantly changing. As new research and experience broaden our understanding, changes in research methods, professional practices, or medical treatment may become necessary.

Practitioners and researchers must always rely on their own experience and knowledge in evaluating and using any information, methods, compounds, or experiments described herein. In using such information or methods they should be mindful of their own safety and the safety of others, including parties for whom they have a professional responsibility.

To the fullest extent of the law, neither the Publisher nor the authors, contributors, or editors, assume any liability for any injury and/or damage to persons or property as a matter of products liability, negligence or otherwise, or from any use or operation of any methods, products, instructions, or ideas contained in the material herein.

ISBN: 978-0-12-804584-8
ISSN: 0076-6879

For information on all Academic Press publications
visit our website at https://www.elsevier.com/

Publisher: Zoe Kruze
Acquisition Editor: Zoe Kruze
Editorial Project Manager: Sarah Lay
Production Project Manager: Magesh Kumar Mahalingam
Cover Designer: Maria Ines Cruz

Typeset by SPi Global, India

CONTENTS

Contributors — xi
Preface — xv

1. Directing Biosynthesis: Practical Supply of Natural and Unnatural Cyanobactins — 1
D. Sardar, M.D. Tianero, and E.W. Schmidt

1. Introduction — 1
2. Discovery of Cyanobactin Pathways — 2
3. Elucidating Natural Rules of Engineering in Cyanobactin Pathways — 3
4. Heterologous Expression of Cyanobactin Pathways in *E. coli* — 5
5. Optimization for Increased Yield of Cyanobactins in *E. coli* — 9
6. Synthesis of Cyanobactins In Vitro — 11
7. Conclusions — 16
8. Outlook — 18

Acknowledgments — 18
References — 18

2. Synthetic Biology Approaches to New Bisindoles — 21
L.M. Alkhalaf, Y.-L. Du, and K.S. Ryan

1. Introduction — 22
2. Identification of New Bisindole Gene Clusters — 25
3. Heterologous Expression — 27
4. Mutational Biosynthesis to Generate New Bisindoles — 30
5. Mixing Genes from Phylogenetically Related Clusters to Generate New Bisindoles — 32
6. Chemical Isolation and Structural Characterization — 32
7. Conclusions — 34

References — 34

3. Enzymatic [4+2] Cycloadditions in the Biosynthesis of Spirotetramates and Spirotetronates — 39
B. Pang, G. Zhong, Z. Tang, and W. Liu

1. Introduction — 40
2. Strategy — 44
3. Methods — 48

4. Discussion and Perspectives	58
References	61

4. Application and Modification of Flavin-Dependent Halogenases 65
K.-H. van Pée, D. Milbredt, E.P. Patallo, V. Weichold, and M. Gajewi

1. Introduction	66
2. Inactivation of Halogenases Under Reaction Conditions	67
3. The Biocatalytic Scope of Flavin-Dependent Tryptophan Halogenases	73
4. Synthesis of PCP-Bound Halogenase Substrates	83
5. Conclusions	88
Acknowledgments	89
References	89

5. Engineering Flavin-Dependent Halogenases 93
J.T. Payne, M.C. Andorfer, and J.C. Lewis

1. Introduction	93
2. Improving the Stability of FDHs	98
3. Altering the Regioselectivity of FDHs	109
4. Expanding the Substrate Scope of FDHs	114
5. Conclusions	122
References	122

6. Heterologous Expression of Fungal Secondary Metabolite Pathways in the *Aspergillus nidulans* Host System 127
J.W.A. van Dijk and C.C.C. Wang

1. Introduction	128
2. Identification of Secondary Metabolite Genes in Fungal Genomes	130
3. Design Primers for Fusion PCR	130
4. Obtain Genomic DNA for PCR Template	131
5. Fusion PCR Construction	134
6. Transformation	136
7. Diagnostic PCR	138
8. Liquid Culturing of Mutant Strains	138
9. Recipes	139
10. Conclusions	140
Acknowledgments	141
References	141

7. Plug-and-Play Benzylisoquinoline Alkaloid Biosynthetic Gene Discovery in Engineered Yeast — 143
J.S. Morris, M. Dastmalchi, J. Li, L. Chang, X. Chen, J.M. Hagel, and P.J. Facchini

1. Introduction — 144
2. Transcriptome Resources and Mining Candidate Genes — 155
3. Building Yeast Platform Strains — 161
4. Functional Testing of Candidate Genes — 166
5. Liquid Chromatography—Tandem Mass Spectrometry — 167
6. Summary and Future Prospects — 172
Acknowledgments — 173
References — 173

8. Optimizing Metabolic Pathways for the Improved Production of Natural Products — 179
J.A. Jones and M.A.G. Koffas

1. Introduction — 180
2. Genetic Optimization — 180
3. Fermentation Optimization — 185
4. Conclusion — 189
Acknowledgments — 189
References — 189

9. Reconstituting Plant Secondary Metabolism in *Saccharomyces cerevisiae* for Production of High-Value Benzylisoquinoline Alkaloids — 195
M.E. Pyne, L. Narcross, E. Fossati, L. Bourgeois, E. Burton, N.D. Gold, and V.J.J. Martin

1. Introduction — 196
2. BIA Biosynthetic Gene Sources and Selection of Candidate Genes — 199
3. Yeast Functional Expression Strategies — 200
4. Pathway Reconstitution and Optimization — 215
5. Challenges and Future Directions — 219
References — 221

10. Engineering Microbes to Synthesize Plant Isoprenoids — 225
K. Zhou, S. Edgar, and G. Stephanopoulos

1. Introduction — 226
2. Improving Production of C20 ISMs by Modular Genetic Engineering of *Escherichia coli* — 228

3.	Improving Production of C20 ISMs by Optimizing Bioreactor Operation	236
4.	Oxygenating C20 ISMs by Using *E. coli–Saccharomyces cerevisiae* Coculture	238
5.	Discussion	244
	References	244

11. Natural Product Biosynthesis in *Escherichia coli*: Mentha Monoterpenoids   247

H.S. Toogood, S. Tait, A. Jervis, A. Ní Cheallaigh, L. Humphreys, E. Takano, J.M. Gardiner, and N.S. Scrutton

1.	Introduction	248
2.	Operon Construction	251
3.	Multigene Protein Expression and Validation	257
4.	Biotransformations	258
5.	Further Studies	261
6.	Conclusions	266
	Acknowledgments	267
	References	267

12. High-Efficiency Genome Editing of *Streptomyces* Species by an Engineered CRISPR/Cas System   271

Y. Wang, R.E. Cobb, and H. Zhao

1.	Introduction	271
2.	Design of pCRISPomyces for Genome Editing in *Streptomyces* Species	273
3.	pCRISPomyces for Single Gene Disruption in *S. lividans*	275
4.	pCRISPomyces for Multiplex Gene Deletion in *S. lividans*	278
5.	Conjugation, Genotype Screening, and pCRISPomyces Clearance	279
6.	Evaluation of pCRISPomyces in Other *Streptomyces* Species	281
7.	Conclusion	282
	Acknowledgment	283
	References	283

13. Rapid Optimization of Engineered Metabolic Pathways with Serine Integrase Recombinational Assembly (SIRA)   285

C.A. Merrick, C. Wardrope, J.E. Paget, S.D. Colloms, and S.J. Rosser

1.	Introduction	286
2.	Mechanism of Serine Integrase Recombination	288
3.	How SIRA Works	291
4.	Applications of SIRA	294
5.	Design Principles for *att* Sites	297

6.	Materials Preparation	300
7.	Example SIRA Protocols	308
8.	The Future of SIRA	312
	Acknowledgments	314
	References	314

14. Rewiring Riboswitches to Create New Genetic Circuits in Bacteria   319

C.J. Robinson, D. Medina-Stacey, M.-C. Wu, H.A. Vincent, and J. Micklefield

1.	Introduction	320
2.	Overview	324
3.	Methods	324
4.	Concluding Remarks	346
	Acknowledgments	346
	References	347

Author Index *349*
Subject Index *371*

CONTRIBUTORS

L.M. Alkhalaf
University of British Columbia, Vancouver, BC, Canada

M.C. Andorfer
University of Chicago, Chicago, IL, United States

L. Bourgeois
Centre for Structural and Functional Genomics, Concordia University, Montréal, QC, Canada

E. Burton
Centre for Structural and Functional Genomics, Concordia University, Montréal, QC, Canada

L. Chang
University of Calgary, Calgary, AB, Canada

X. Chen
University of Calgary, Calgary, AB, Canada

R.E. Cobb
University of Illinois at Urbana-Champaign, Urbana, IL, United States; Carl R. Woese Institute for Genomic Biology, University of Illinois at Urbana-Champaign, Urbana, IL, United States

S.D. Colloms
University of Glasgow, Glasgow, United Kingdom

M. Dastmalchi
University of Calgary, Calgary, AB, Canada

Y.-L. Du
University of British Columbia, Vancouver, BC, Canada

S. Edgar
Massachusetts Institute of Technology, Cambridge, MA, United States

P.J. Facchini
University of Calgary, Calgary, AB, Canada

E. Fossati
Centre for Structural and Functional Genomics, Concordia University, Montréal, QC, Canada

M. Gajewi
Allgemeine Biochemie, TU Dresden, Dresden, Germany

J.M. Gardiner
School of Chemistry, Manchester Institute of Biotechnology, University of Manchester, Manchester, United Kingdom

N.D. Gold
Centre for Structural and Functional Genomics, Concordia University, Montréal, QC, Canada

J.M. Hagel
University of Calgary, Calgary, AB, Canada

L. Humphreys
GlaxoSmithKline, Medicines Research Centre, Stevenage, United Kingdom

A. Jervis
Manchester Institute of Biotechnology, SYNBIOCHEM, University of Manchester, Manchester, United Kingdom

J.A. Jones
Rensselaer Polytechnic Institute, Troy, NY, United States; Center for Biotechnology and Interdisciplinary Sciences, Rensselaer Polytechnic Institute, Troy, NY, United States

M.A.G. Koffas
Rensselaer Polytechnic Institute, Troy, NY, United States; Center for Biotechnology and Interdisciplinary Sciences, Rensselaer Polytechnic Institute, Troy, NY, United States

J.C. Lewis
University of Chicago, Chicago, IL, United States

J. Li
University of Calgary, Calgary, AB, Canada

W. Liu
State Key Laboratory of Bioorganic and Natural Products Chemistry, Shanghai Institute of Organic Chemistry, Chinese Academy of Sciences, Shanghai; Huzhou Center of Bio-Synthetic Innovation, Huzhou, PR China

V.J.J. Martin
Centre for Structural and Functional Genomics, Concordia University, Montréal, QC, Canada

D. Medina-Stacey
School of Chemistry, Manchester Institute of Biotechnology, The University of Manchester, Manchester, United Kingdom

C.A. Merrick
University of Edinburgh, Edinburgh, United Kingdom

J. Micklefield
School of Chemistry; Centre for Synthetic Biology of Fine and Speciality Chemicals (SYNBIOCHEM), Manchester Institute of Biotechnology, The University of Manchester, Manchester, United Kingdom

D. Milbredt
Allgemeine Biochemie, TU Dresden, Dresden, Germany

J.S. Morris
University of Calgary, Calgary, AB, Canada

A. Ní Cheallaigh
School of Chemistry, Manchester Institute of Biotechnology, University of Manchester, Manchester, United Kingdom

L. Narcross
Centre for Structural and Functional Genomics, Concordia University, Montréal, QC, Canada

J.E. Paget
University of Edinburgh, Edinburgh, United Kingdom

B. Pang
State Key Laboratory of Bioorganic and Natural Products Chemistry, Shanghai Institute of Organic Chemistry, Chinese Academy of Sciences, Shanghai, PR China

E.P. Patallo
Allgemeine Biochemie, TU Dresden, Dresden, Germany

J.T. Payne
University of Chicago, Chicago, IL; Stanford University, Stanford, CA, United States

M.E. Pyne
Centre for Structural and Functional Genomics, Concordia University, Montréal, QC, Canada

C.J. Robinson
School of Chemistry; Centre for Synthetic Biology of Fine and Speciality Chemicals (SYNBIOCHEM), Manchester Institute of Biotechnology, The University of Manchester, Manchester, United Kingdom

S.J. Rosser
University of Edinburgh, Edinburgh, United Kingdom

K.S. Ryan
University of British Columbia, Vancouver, BC, Canada

D. Sardar
University of Utah, Salt Lake City, UT, United States

E.W. Schmidt
University of Utah, Salt Lake City, UT, United States

N.S. Scrutton
Manchester Institute of Biotechnology, Faculty of Life Sciences, University of Manchester, Manchester, United Kingdom

G. Stephanopoulos
Massachusetts Institute of Technology, Cambridge, MA, United States

S. Tait
Manchester Institute of Biotechnology, Faculty of Life Sciences, University of Manchester, Manchester, United Kingdom

E. Takano
Manchester Institute of Biotechnology, Faculty of Life Sciences, University of Manchester, Manchester, United Kingdom

Z. Tang
State Key Laboratory of Bioorganic and Natural Products Chemistry, Shanghai Institute of Organic Chemistry, Chinese Academy of Sciences, Shanghai, PR China

M.D. Tianero
University of Utah, Salt Lake City, UT, United States

H.S. Toogood
Manchester Institute of Biotechnology, Faculty of Life Sciences, University of Manchester, Manchester, United Kingdom

J.W.A. van Dijk
School of Pharmacy, University of Southern California, Los Angeles, CA, United States

K.-H. van Pée
Allgemeine Biochemie, TU Dresden, Dresden, Germany

H.A. Vincent
School of Chemistry, Manchester Institute of Biotechnology, The University of Manchester, Manchester, United Kingdom

C.C.C. Wang
School of Pharmacy; Dornsife College of Letters, Arts, and Sciences, University of Southern California, Los Angeles, CA, United States

Y. Wang
University of Illinois at Urbana-Champaign, Urbana, IL, United States

C. Wardrope
University of Edinburgh, Edinburgh, United Kingdom

V. Weichold
Allgemeine Biochemie, TU Dresden, Dresden, Germany

M.-C. Wu
School of Chemistry, Manchester Institute of Biotechnology, The University of Manchester, Manchester, United Kingdom

H. Zhao
University of Illinois at Urbana-Champaign, Urbana, IL, United States; Carl R. Woese Institute for Genomic Biology, University of Illinois at Urbana-Champaign, Urbana, IL, United States

G. Zhong
State Key Laboratory of Bioorganic and Natural Products Chemistry, Shanghai Institute of Organic Chemistry, Chinese Academy of Sciences, Shanghai, PR China

K. Zhou
National University of Singapore, Singapore, Singapore

PREFACE

Advances in sequencing, bioinformatics, and genome editing now enable us to access the rich chemistry encoded within the metabolic pathways of plants and microbes. A major focus in metabolism is the secondary, or specialized, pathways that produce small, biologically active molecules with applications in pharmaceutical, agrochemical, or other biotechnological sectors. In recent years, metabolic engineering/synthetic biology approaches have shown remarkable promise for the exploitation of these pathways for human use. In these two volumes, we highlight some of the most important approaches that have been used to harness microbial and plant metabolic pathways.

In Volume 1, we focus on advances that have been made in microbial-based systems. The discovery some three decades ago that bacterial specialized metabolic pathways are clustered on the genome has greatly facilitated the identification and characterization of these pathway genes. Coupled with the fact that bacterial genomes can now be sequenced rapidly and inexpensively, the last decade has seen a staggering increase in our knowledge of bacterial specialized metabolism. Additionally, it is now known that fungal specialized pathways also cluster on the genome. While fungal genomes are larger than those from bacteria, these genomes can still can be easily sequenced, and substantial advances in elucidating fungal metabolism have been made. Consequently, a wealth of new opportunities in metabolic engineering have been opened. In this volume, we highlight how better production of these compounds can be achieved, and how these biosynthetic enzymes can be engineered to generate new biocatalysts and new products. We also discuss how microbial species can be manipulated to serve as a host for reconstitution of plant pathways. The volume concludes with several representative examples of new tools that allow us to rapidly manipulate the genetic material of the microbial host.

Volume 2 focuses on the metabolism of plants. Historically, elucidating plant metabolism has been challenging due to the lack of tightly genome-clustered pathways that are observed in microbial systems, along with the large size of plant genomes and transcriptomes. The first set of articles in this volume describe a variety of strategies to elucidate plant-specialized metabolism. Notably, the specialized metabolism of plants is controlled by complex regulatory processes. Furthermore, plant biosynthetic processes are also complicated by the fact that the metabolic reactions occur in a variety of different cell types and subcellular compartments. Therefore, several articles in this volume also describe efforts to control the regulatory networks that maintain the levels of metabolism production in plants, along with methods to understand the mechanisms of transport and localization of specialized metabolic intermediates. Finally, we highlight emerging tools to harness plant metabolism: new plant-based expression platforms and expression tools for production of metabolites are discussed.

Metabolic engineering has progressed rapidly in the last several years. The advent of genome editing, the ability to sequence complex genomes quickly and inexpensively, and the successful manipulation of plant and microbial hosts for more effective pathway reconstitution have collectively demonstrated that metabolic engineering holds substantial promise for improving our access to the end products of specialized metabolism. I note that these two volumes scratch the surface of this field, providing only a survey of some of the efforts being made in this area. I am deeply indebted to all of the contributors to this volume who graciously provided their time and effort to make a contribution to this work.

S.E. O'CONNOR
The John Innes Centre
Department of Biological Chemistry

CHAPTER ONE

Directing Biosynthesis: Practical Supply of Natural and Unnatural Cyanobactins

D. Sardar, M.D. Tianero, E.W. Schmidt[1]
University of Utah, Salt Lake City, UT, United States
[1]Corresponding author: e-mail address: ews1@utah.edu

Contents

1. Introduction — 1
2. Discovery of Cyanobactin Pathways — 2
3. Elucidating Natural Rules of Engineering in Cyanobactin Pathways — 3
4. Heterologous Expression of Cyanobactin Pathways in *E. coli* — 5
5. Optimization for Increased Yield of Cyanobactins in *E. coli* — 9
6. Synthesis of Cyanobactins In Vitro — 11
7. Conclusions — 16
8. Outlook — 18
Acknowledgments — 18
References — 18

Abstract

The increasingly rapid accumulation of genomic information is revolutionizing natural products discovery. However, the translation of sequence data to chemical products remains a challenge. Here, we detail methods used to circumvent the supply problem of cyanobactin natural products, both by engineered synthesis in *Escherichia coli* and by using purified enzymes in vitro. Such methodologies exploit nature's strategies of combinatorial chemistry in the cyanobactin class of RiPP natural products. As a result, it is possible to synthesize a wide variety of natural and unnatural compounds.

1. INTRODUCTION

Natural products remain a major component of drug discovery efforts (Newman & Cragg, 2012), but such efforts are still hindered by the need to synthesize a sufficient supply and to create analogs. Mimicking biosynthesis has emerged as a powerful tool to overcome these problems, providing a tractable alternative to traditional chemical synthesis. The key advance that has aided this is the explosion in genomic data, which has allowed connecting natural products to their corresponding genes (Walsh & Fischbach, 2010). In turn, rapid gene identification enables heterologous expression of enzymes to perform total synthesis in vivo or in vitro. Connecting genes to molecules has thus aided discovery based on sequencing, overcoming limitations of conventional natural products discovery, which relies on actual physical isolation of compounds, often from far-reaching locations of the earth. In addition, assigning genes to specific chemical modifications allows exploitation of enzyme function to modify nature's arsenal of chemistry far beyond, by technologies such as directed evolution (Renata, Wang, & Arnold, 2015).

This chapter focuses on directing biosynthesis using the cyanobactin biosynthetic machinery to engineer synthesis. Cyanobactins are natural products found in marine animals or in relatively slow-growing cyanobacteria. Due to the rare and variable distribution of the producing organisms, supply is an issue hindering cyanobactin development. An advantage is that several cyanobactin pathways are exceptionally broad-substrate tolerant, which allows cyanobactin enzymes and pathways to be used in the synthesis of thousands of derivatives. Recent advances enable such synthesis at scale in *Escherichia coli* and in vitro using purified enzymes. Later, we describe methods enabling the practical synthesis and engineering of cyanobactin pathways.

2. DISCOVERY OF CYANOBACTIN PATHWAYS

The cyanobactins belong to RiPP (*ri*bosomally synthesized and *p*osttranslationally modified *p*eptides) class of natural products and are present in about 30% of all cyanobacteria (Arnison et al., 2013; Sivonen, Leikoski, Fewer, & Jokela, 2010). The first cyanobactins, patellamides, were isolated from marine ascidian animals (Ireland & Scheuer, 1980), and related compounds have also been found in both ascidians and free-living cyanobacteria. Subsequently, biosynthetic genes for cyanobactins were discovered. A small sample (several grams) of the ascidian *Lissoclinum patella* was obtained near breaking surf on the reef flat above Blue Corner, Palau. Metagenome sequencing of the marine animal and its symbiotic bacteria led to identification of a RiPP pathway, *pat* in symbiotic cyanobacteria, *Prochloron*. *pat* carried a gene encoding the amino acid sequence of patellamide. Heterologous expression of *pat* in *E. coli* confirmed the symbiont *Prochloron* as the source of the cyanobactin. This was an early application of whole (meta)genome sequencing to identify a natural product source, leading to successful transfer of the producing pathway from a noncultivable producer in the ocean to a model host in the laboratory (Schmidt et al., 2005).

Due to the ribosomal nature of the cyanobactin pathway, subsequent genome mining efforts were focused on discovery through screening of homologous sequences, leading to identification of a new cyanobactin, trichamide (Sudek, Haygood, Youssef, & Schmidt, 2006), heralding the era of genome-based RiPP discovery. In RiPP pathways, a precursor peptide is modified by enzymes to yield natural products. In the case of cyanobactin pathways, further analysis revealed that the precursor peptides exist as natural combinatorial libraries (Donia et al., 2006; Donia, Ravel, & Schmidt, 2008). Many more new pathways were discovered in this way by genome mining (Donia & Schmidt, 2011; Leikoski, Fewer, & Sivonen, 2009; Martins, Leao, Ramos, & Vasconcelos, 2013; Ziemert et al., 2008), including the noteworthy discovery of a new class of cyanobactins that were linear with the ends protected by N-terminal prenylation and C-terminal methylation (Leikoski et al., 2013).

Ascidians contain a wide array of different cyanobactins with different structures and posttranslational modifications. However, across time and space in the oceans the ascidian-derived cyanobactin pathways are very closely related, being nearly 100% gene-sequence identical across their biosynthetic pathways. The exception is in precise regions that encode new sequence variants or new posttranslational modifications. This natural precision mutation has greatly aided studies of biosynthetic mechanism and engineering (Donia et al., 2008; Fig. 1). The most conserved genes are the N- and C-terminal proteases, a feature that can be exploited to discover new cyanobactin pathways by blast searching (Donia & Schmidt, 2011). By contrast, more variable regions and enzymes have allowed us to understand the rules that govern combinatorial biosynthesis (Donia et al., 2006; Sardar & Schmidt, 2015), which was subsequently exploited for engineering as described later.

3. ELUCIDATING NATURAL RULES OF ENGINEERING IN CYANOBACTIN PATHWAYS

Across multiple cyanobactin pathways, two kinds of genetic recombination events can be observed. First, the precursor peptide substrate is shuffled. Only a small number of amino acids in the precursor peptide

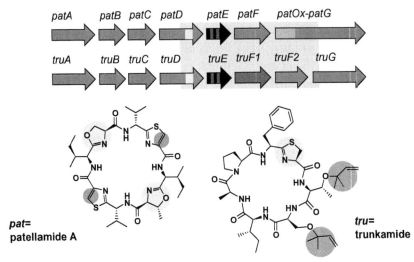

Fig. 1 Evolution of cyanobactin pathways in ascidians. The *pat* and *tru* cyanobactin gene clusters are shown, where the precursor peptide gene *patE/truE* is in *black*. The *red bars* within the precursor peptide gene represent the variable core sequences that encode the final natural products, whereas the remaining sequence (*black*) share >80% identity. The genes flanking the precursor code for posttranslational enzymes and other functions. Outside the precursor gene, regions in *gray* are similar in sequence, whereas the colored segments represent variation in sequence. The most variability apart from the core sequence corresponds to heterocyclization (*yellow*), prenylation (*green*), and oxidation (*blue*) posttranslational chemistry (*purple box*). This variation translates clearly to structural variability in the *pat* and *tru* natural products as shaded in the same corresponding color as the genes. The *pat* pathway products carry both thiazolines and oxazoline/methyloxazolines (*yellow circles*), whereas *tru* products carry only thiazoline (*yellow circle*). In addition, *tru* products are prenylated (*green circles*) corresponding to presence of the TruF1 prenyltransferase that is absent in *pat*. Similarly, *tru* products lack oxidation of thiazolines, since the corresponding oxidase domain (*blue*) is absent in them, in contrast to *pat* products. (See the color plate.)

(the core sequence) encode the final natural product; these are hypervariable. By contrast, the core peptide is flanked by highly conserved sequences that largely serve as recognition sequences (RSs) for enzymes (Sardar, Pierce, McIntosh, & Schmidt, 2015; Schmidt et al., 2005). Often, multiple copies of the core or multiple precursors with different cores are present in the same biosynthetic gene cluster. This results in the observed natural combinatorial chemistry in the cyanobactins (Donia et al., 2008). Second, the modifying enzymes are shuffled. New enzymes encoded in a particular cluster directly correspond to new chemistry in the final natural product. For example, the *tru* pathway produces prenylated natural products, a feature endowed by the prenyltransferase TruF1 (Sardar, Lin, & Schmidt, 2015). In contrast, the *pat* pathway that lacks prenylated natural product does not carry a functional TruF1 homolog in its cluster. These principles of nature to introduce new sequences into the natural peptide backbone (by shuffling precursor peptides) followed by introducing new chemistry on the peptide scaffold (by shuffling posttranslational modification enzymes) have been recapitulated in the laboratory to produce desired novel peptide derivatives.

The question that arises is how do the same posttranslational modification enzymes deal with this immense precursor peptide substrate diversity? The enzymes must be extremely promiscuous. This promiscuity is enabled by highly conserved sequences within the precursor peptide, the RSs that serve to guide the posttranslational enzymes (Donia et al., 2008; Sardar, Pierce, et al., 2015). This enables the core peptides, encoding the natural product, to vary without losing affinity for the enzyme. This evolutionary mechanism creates a unique natural engineering strategy that allows synthesis of novel derivatives in the laboratory as detailed later (Fig. 2).

4. HETEROLOGOUS EXPRESSION OF CYANOBACTIN PATHWAYS IN *E. coli*

The patellamides were the first cyanobactins to be heterologously expressed (Long, Dunlap, Battershill, & Jaspars, 2005; Schmidt et al., 2005). Due to their ribosomal origins, and the unique engineering rules of the cyanobactin biosynthetic route, specific mutations could be easily made within the pathway to encode nonnative cyanobactin derivatives (Donia et al., 2006). For cyanobactin production, we designed an *E. coli* expression platform that now uses a pUC-based vector pTru-SD (Symbion Discovery, Inc.), which carries the *tru* pathway under control of the *lac* promoter. This includes the biosynthetic enzymes TruA (N-terminal protease), TruB/TruC (hypothetical proteins), TruD (heterocyclase), TruF1 (prenyltransferase), TruF2 (prenyltransferase of unknown function), TruG (C-terminal protease/macrocyclase), and the precursor peptide substrate TruE that encodes the natural products patellins 3 and 2 (patellin 3 = diprenylated cyc-TVPVPTLC★ and patellin 2 = diprenylated cyc-TVPTLC★, where C★ is thiazoline). A typical expression experiment involves the following steps (Fig. 3):

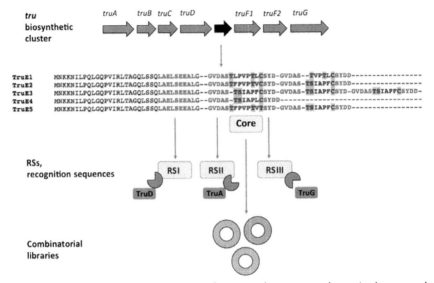

Fig. 2 Combinatorial libraries using *tru*. The *tru* pathway gene cluster is shown, and the precursor peptide variant TruE (*black*) is magnified to its translated amino acid sequence. Observation of this sequence clearly demarcates the hypervariable regions (*blue*) that represent the core encoding the final natural products. In contrast, the rest of the precursor is highly conserved and correspond to recognition sequences (RSs, *red*) that direct specific posttranslational enzymes. This phenomenon of substrate evolution, wherein the substrates evolve to maintain a balance between variations in the core (creates diversity) and conservation of the flanking RSs (maintains modification chemistry) allows the creation of natural combinatorial libraries. (See the color plate.)

- Day 1: Transform vector pTru-SD into *E. coli* DH10β and plate on LB agar supplemented with ampicillin (50 μg mL^{-1}).[1]
- Day 2: Pick six colonies into six wells in a 24-well plate containing liquid 2xYT medium (3 mL) with ampicillin (50 μg mL^{-1}). Grow overnight at 30°C with shaking at 150 rpm.
- Day 3: Pool the cultures. Use the pooled cultures (20 μL of pooled culture per 6 mL final volume) to inoculate wells in a 24-well plate (rounded ends). Each well contains 2xYT (6 mL) and ampicillin (50 μg mL^{-1}). In addition, several other media components can be added (described later as "yield optimizing additives"). Cover each 24-well plate with foil (sealing), and use a sterile needle to poke a small

---

[1] *Note*: Other cyanobactin vectors, or even libraries containing multiple cyanobactin precursor variants, use roughly the same protocol. Some details vary. For example, the yield is much lower with vectors that are not codon-optimized.

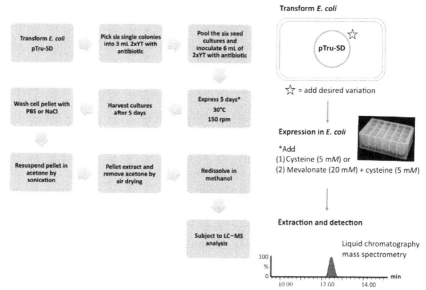

Fig. 3 Schematic for heterologous expression in *E. coli*. There are three basic steps: (1) transformation of *E. coli* with the pTru-SD vector that carries the *tru* pathway. At this step, the sequence of the precursor can be manipulated to add desired motifs to the final product; (2) expression in *E. coli* for an optimum of 5 days. At this step, addition of cysteine or cysteine with mevalonate results in higher yields of compounds; and (3) extraction of compounds from the *E. coli* cell pellet and mass spectrometry-based detection. (See the color plate.)

hole directly above each well to enable gas exchange. Grow cultures at 30°C with shaking at 150 rpm. Optimum harvest time varies by condition, but as described here 5 days is usually best.

- Day 8: Harvest the cells by centrifuge using either the 24-well plates or combined cultures from multiple wells.[2] Wash the resulting pellet twice with either 100 mM NaCl or phosphate-buffered saline. At this stage, the cell pellets can be extracted immediately or stored at −80°C for later extraction.
- To extract cells, resuspend the pellets (fresh or frozen) in acetone (2–3 mL) and place in a sonicating bath for 30 min. Remove cell debris

---

[2] *Note*: The yield is best if cells are grown in individual wells rather than combined in a flask. It is trivial to aliquot seeded media into multiple wells, which can then be combined and extracted at the end of the experiment. Because yield is not affected by stacking 24-well plates, this method optimizes use of shaker space, and the entire volume of a shaking incubator can be occupied. Yield is reasonable in flasks or in a fermentor under some conditions when using "yield optimizing additives," but it is still less reliable than 24-well plates.

by centrifugation, and the acetone extract is processed (later) for further analyses.
- Example use 1: The acetone layer is extracted, air-dried, and resuspended in liquid chromatography mass spectrometry (LC–MS) grade methanol, which is directly injected into the HPLC–mass spectrometer for analysis.
- Example use 2: The acetone layer is passed through a plug containing C18 resin, and fractions containing desired cyanobactins are further purified by HPLC.
- The yield of cyanobactins varies, but is generally $\sim 10\ \mu g\ L^{-1}$ under these conditions. "Yield optimizing additives" described later improve yield by 3000-fold but add complexity.

Over the years, this platform has been extensively optimized to engineer both for production of derivatives and improve yields. The synthesis of non-native cyanobactin was reported first in the creation of eptidemnamide, an analog of the rattlesnake derived anticoagulant eptifibatide. This was achieved by replacing the core sequence of patellamide precursor peptide with the eptifibatide sequence designed to carry an amide bond instead of a disulfide linkage (Donia et al., 2006). This feat was furthered by synthesis of cyanobactin derivatives containing nonproteinogenic amino acids like chlorinated and brominated tyrosine residues. Using methodologies developed by the Schultz group (Young et al., 2011), incorporation of nonproteinogenic amino acids in the core sequence was carried out (Tianero, Donia, Young, Schultz, & Schmidt, 2012). Production of patellins 3 and 2 from the parent precursor pTru-SD serves as internal controls of expression ensuring that the *tru* pathway is functional. A typical procedure for the incorporation of unnatural amino acids in cyanobactins is as follows:
- Clone the precursor peptide *truE* into pRSF vector under the control of a *lac* promoter and create mutants replacing the codons of amino acids in one or multiple positions with the amber stop codon, TAG. Select for the mutants using kanamycin (50 mg mL^{-1}) and purify the plasmids.
- Cotransform the resulting pRSF plasmid containing the *truE* with TAG codon(s) with the pTru-SD containing the *tru* operon and pEvol/pCNF[3] which contains the tRNA/aaRS pair to incorporate the unnatural amino acids into the TAG/stop codon position. Plate in LB agar

---

[3] pEvol/pCNF plasmid contains arabinose-inducible copy of the aaRS. In our hands, successful incorporation of the unnatural amino acids was achieved without induction.

containing kanamycin (50 µg mL^{-1}), ampicillin (50 µg mL^{-1}), and chloramphenicol (25 µg mL^{-1}). Grow at 30°C and 150 rpm overnight.
- Pick colonies and start seed cultures in 2xYT broth containing the three antibiotics (kanamycin, ampicillin, and chloramphenicol) as described earlier. After overnight growth, inoculate a fresh 2xYT broth (6 mL) containing the antibiotics with 20 µL of the pooled seed culture. To this, add the prepared unnatural amino acid (2 m*M*) and "yield optimizing additive."
- Grow the cultures for 5 days in a shaking incubator at 30°C and 150 rpm. Extract the cultures as described earlier and analyze the extracts for both the cyanobactin(s) containing the unnatural amino acids and patellins 2 and 3.
- This procedure can also be performed using amber mutations within the *tru* operon, rather than having a second copy encoding TruE.

In addition to creating nonnative cyanobactin derivatives, recently a *tru* pathway library was made in *E. coli*, revealing an ability to make potentially millions of compounds. The same method as described earlier was followed, and the library helped elucidate rules of core sequence selectivity and amino acid residue preferences at each position in the core sequence, leading to creation of >300 new compounds (Ruffner, Schmidt, & Heemstra, 2015).

5. OPTIMIZATION FOR INCREASED YIELD OF CYANOBACTINS IN *E. coli*

Although the strategies described earlier allowed production of both natural and unnatural cyanobactins in *E. coli*, a significant gap still existed in that heterologous expression often led to low compound yield. Although we had the technological expertise to create combinatorial libraries in the *E. coli* factory for drug discovery, downstream applications such as drug development were hindered by the yield problem.

Conventional protocols such as transcriptional activation, varying media conditions, or using different strains led to no significant improvement to yield (Donia & Schmidt, 2011). We then resorted to a novel metabolite-directed approach with the hypothesis that certain reagents can affect the metabolic flux in *E. coli*, which may result in increased compound yields. A screen of such metabolites led to the identification of the amino acid cysteine (yield optimizing additive), which provided a 150-fold increase in compound production (Tianero et al., 2016).

The addition of cysteine is identical to that described earlier except on Day 3:
- Day 3: Pool the cultures. Use the pooled cultures (200 μL of pooled culture per 6 mL final volume) to inoculate wells in a 24-well plate. Each well contains 2xYT (6 mL), cysteine (5 mM),[4] and ampicillin (50 μg mL^{-1}). Cover with foil, and poke a small hole directly above each well to enable gas exchange. Cultures are grown at 30°C with shaking at 150 rpm. Harvest after 5 days.
- This condition provides up to about 2 mg L^{-1} of cyanobactins.

Further investigation revealed that hydrogen sulfide, the breakdown product of cysteine in *E. coli*, was the actual mediator of increased cyanobactin production. The cysteine effect could be recapitulated by introducing hydrogen sulfide to the media as described later. In addition, the mechanism of sulfide action has been elucidated (Tianero et al., 2016).

The protocol is identical to that described earlier, except for Day 3:
- Day 3: Pool the cultures. Use the pooled cultures (20 μL of pooled culture per 6 mL final volume) to inoculate a glass test tube containing 2xYT (6 mL) and ampicillin (50 μg mL^{-1}). Place the test tube inside a 50 mL conical Falcon tube containing sodium phosphate buffer pH 8 (0.2 M, 10 mL) and Na$_2$S (10 mM). Seal the Falcon tube with a rubber cap allowing a little open headspace above the glass culture tube. This generates low doses of hydrogen sulfide that can diffuse into the culture tube through the headspace.
- Sulfide can also be applied at other scales and other vessel combinations.

A second metabolite was also identified along with cysteine that could drive increased compound titers: the isoprene precursor mevalonate, which is converted into dimethylallylpyrophosphate (DMAPP; Tianero et al., 2016). Addition of the *mev* pathway and mevalonate to *E. coli* cultures not only increases the degree of prenylated products but also increases total compound production, although the effect of mevalonate was only pronounced in the presence of cysteine. A typical expression platform involving mevalonate is as follows:
- Day 1: Transform vector pTru-SD[5] and pMBI[6] (Martin, Pitera, Withers, Newman, & Keasling, 2003) carrying the mevalonate pathway into

---

[4] *Note*: Optimum cysteine concentration varies between 1 and 20 mM depending upon the exact condition, but 5 mM provides a good, standardized dose that works.
[5] *Note*: Other cyanobactin-encoding vectors, including those for libraries of compounds or for other pathways such as *pat*, are effectively produced in this protocol.
[6] *Note*: In addition to pMBI (Martin et al., 2003), it is possible to use vectors that encode mevalonate synthesis from acetate instead, but cyanobactin yields are lower and more variable.

E. coli DH10β and plate on LB agar supplemented with ampicillin (50 µg mL^{-1}) and tetracycline (5 µg mL^{-1}).
- Day 2: Pick six colonies into six wells in a 24-well plate containing liquid 2xYT medium (3 mL) with ampicillin (50 µg mL^{-1}) and tetracycline (5 µg mL^{-1}). Grow overnight at 30°C with shaking at 150 rpm.
- Day 3: Pool the cultures. Use the pooled cultures (20 µL of pooled culture per 6 mL final volume) to inoculate wells in a 24-well plate (rounded ends). Each well contains 2xYT (6 mL), cysteine (5 mM), mevalonate (20 mM)[7], ampicillin (50 µg mL^{-1}), and tetracycline (5 µg mL^{-1}).
- Cultures are otherwise treated as described earlier. This condition provides up to about 30 mg L^{-1} of cyanobactins.

6. SYNTHESIS OF CYANOBACTINS IN VITRO

In many events, certain sequences exist that can be fully processed in vitro, yet fail to be produced in E. coli. This can be caused by any number of reasons, such as toxicity and/or degradation (Sardar, Lin, et al., 2015). In such cases, having a robust in vitro synthetic platform at hand is necessary for drug discovery.

Since discovery of the first cyanobactin gene cluster (Schmidt et al., 2005), a series of biochemical studies characterized representatives of the most ubiquitous cyanobactin enzymes, including the heterocyclase (TruD and homologs; Koehnke et al., 2013; McIntosh, Donia, & Schmidt, 2010; McIntosh & Schmidt, 2010; Sardar, Pierce, et al., 2015), the N-terminal and the C-terminal protease/macrocyclase (TruA and TruG homologs; Agarwal, Pierce, McIntosh, Schmidt, & Nair, 2012; Houssen et al., 2012; Koehnke et al., 2012; Lee, McIntosh, Hathaway, & Schmidt, 2009; McIntosh, Robertson, et al., 2010), and the prenyltransferase (TruF1 and homologs; McIntosh, Donia, Nair, & Schmidt, 2011; Sardar, Lin, et al., 2015). An elegant alternative in vitro approach to probing enzyme activity was recently reported by Goto et al., using a cell-free translation system. Interestingly, unprecedented heterocyclic motifs were generated, including a sequence with tandem thiazoline rings, a feature not observed in nature (Goto, Ito, Kato, Tsunoda, & Suga, 2014).

---

[7] *Note*: In general, the more mevalonate added, the better. However, mevalonate toxicity is apparent above about 40 mM, so that 20 mM as the upper limit is usually safe. Doubling time of E. coli is delayed to ~400 min in this condition.

Despite extensive exploitation of individual cyanobactin enzymes to create unnatural derivatives, the use of a combination of these enzymes for recapitulation of the entire multistep biosynthetic route was achieved only recently (Sardar, Lin, et al., 2015). A key finding enabling this event was that the N-terminal protease is inhibited by the reducing agent dithiothreitol (DTT). This required that if the preceding heterocyclization required DTT, then the heterocyclized product had to be purified for subsequent proteolysis. Alternatively, the protease cleavage site could be replaced to include a commercial protease cleavage site (Houssen et al., 2014), although this is not advantageous for one-pot synthesis. In addition, certain substrates that did not carry intramolecular disulfides could be easily processed by the heterocyclase without the need for reduction, and such substrates could be used in one-pot reaction schemes that could be modified to the final natural product carrying up to at least four posttranslational modifications (Sardar, Lin, et al., 2015). A typical in vitro pathway reconstitution method is described later (Fig. 4).

The cyanobactin proteins are expressed and purified as follows:
- All expression constructs are cloned into pET-28(b) vector backbone within the NdeI and XhoI restriction sites, which maintains an N-terminal His-tag.
- Day 1: Transform desired construct into BL21(DE3) or R2D-BL21 cells, and plate on LB agar supplemented with kanamycin (50 µg mL^{-1}). Add chloramphenicol (25 µg mL^{-1}) if R2D-BL21 cells are used.
- Day 2: Pick 1–5 colonies into LB broth (10 mL) supplemented with kanamycin (50 µg mL^{-1}), with addition of chloramphenicol (25 µg mL^{-1}) if R2D-BL21 cells were used, for an overnight seed culture.
- Day 3: Inoculate either LB or 2xYT media supplemented with the necessary antibiotics as above with the overnight culture from day 2, using 10 mL L^{-1}. Incubate at 30°C with shaking at 200–225 rpm, until the OD$_{600}$ reaches 0.4–0.6 units. For precursor peptide expression, induce cultures with IPTG (1 mM) and raise temperature to 37°C for an additional 3 h. This drives the precursor peptide into the pellet and improves expression. For expression of enzymes, lower the expression temperature to 18°C and induce cultures with 0.1 mM IPTG for 18 h. Typically, 6–8 L culture scales are used for each protein.
- After completion of induction time, harvest cells by spinning at 4000 rpm for 10 min. The pellets are collected and stored at −80°C till processed for purification.

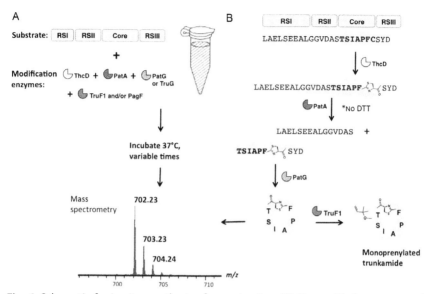

Fig. 4 Schematic for in vitro synthesis of cyanobactins. (A) The purified enzymes and substrate are mixed in a reaction tube under optimum conditions, and the products detected by mass spectrometry. It is essential to maintain the necessary recognition sequences (RSs) in the substrate for posttranslational chemistry. (B) A detailed reaction scheme is shown with each modification step. The representative substrate carries the trunkamide core sequence, flanked by the required RSs. The heterocyclase ThcD (directed by RSI) modifies the cysteine residue in the core to thiazoline. This is followed by N-terminal proteolysis by the protease PatA (directed by RSII). It is helpful to keep the reaction medium free of reducing agents for PatA action. The subsequent protease/macrocyclase PatG (directed by RSIII) cleaves off the RSIII and joins the ends to generate the cyclic product. Further modification of prenylation is appended on the backbone by the enzyme TruF1. (See the color plate.)

- All precursor peptides are purified using denaturing conditions and all enzymes are purified under native conditions using nickel column affinity chromatography, following standard purification protocols. All proteins are aliquoted and flash frozen (the enzymes are stored with 5% glycerol). The precursor peptides are stable and can be thawed multiple times or stored at 4°C for long time periods, whereas with the enzymes, a fresh aliquot is used every time.

A typical in vitro reaction setup is as follows:
- We use the enzyme ThcD from the *thc* pathway for introducing thiazolines. TruD from *tru* pathway also exhibits the same chemoselectivity as ThcD (Sardar, Pierce, et al., 2015). Heterocyclization reactions are set up with enzyme (2 μM), substrate (typically 50–100 μM), MgCl$_2$

(5 mM), DTT (if present, 7.5 mM), and ATP (1 mM) in Tris buffer pH 7.5 (50 mM). If DTT is present, the reaction is purified for subsequent steps by HPLC. Alternatively, a similar method can be used with PatD to simultaneously heterocyclize cysteine and serine/threonine.
- The second step after heterocyclization is N-terminal proteolysis. We use the enzyme PatA from the *pat* pathway for this purpose. The purified heterocyclized product is subjected to N-terminal proteolysis with enzyme (2–5 μM), MgCl$_2$ (5 mM), and CaCl$_2$ (10 mM, this is not necessary and can be left out) in Tris buffer pH 7.5 (50 mM). If no DTT was used in the previous step, these components can be directly added to the heterocyclization reaction mix.
- N-terminal proteolysis is followed by C-terminal cleavage in tandem with macrocyclization. We have used both PatG protease domain (*pat* pathway) and TruG full-length construct (*tru* pathway) for this. Addition of the C-terminal protease/macrocyclase can be done either with or without purification of the N-terminal proteolytic product, although HPLC purification usually yields a cleaner subsequent reaction. The C-terminal protease/macrocyclase (10–20 μM) is added to the previous reaction maintaining MgCl$_2$ (5 mM) in Tris buffer pH 7.5 (50 mM).
- In certain pathways, isoprene is appended on the cyclic peptide backbone. We have used the prenyltransferases TruF1 (a Ser/Thr *O*-prenyltransferase from the *tru* pathway), PagF (a Tyr *O*-prenyltransferase from the *pag* pathway) and LynF (a reverse Tyr *O*-prenyltransferase from the *lyn* pathway). For prenylation, the enzyme (10 μM) is added to the macrocyclization reaction mix with DMAPP (10 mM) as the prenyl donor. Due to the high cost of commercially available DMAPP, we chemically synthesize it in the laboratory using previously established protocols (Davisson et al., 1986; Woodside, Huang, & Poulter, 1988).
- All reactions are carried out at 37°C in a PCR cycler with a heated lid to prevent evaporation within the reaction tubes. Typically, both the heterocyclization and N-terminal proteolysis reactions are complete within 2 h, whereas the C-terminal proteolysis/macrocyclization usually runs for 24 h or more, and the subsequent prenylation step even longer and never reaches completion in our hands.
- A one-pot synthetic route to cyanobactin derivatives using the enzymes ThcD, PatA, and PatG (or TruG) was also reported (Sardar, Lin, et al., 2015). For one-pot reaction schemes, the same concentrations and conditions as detailed earlier are maintained in a single reaction tube for 7.5–10 h.

Purification and characterization of reaction intermediates and products is done as follows:
- If it is desirable to stop the reaction at an earlier time point, boiling for 15 min quenches the reactions. The tubes are then centrifuged briefly to pellet precipitated material, and the supernatant is analyzed.
- HPLC purification: A semipreparative C18 column is used with a mobile phase comprising H_2O/ACN gradient from 1% to 99% ACN over 20 min. Note that no acid is used in the HPLC mobile phase since it renders the thiazoline prone to ring opening. Typically, fractions are collected every minute, dried under vacuum, and the fractions with UV absorbance are analyzed by mass spectrometry to detect the desired species. A 255-nm shoulder in the UV spectrum indicates presence of thiazoline ring.
- C18 purification: In certain cases, an alternative to HPLC purification is a rapid purification by C18 resin (Sigma). A small C18 plug is made in a Pasteur pipette. The resin is equilibrated with 100% ACN followed by H_2O. The reaction solution is added to the resin, which is washed twice with water, and elutions are collected at 25%, 50%, and 100% ACN. Typically, the desired products elute at 50% and/or 100% ACN.
- Product characterization: All intermediate and product species are characterized by LC–MS analysis. A C18 or C4 analytical column is used in all cases, and a C4 column results in negligible column bleed-through in case of larger peptides >1000 Da mass. A mobile phase of H_2O (0.1% formic acid)/ACN is used, with an ACN gradient of 1–99% over 20 min.
- If reactions are done in large scale, purity of products is further assessed by NMR spectroscopy. Typically, the cyclic peptides are dissolved in 3:2 D_2O/ACN-d_3 for best results.

Alternative routes to generate products and derivatives in vitro have also been reported, such as use of commercial protease cleavage sites in addition to cyanobactin enzymes (Houssen et al., 2014), use of posttranslational enzymes fused to substrate leader sequence (Oueis et al., 2015), and use of an in vitro translation platform (Goto et al., 2014). A unique feature of cyanobactin RiPP pathways that has enabled in vitro synthesis of cyanobactin derivatives is their modularity (Sardar, Lin, et al., 2015). This implies that parts of the pathways can be mixed and matched with parts of other pathways to create hybrid natural products. Thus, different enzymes from varied cyanobactin families can successfully process nonnative substrates from other families, or even chimeric substrates that carry cores from

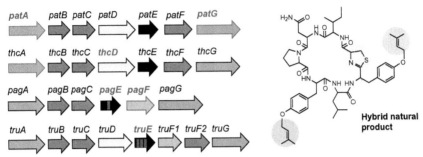

Fig. 5 Modularity of cyanobactin pathways. Enzymes and substrates from different pathways can be mixed and matched to create hybrid natural products for combinatorial chemistry. For example, the unnatural cyanobactin shown on the *right* is derived from elements belonging to four different pathways *pat*, *thc*, *pag*, and *tru*. The specific elements involved from each gene cluster for the creation of this hybrid are highlighted in *red*. The cyclic peptide sequence is derived from *pagE* precursor sequence INPYLYP (*red bar* in *pagE*), wherein the proline is mutated to cysteine to allow introduction of a heterocycle. The recognition sequences in the precursor substrate were derived from *truE* (*red bars* in *truE*) to direct the enzymes heterocyclase (*thcD*, from the *thc* pathway), the N-terminal protease (*patA*, from the *pat* pathway), and the C-terminal protease/macrocyclase (*patG*, from the *pat* pathway). Additional modification of prenylation was introduced by *pagF* (from the *pag* pathway). This resulted in a derivative of the natural product prenylagaramide B, carrying chemistry not found in nature such as thiazoline ring (*yellow*) and double prenylation (*blue*). (See the color plate.)

different families. The only requirement necessary to allow such hybridization is the maintenance of the conserved RSs that act as docking sites for the respective posttranslational modification reaction (Fig. 5; Sardar, Pierce, et al., 2015).

7. CONCLUSIONS

This chapter details the methods that demonstrate the powerful ability of rerouting biosynthetic routes for synthesis. Since the cyanobactins carry a peptide backbone, research in this area provides tools for the synthesis of desirable peptide motifs. An advantage of such tools is that they require simple manipulations at the genetic level for the creation of derivatives. In contrast, a chemical synthetic route is tedious with complicated scope for combinatorial chemistry. For example, the total synthesis of trunkamide involves 14 steps (Wipf & Uto, 1999). In contrast, production of trunkamide and derivatives with the use of our optimized *E. coli* expression system and in vitro platform requires far lesser effort with maximized yields.

We have detailed methods for synthesis in vivo and in vitro. Such synthesis is possible based on an understanding of nature's engineering rules of modularity in the cyanobactin RiPPs. This has allowed creation of unique peptide scaffolds, a few examples of which include cyclic peptides with halogenation, azide functionality, polyketide insertions, and large macrocyclic ring size. In addition, pathway hybrids with nonnative patterns of heterocyclization and prenylation and linear peptides with heterocycles at unexpected positions are among few of the nonnative chemical motifs that have been captured in the laboratory, expanding nature's array of chemistry using nature's tools (Fig. 6).

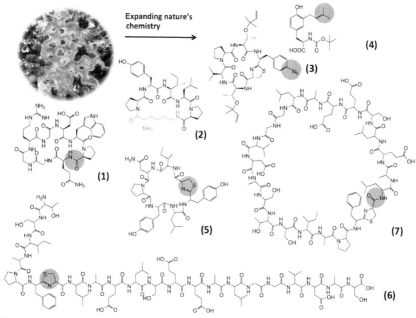

Fig. 6 Representative examples of modified peptide motifs not found in nature created using the cyanobactin RiPP machinery both in vivo and in vitro. (1) Eptidemnamide that resembles the rattlesnake derived anticoagulant eptifibatide, (2) macrocyclic peptides with polyketide insertions, (3) macrocyclic peptides with nonproteinogenic amino acid insertions, (4) small molecules carrying isoprene units, (5) thiazoline containing cyclic peptides, (6) linear peptide with the thiazoline heterocycle at desirable positions, and (7) an unusually large macrocycle of 22 ring size. (See the color plate.)

8. OUTLOOK

Although this chapter is focused primarily on the cyanobactin story, similar features of promiscuity that allows combinatorial biosynthesis are also beginning to be observed in other RiPP families (Sardar & Schmidt, 2015). The most notable of this is the lanthipeptide family of RiPPs, which include the marine compounds called the prochlorosins (Zhang, Yu, Velasquez, & van der Donk, 2012). Apart from the cyanobactins and the lanthipeptides, it is likely that similar stories and strategies will be found among the ubiquitous and widespread families of RiPPs.

ACKNOWLEDGMENTS
Our work on RiPPs is funded by NIH R01 GM102602.

REFERENCES

Agarwal, V., Pierce, E., McIntosh, J., Schmidt, E. W., & Nair, S. K. (2012). Structures of cyanobactin maturation enzymes define a family of transamidating proteases. *Chemistry & Biology*, 19, 1411–1422.

Arnison, P. G., Bibb, M. J., Bierbaum, G., Bowers, A. A., Bugni, T. S., Bulaj, G., et al. (2013). Ribosomally synthesized and post-translationally modified peptide natural products: Overview and recommendations for a universal nomenclature. *Natural Product Reports*, 30, 108–160.

Davisson, V. J., Woodside, A. B., Neal, T. R., Stremler, K. E., Muehlbacher, M., & Poulter, C. D. (1986). Phosphorylation of isoprenoid alcohols. *The Journal of Organic Chemistry*, 51, 4768–4779.

Donia, M. S., Hathaway, B. J., Sudek, S., Haygood, M. G., Rosovitz, M. J., Ravel, J., et al. (2006). Natural combinatorial peptide libraries in cyanobacterial symbionts of marine ascidians. *Nature Chemical Biology*, 2, 729–735.

Donia, M. S., Ravel, J., & Schmidt, E. W. (2008). A global assembly line for cyanobactins. *Nature Chemical Biology*, 4, 341–343.

Donia, M. S., & Schmidt, E. W. (2011). Linking chemistry and genetics in the growing cyanobactin natural products family. *Chemistry & Biology*, 18, 508–519.

Goto, Y., Ito, Y., Kato, Y., Tsunoda, S., & Suga, H. (2014). One-pot synthesis of azoline-containing peptides in a cell-free translation system integrated with a posttranslational cyclodehydratase. *Chemistry & Biology*, 21, 766–774.

Houssen, W. E., Bent, A. F., McEwan, A. R., Pieiller, N., Tabudravu, J., Koehnke, J., et al. (2014). An efficient method for the in vitro production of azol(in)e-based cyclic peptides. *Angewandte Chemie (International Ed. in English)*, 53, 14171–14174.

Houssen, W. E., Koehnke, J., Zollman, D., Vendome, J., Raab, A., Smith, M. C., et al. (2012). The discovery of new cyanobactins from Cyanothece PCC 7425 defines a new signature for processing of patellamides. *Chembiochem*, 13, 2683–2689.

Ireland, C., & Scheuer, J. (1980). Ulicyclamide and ulithiacyclamide, two new small peptides from a marine tunicate. *Journal of the American Chemical Society*, 102, 5688–5691.

Koehnke, J., Bent, A., Houssen, W. E., Zollman, D., Morawitz, F., Shirran, S., et al. (2012). The mechanism of patellamide macrocyclization revealed by the characterization of the PatG macrocyclase domain. *Nature Structural & Molecular Biology*, 19, 767–772.

Koehnke, J., Bent, A. F., Zollman, D., Smith, K., Houssen, W. E., Zhu, X., et al. (2013). The cyanobactin heterocyclase enzyme: A processive adenylase that operates with a defined order of reaction. *Angewandte Chemie (International Ed in English), 52*, 13991–13996.

Lee, J., McIntosh, J., Hathaway, B. J., & Schmidt, E. W. (2009). Using marine natural products to discover a protease that catalyzes peptide macrocyclization of diverse substrates. *Journal of the American Chemical Society, 131*, 2122–2124.

Leikoski, N., Fewer, D. P., & Sivonen, K. (2009). Widespread occurrence and lateral transfer of the cyanobactin biosynthesis gene cluster in cyanobacteria. *Applied and Environmental Microbiology, 75*, 853–857.

Leikoski, N., Liu, L., Jokela, J., Wahlsten, M., Gugger, M., Calteau, A., et al. (2013). Genome mining expands the chemical diversity of the cyanobactin family to include highly modified linear peptides. *Chemistry & Biology, 20*, 1033–1043.

Long, P. F., Dunlap, W. C., Battershill, C. N., & Jaspars, M. (2005). Shotgun cloning and heterologous expression of the patellamide gene cluster as a strategy to achieving sustained metabolite production. *Chembiochem, 6*, 1760–1765.

Martin, V. J., Pitera, D. J., Withers, S. T., Newman, J. D., & Keasling, J. D. (2003). Engineering a mevalonate pathway in *Escherichia coli* for production of terpenoids. *Nature Biotechnology, 21*, 796–802.

Martins, J., Leao, P. N., Ramos, V., & Vasconcelos, V. (2013). N-terminal protease gene phylogeny reveals the potential for novel cyanobactin diversity in cyanobacteria. *Marine Drugs, 11*, 4902–4916.

McIntosh, J. A., Donia, M. S., Nair, S. K., & Schmidt, E. W. (2011). Enzymatic basis of ribosomal peptide prenylation in cyanobacteria. *Journal of the American Chemical Society, 133*, 13698–13705.

McIntosh, J. A., Donia, M. S., & Schmidt, E. W. (2010). Insights into heterocyclization from two highly similar enzymes. *Journal of the American Chemical Society, 132*, 4089–4091.

McIntosh, J. A., Robertson, C. R., Agarwal, V., Nair, S. K., Bulaj, G. W., & Schmidt, E. W. (2010). Circular logic: Nonribosomal peptide-like macrocyclization with a ribosomal peptide catalyst. *Journal of the American Chemical Society, 132*, 15499–15501.

McIntosh, J. A., & Schmidt, E. W. (2010). Marine molecular machines: Heterocyclization in cyanobactin biosynthesis. *Chembiochem, 11*, 1413–1421.

Newman, D. J., & Cragg, G. M. (2012). Natural products as sources of new drugs over the 30 years from 1981 to 2010. *Journal of Natural Products, 75*, 311–335.

Oueis, E., Adamson, C., Mann, G., Ludewig, H., Redpath, P., Migaud, M., et al. (2015). Derivatisable cyanobactin analogues: A semisynthetic approach. *Chembiochem, 16*, 2646–2650.

Renata, H., Wang, Z. J., & Arnold, F. H. (2015). Expanding the enzyme universe: Accessing non-natural reactions by mechanism-guided directed evolution. *Angewandte Chemie (International Ed. in English), 54*, 3351–3367.

Ruffner, D. E., Schmidt, E. W., & Heemstra, J. R. (2015). Assessing the combinatorial potential of the RiPP cyanobactin tru pathway. *ACS Synthetic Biology, 4*, 482–492.

Sardar, D., Lin, Z., & Schmidt, E. W. (2015). Modularity of RiPP enzymes enables designed synthesis of decorated peptides. *Chemistry & Biology, 22*, 907–916.

Sardar, D., Pierce, E., McIntosh, J. A., & Schmidt, E. W. (2015). Recognition sequences and substrate evolution in cyanobactin biosynthesis. *ACS Synthetic Biology, 4*, 167–176.

Sardar, D., & Schmidt, E. W. (2015). Combinatorial biosynthesis of RiPPs: Docking with marine life. *Current Opinion in Chemical Biology, 31*, 15–21.

Schmidt, E. W., Nelson, J. T., Rasko, D. A., Sudek, S., Eisen, J. A., Haygood, M. G., et al. (2005). Patellamide A and C biosynthesis by a microcin-like pathway in Prochloron didemni, the cyanobacterial symbiont of Lissoclinum patella. *Proceedings of the National Academy of Sciences of the United States of America, 102*, 7315–7320.

Sivonen, K., Leikoski, N., Fewer, D. P., & Jokela, J. (2010). Cyanobactins-ribosomal cyclic peptides produced by cyanobacteria. *Applied Microbiology and Biotechnology, 86,* 1213–1225.

Sudek, S., Haygood, M. G., Youssef, D. T., & Schmidt, E. W. (2006). Structure of trichamide, a cyclic peptide from the bloom-forming cyanobacterium Trichodesmium erythraeum, predicted from the genome sequence. *Applied and Environmental Microbiology, 72,* 4382–4387.

Tianero, M. D., Donia, M. S., Young, T. S., Schultz, P. G., & Schmidt, E. W. (2012). Ribosomal route to small-molecule diversity. *Journal of the American Chemical Society, 134,* 418–425.

Tianero, M. D., Pierce, E., Raghuraman, S., Sardar, D., McIntosh, J. A., Heemstra, J. R., et al. (2016). Metabolic model for diversity-generating biosynthesis. *Proceedings of the National Academy of Sciences of the United States of America, 113,* 1772–1777.

Walsh, C. T., & Fischbach, M. A. (2010). Natural products version 2.0: Connecting genes to molecules. *Journal of the American Chemical Society, 132,* 2469–2493.

Wipf, P., & Uto, Y. (1999). Total synthesis of the putative structure of the marine metabolite trunkamide A. *Tetrahedron Letters, 40,* 5165–5169.

Woodside, A. B., Huang, Z., & Poulter, C. D. (1988). Trisammonium geranyl diphosphate. *Organic Syntheses, 66,* 211.

Young, T. S., Young, D. D., Ahmad, I., Louis, J. M., Benkovic, S. J., & Schultz, P. G. (2011). Evolution of cyclic peptide protease inhibitors. *Proceedings of the National Academy of Sciences of the United States of America, 108,* 11052–11056.

Zhang, Q., Yu, Y., Velasquez, J. E., & van der Donk, W. A. (2012). Evolution of lanthipeptide synthetases. *Proceedings of the National Academy of Sciences of the United States of America, 109,* 18361–18366.

Ziemert, N., Ishida, K., Quillardet, P., Bouchier, C., Hertweck, C., de Marsac, N. T., et al. (2008). Microcyclamide biosynthesis in two strains of Microcystis aeruginosa: From structure to genes and vice versa. *Applied and Environmental Microbiology, 74,* 1791–1797.

CHAPTER TWO

Synthetic Biology Approaches to New Bisindoles

L.M. Alkhalaf[1], Y.-L. Du[1], K.S. Ryan[2]
University of British Columbia, Vancouver, BC, Canada
[2]Corresponding author: e-mail address: ksryan@chem.ubc.ca

Contents

1. Introduction — 22
2. Identification of New Bisindole Gene Clusters — 25
3. Heterologous Expression — 27
   3.1 Host Strains Commonly Used — 27
   3.2 Introduction of the Bisindole Gene Cluster into a Chassis Host — 28
4. Mutational Biosynthesis to Generate New Bisindoles — 30
5. Mixing Genes from Phylogenetically Related Clusters to Generate New Bisindoles — 32
6. Chemical Isolation and Structural Characterization — 32
   6.1 Strain Growth and Product Extraction — 32
   6.2 Analysis and Purification — 33
   6.3 Metabolite Characterization — 33
7. Conclusions — 34
References — 34

Abstract

Bisindoles are a class of natural products derived from oxidative dimerization of tryptophan, and many of these molecules have potential use as anticancer agents. The recent isolation of new bisindoles and their corresponding gene clusters has greatly expanded the repertoire of biosynthetic genes available to synthetic biologists. This chapter describes methods to exploit the biosynthetic pathways leading to bisindoles, using cladoniamides as a representative example. Specifically, we describe how to identify and heterologously express gene clusters and how to manipulate pathways in order to generate new bisindoles. We also discuss methods for cultivating, extracting, purifying, and characterizing these new metabolites.

---

[1] Equal contributors

1. INTRODUCTION

Microbial bisindoles are a natural product family derived from oxidation and dimerization of two tryptophan monomers. These molecules have shown promise as anticancer compounds, and analogs of bisindoles staurosporine and rebeccamycin have gone through advanced clinical trials (Bharate, Sawant, Singh, & Vishwakarma, 2013; Schwandt et al., 2012; Sherer & Snape, 2015). Bisindoles are also active as antibacterials (Fernandez et al., 2006) and antivirals (Marschall et al., 2002). The recent isolation of a large number of new bisindoles has further expanded this natural product class (Du & Ryan, 2016). For instance, cladoniamide A (Williams et al., 2008) is a cytotoxic compound thought to target the proteolipid subunit of the vacuolar H^+-ATPase (Chang, Kawashima, & Brady, 2014; Kimura et al., 2012) and was recently shown to have potent antimalarial activity (Deng et al., 2015).

The first two steps in the biosynthesis of microbial bisindoles from L-tryptophan are shared among different pathways (Ryan & Drennan, 2009) (Fig. 1), with tryptophan sometimes initially converted to a chlorinated derivative by an $FADH_2$-dependent halogenase such as RebH (Yeh, Garneau, & Walsh, 2005). The first shared step in the pathways is oxidation of tryptophan (or chlorotryptophan) by a flavoprotein oxidase (ClaO and homologs) into the corresponding indole pyruvate (IPA) imine (Nishizawa, Aldrich, & Sherman, 2005). Second, two molecules of IPA enamine are coupled by a heme-containing enzyme (ClaD and homologs) to form chromopyrrolic acid (CPA), which is a common intermediate in the biosynthetic pathways (Asamizu et al., 2012). From here, the pathways branch in several possible directions. In methylarcyriarubin biosynthesis, a Rieske dioxygenase MarC, and methyltransferase MarM, catalyzes oxidative decarboxylation and methylation of CPA to give methylarcyriarubin (Chang & Brady, 2014). In pathways to molecules such as the cladoniamides, rebeccamycin, staurosporine, and erdasporine, a cytochrome P450 (ClaP and homologs) catalyzes the formation of an aryl–aryl bond between the C2 indole carbons (Makino et al., 2007), followed by nonenzymatic transformations (Howard-Jones & Walsh, 2007). A series of different oxidations, reductions, and decarboxylations are then responsible for producing the diverse upper ring structures found in different microbial bisindoles. The pyrrolinium ring in reductasporine is thought to form via reduction and dimethylation of an unstable carboxylate intermediate by reductase RedE and methyltransferase RedM (Chang, Ternei, Calle, & Brady, 2015),

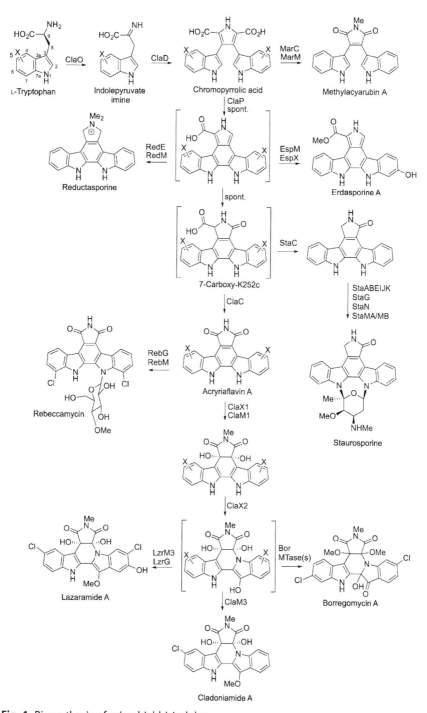

Fig. 1 Biosynthesis of microbial bisindoles.

whereas the pyrrole carboxylic acid found in erdasporine is thought to form from the same intermediate by O-methylation by EspM, followed by hydroxylation by EspX to give erdasporine (Chang, Ternei, Calle, & Brady, 2013). Formation of the pyrrolinone in staurosporine and the maleimide in rebeccamycin is catalyzed by flavoproteins (StaC or RebC) from tautomers of 7-carboxy-K252c (Goldman et al., 2012; Howard-Jones & Walsh, 2006; Ryan et al., 2007).

One further alteration to the core structure is achieved by "flipping" one of the indole rings of the indolocarbazole to generate an indolotryptoline, as in cladoniamide, BE-54017, lazaramide, and borregomycin biosynthesis (Chang & Brady, 2011, 2013; Montiel, Kang, Chang, Charlop-Powers, & Brady, 2015; Williams et al., 2008). This reaction is carried out by two flavoenzymes. The first flavoenzyme (ClaX1 and homologs) forms a cis-epoxide across the maleimide C=C double bond. Then, following N-methylation of the upper ring, a second flavoenzyme (ClaX2 and homologs) catalyzes oxidative destruction of the indolocarbazole, putatively via epoxidation of the pyrrole C=C double bond of the indole (Du, Williams, Patrick, Andersen, & Ryan, 2014).

A variety of tailoring reactions can occur to generate further diversity. One key example is the chlorination that can occur at the start of the biosynthesis by the action of $FADH_2$-dependent chlorinases. Chlorination at the C7 (Yeh et al., 2005), C6 (Chang & Brady, 2013), and C5 (Du & Ryan, 2015) positions has been reported in microbial bisindole pathways. Additionally, further modifications can occur once the core scaffolds have been put together. N- and O-methylation is observed in cladoniamide (Du, Ding, & Ryan, 2013), BE-54017 (Chang & Brady, 2011), borregomycin (Chang & Brady, 2013), and lazarimide (Montiel et al., 2015) pathways, while hydroxylation of the indole is observed in both erdasporine (Chang et al., 2013) and lazarimide biosynthesis. More complex additions are present in both rebeccamycin and staurosporine, in which glycosylation occurs on the indole nitrogen(s). In rebeccamycin biosynthesis, RebG catalyzes formation of the new glycoside bond with glucose, which is subsequently methylated by RebM (Onaka, Taniguchi, Igarashi, & Furumai, 2003; Zhang et al., 2006). In staurosporine biosynthesis glucose is first converted by staurosporine enzymes (StaA, StaB, StaE, StaI, StaJ, StaK) to L-ritosamine, which is then attached to the first indole nitrogen by StaG, then to the second indole nitrogen by StaN. Methyltransferases StaMA and StaMB then methylate the amino and hydroxyl groups of the sugar, respectively (Onaka, Taniguchi, Igarashi, & Furumai, 2002; Salas et al., 2005).

The close relationship of the biosynthetic pathways, as well as some flexibility in enzyme substrate tolerance, has allowed for the construction of new bisindoles. The pioneering work by Salas and coworkers focused on combinatorial biosynthesis with rebeccamycin and staurosporine biosynthetic genes (Salas et al., 2005; Sánchez, Méndez, & Salas, 2006; Sánchez et al., 2009, 2005). In their approach they coexpressed different combinations of *reb/sta* genes, along with genes for the biosynthesis of different sugars, in order to generate novel bisindoles. One of the resulting "nonnatural" bisindole products, EC-70124, has been shown to be a promising molecule for treatment of triple-negative breast cancer (Cuenca-López et al., 2015). With the recent isolation of new bisindoles and their corresponding gene clusters, the genetic toolbox available to synthetic biologists has greatly expanded. Our lab has applied a similar strategy as the Salas group to generate new bisindole metabolites (Fig. 2) (Du & Ryan, 2015). However, unlike the previous studies in which individual genes were stitched together for coexpression, we start with the intact cladoniamide (*cla*) biosynthetic gene cluster (Ryan, 2011) to minimize the disruption of natural operons and regulatory elements.

Here, we describe methods for manipulating biosynthetic pathways to produce novel bisindoles (Fig. 3). The first step is identification of bisindole gene clusters by utilizing degenerate primers or examination of a sequenced genome. Once a gene cluster has been identified, transferring this gene cluster into a heterologous expression host, commonly a modified *Streptomyces* strain, allows for production of the natural product. Different methods can then be employed to produce novel analogs: deleting genes within the gene cluster can produce pathway intermediates and shunt products, whereas expressing other, phylogenetically related genes can generate molecules with new diversity. Methods to cultivate, extract, purify, and characterize new metabolites are also described.

2. IDENTIFICATION OF NEW BISINDOLE GENE CLUSTERS

As described in Section 1, most microbial bisindoles derive from the oxidative dimerization of L-Trp by the action of core indolocarbazole enzymes. Furthermore, as for most bacterial natural product genes, bisindole biosynthetic genes are generally colocalized in the genome. Thus, primers specific for the genes encoding the core indolocarbazole enzymes can be used as probes to isolate new bisindole biosynthetic gene clusters. Two sets

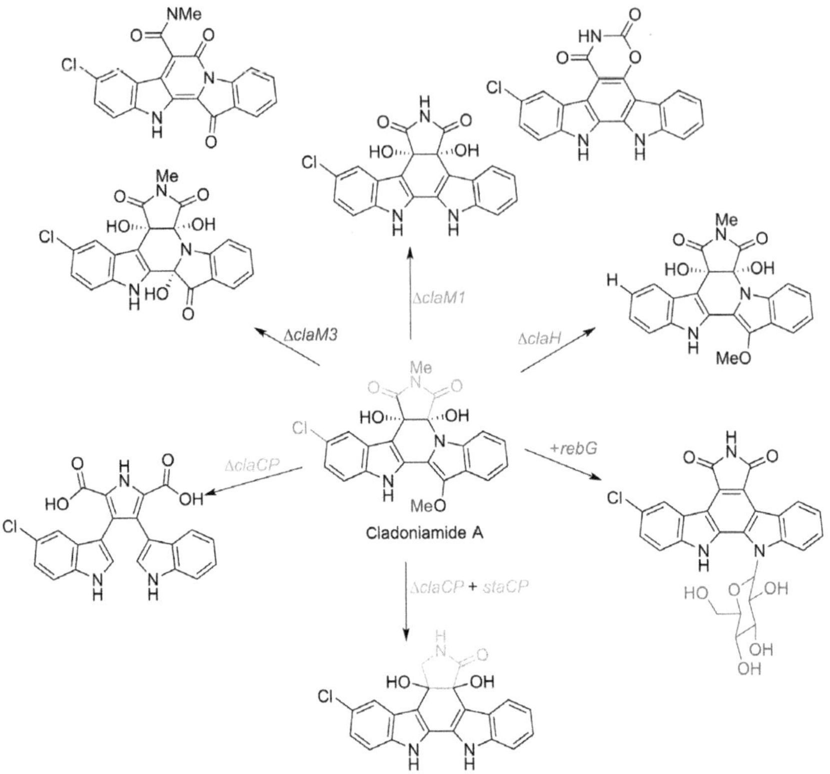

Fig. 2 Analogs of cladoniamide A obtained through combinatorial biosynthesis. Colors link biosynthetic enzymes with the corresponding fragment of the cladoniamide structure. Major metabolites are shown, although metabolites with distinct chlorination patterns are frequently coisolated (Du, Ding, & Ryan, 2013; Du & Ryan, 2015; Du et al., 2014). (See the color plate.)

of indolocarbazole-specific primers have been reported. The first set consists of degenerate primers for *rebD/staD/vioB* (Chang & Brady, 2011), and the second set of primers are specific to conserved regions in the flavoprotein *rebC/staC* genes (Ryan, 2011). Since RebC is employed after a major branch point in the biosynthesis, many bisindole gene clusters lack *rebC* homologs, thus the primers for *rebC* homologs are less general than those for *rebD*. This method of employing degenerate primers is particularly useful for the screening of environmental DNA libraries (Chang et al., 2013) or large microbial libraries (Zhang et al., 2012). However, for microbial strains already known to produce bisindoles, the current approach is genome scanning, which is a faster and potentially cheaper way to identify bisindole gene clusters. Once gene clusters are identified, they are annotated using standard bioinformatics tools. In the case of *Streptomyces*-derived gene

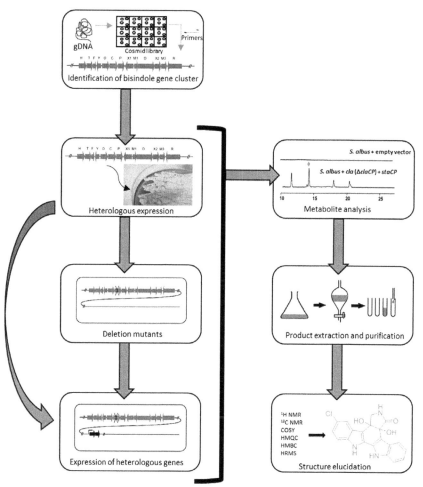

Fig. 3 Workflow for isolation and combinatorial biosynthesis of tryptophan-derived microbial bisindoles, from identification of a gene cluster to characterization of novel metabolites. (See the color plate.)

clusters, the biased codon usage by *Streptomyces* species, which leads to a very high G+C content, can be used to predict coding regions of DNA with FramePlot (Ishikawa & Hotta, 1999).

3. HETEROLOGOUS EXPRESSION

3.1 Host Strains Commonly Used

The same codon bias that is useful in assigning coding regions of DNA often makes heterologous expression of *Streptomyces* genes in traditional hosts such as *Escherichia coli* and yeast challenging, although these hosts are sometimes used (Chang et al., 2013). Consequently a number of *Streptomyces* hosts have been developed. Our group commonly uses *Streptomyces coelicolor* M1146 and *Streptomyces albus* J1074. *S. coelicolor* M1146 is a modified host in which the actinorhodin, prodiginine, calcium-dependent antibiotic, and coelimycin gene clusters have been deleted (Gomez-Escribano & Bibb, 2011; Gomez-Escribano et al., 2012). This host has the benefit of being a proficient natural product producer, while not producing high levels of native natural products that might compete for biosynthetic precursors. *S. albus* J1074, by contrast, is a naturally minimized *Streptomyces* strain (Baltz, 2010). A number of other *Streptomyces* hosts are available, which reduce production of unwanted metabolites, improve production of expressed enzymes, and/or provide a ready supply of necessary precursors (Gomez-Escribano & Bibb, 2012). The genetic manipulation techniques are largely the same for all these strains.

3.2 Introduction of the Bisindole Gene Cluster into a Chassis Host

Transfer of genes into a predesigned plasmid can be accomplished using approaches such as transformation-associated recombination (Kim et al., 2010; Kouprina & Larionov, 2008; Yamanaka et al., 2014), linear-plus-linear homologous recombination (Fu et al., 2012), or in vitro Gibson assembly (Temme, Zhao, & Voigt, 2012). However, in the case of most reported microbial bisindoles, gene clusters are small (<30 kbp) and the genes are already contained on a single cosmid. Thus, it is often simpler to adapt the existing cosmid to allow conjugation and maintenance into a heterologous *Streptomyces* host. In the case of cladoniamide, we retrofitted the SuperCos1-derived cosmid 7H1 containing the complete cladoniamide (*cla*) gene cluster for heterologous expression (Du, Ding, & Ryan, 2013). First, the *bla* gene in the backbone of the cosmid, which confers ampicillin resistance, was replaced with a 4.9 kb, gel-purified *Dra*I-*Bsa*I fragment from pIJ787 (containing an *oriT* site, *tet* gene, *attP* site, and the integrase cassette from phage ΦC31; Eustáquio et al., 2005) using λ-Red-mediated recombination (described in Section 4), to give pYD1 (Fig. 4). The *tet* gene, conferring tetracycline resistance, was then replaced with a 1 kb fragment containing the apramycin resistance gene (*aac(3)IV*), amplified from pHY773, using the λ-Red-mediated recombination. The resulting plasmid pYD3 was then conjugated into a *Streptomyces* host via passage through nonmethylating *E. coli* ET12567/pUZ8002 using standard protocols

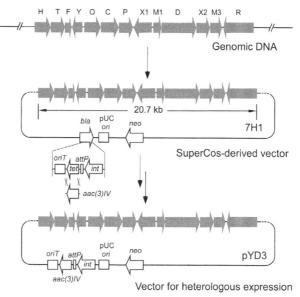

Fig. 4 Strategy of constructing a vector for heterologous expression of cladoniamide gene cluster.

(Kieser, Bibb, Buttner, Chater, & Hopwood, 2000). Exconjugates were grown in liquid medium (such as tryptic soy broth (TSB)) for isolation of genomic DNA. PCR was performed to verify the successful integration of pYD3 into the chromosome of the host strain, with primers specifically targeting bisindole biosynthetic genes. Two to three verified exconjugates were then used to analyze for metabolite production. As a negative control, pIJ787 was similarly retrofitted then conjugated into the heterologous host.

If metabolite production is not observed following such standard heterologous expression techniques, additional methods are available. For example, transcriptional upregulation of gene clusters can be accomplished in *Streptomyces* hosts by overexpression of the pathway-specific positive regulator. For example, following heterologous expression of the borregomycin gene cluster only very low levels of metabolite were observed (Chang & Brady, 2013). To improve metabolite production the transcriptional regulator from the *bor* cluster, *borR* was separately cloned under control of the constitutive, strong *ermE** promoter. Cointroduction of this plasmid with the *bor* cluster resulted in significantly improved production of clone-specific metabolites and led to the isolation of borregomycins. Alternatively, promoter engineering can be applied. Transformation-associated recombination was used to replace the natural promoter regions of the gene cluster of lazarimide with synthetic promoters (Montiel et al., 2015). Combined with replacement of a gene thought to be inactive (*lxrX1*) with a homolog (*abeX1*) from the BE-54017 pathway, successful production of lazarimide was observed in the heterologous host. In another case, isolation of erdasporine was not possible due to its rapid degradation under the conditions required for bacterial growth (Chang et al., 2013). To overcome this problem, each individual gene from the cluster was provided with its own T7-promoter, and all the genes, except the final cytochrome P450 *espX*, were expressed in *E. coli*. The intermediate produced could then be fed to *E. coli* expressing only *espX* in a more time-controlled manner in order to successfully isolate erdasporine. Similarly, heterologous expression of the reductasporine gene cluster in both *E. coli* and *Streptomyces* hosts showed no clone-specific metabolites (Chang et al., 2015). In this case, it was believed to be due to improper protein folding of the core bisindole biosynthetic genes *redODP*. The problem was overcome by swapping the inactive enzymes for homologous enzymes from the erdasporine pathway (*erdODP*), which had previously been shown to be active in *E. coli*. This approach resulted in successful production of reductasporine.

4. MUTATIONAL BIOSYNTHESIS TO GENERATE NEW BISINDOLES

With an efficient producer of bisindoles in hand, targeted gene inactivation can be carried out to generate strains to accumulate pathway intermediates or shunt products, which not only diversify bisindole structures but also give insight into the biosynthetic pathways of new bisindoles. A straightforward method is to introduce a premodified gene cluster into the chassis host. Generation of a premodified gene cluster can be easily achieved by targeted inactivation of gene(s) using *E. coli*-based λ-Red-mediated recombination (Gust, Challis, Fowler, Kieser, & Chater, 2003). However, newer techniques, such as the clustered regularly interspaced short palindromic repeats (CRISPR)/CRISPR-associated (Cas) protein (CRISPR/Cas) system (Tong, Charusanti, Zhang, Weber, & Lee, 2015) or antisense RNA-mediated gene silencing (Uguru et al., 2013), could also be employed to perform genome editing or mediate gene expression levels in the heterologous host carrying the intact bisindole gene cluster.

For generation of modified *cla* gene clusters, cosmid pYD1 was used for targeted gene inactivation. Gene disruption was carried out by insertion of an *aac(3)IV* gene into the coding region of the *cla* gene, using λ-Red-mediated recombination, which is a minor adaption from that described by Gust et al. (2003). Specifically, vector pYD1 was first introduced into *E. coli* BW25113/pIJ790 (note: pIJ790 carries the λ *red* genes). A PCR-amplified *aac(3)IV* disruption cassette (from pHY773) containing 39-nt extensions for λ-Red-mediated recombination was then introduced into *E. coli* BW25113/pIJ790/pYD1 to disrupt a specific *cla* gene. In the case where in-frame deletion is required for avoiding a polar effect on downstream genes, the *aac(3)IV* gene of targeted pYD1 (eg, pYD1(Δ*claM1::aac(3)IV*)) is deleted by transforming the vector into *E. coli* DH5α/BT340, which harbors enzymes for FLP-mediated recombination (Fig. 5). A series of engineered strains were then generated by separately introducing modified *cla* gene clusters into a chassis strain by intergeneric conjugation. Comparative metabolic profiling by HPLC was carried out to identify strain-specific metabolites.

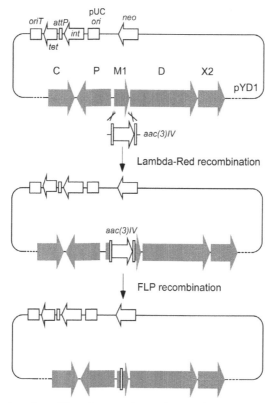

Fig. 5 Construction of modified *cla* gene clusters by λ-Red- and FLP-mediated targeted gene inactivation.

5. MIXING GENES FROM PHYLOGENETICALLY RELATED CLUSTERS TO GENERATE NEW BISINDOLES

To further diversify bisindoles, components from phylogenetically related biosynthetic gene clusters were introduced to the above engineered strains to mimic the evolution of naturally occurring gene clusters. In order to transfer gene(s) into these strains, either replicative or integrative vectors with different selection markers can be used to host the targeted gene(s). The expression levels of these genes can be further tuned by gene copy number and the strength of promoters. In *Streptomyces*, libraries of both natural and synthetic promoters with different strengths have been reported (Bai et al., 2015; Luo, Zhang, Barton, & Zhao, 2015; Seghezzi, Amar, Koebmann, Jensen, & Virolle, 2011; Siegl, Tokovenko, Myronovskyi, & Luzhetskyy, 2013).

In our lab a modified pIJ8600 plasmid is used that confers thiostrepton resistance as well as containing the *oriT* site for replication and ΦC31 *int* gene and *attP* site (Bibb, Kelemen, Fernández-Abalos, & Sun, 1999). This plasmid also contains an *ermE** promoter, which improves expression of the desired genes. The genes of interest are amplified either from genomic DNA or from previously synthesized plasmids by PCR, then appropriately digested. Ligation into the corresponding sites of the plasmid is confirmed by sequencing before conjugation. The heterologous genes can be added to the strains containing the full gene cluster, for example, to incorporate rebeccamycin glucosyltransferase gene *rebG* into the complete cladoniamide gene cluster. Heterologous genes can also be added to strains containing the gene cluster in which enzymes have been disrupted to produce a wider variety of metabolites. A successful example of this is in the incorporation of *staC* and *staP* into *S. albus* containing the cladoniamide gene cluster that lacks functional *claC* and *claP* genes (Du & Ryan, 2015).

6. CHEMICAL ISOLATION AND STRUCTURAL CHARACTERIZATION

6.1 Strain Growth and Product Extraction

Streptomyces strains can be cultured in a baffled flask, or with a spring, to increase aeration. Starter cultures (25 mL) are initially grown from spore stocks for 12–18 h at 30°C, 200 RPM, in the nutrient-rich TSB (Kieser et al., 2000), in order to obtain uniform growth across all flasks. This starter culture is then used to inoculate 200 mL modified R5 medium (Kieser et al., 2000), which is incubated at 30°C, 200 RPM for 12 days. Both culture medium and incubation time may require optimization for observation of different metabolites. Following fermentation, the cultures are extracted twice with one volume of ethyl acetate, the combined extracts are dried under vacuum, and the crude extract is redissolved in methanol. Depending on the target compound, it is sometimes useful to acidify or alkalize the cultures prior to extraction.

6.2 Analysis and Purification

The crude samples are analyzed by comparative HPLC metabolic profiling with appropriate negative controls (Fig. 3). HPLC is typically carried out on a C18 column with a linear gradient of water and acetonitrile, both containing 0.05% (v/v) trifluoroacetic acid. Due to the presence of the indole heterocycle in the metabolites, absorption can be monitored at 280 nm. Following identification of initial peaks of interest, LC–MS is used to identify the molecular weight of the new metabolites using both positive and negative ion electrospray. Once a new metabolite is identified, a larger-scale cultivation and extraction (3–6 L) is carried out. The crude material obtained is fractionated first on a reversed-phase C8 flash column with a methanol/water gradient elution and then on Sephadex LH-20 with methanol elution. Fractions containing metabolites of interest are further purified using semipreparative HPLC on a C18 column with an isocratic elution of acetonitrile:water, which is commonly between 30:70 and 70:30 (v:v) acetonitrile:water for cladoniamide analogs.

6.3 Metabolite Characterization

The structures of the purified metabolites are confirmed by 1D and 2D NMR, commonly ^{1}H, ^{13}C, COSY, HMQC, and HMBC. Usually two isolated spin systems are observed for each substituted benzene ring, and HMBC correlations can be used to connect these systems with the "upper" ring of bisindoles. In cases where similar structures have been identified, comparative NMR can be used to assign chemical shifts. For example, the sugar in glucosyl-3-chloroarcyriaflavin A (Fig. 2) (Du & Ryan, 2015) can be assigned based on similar sugars in rebeccamycin (Sánchez et al., 2002). In some cases, such as for xenocladoniamide F, the lack of protons surrounding the central core of the bisindoles mean that crystallographic data are necessary in order to unambiguously establish chemical structure (Du, Ding, Patrick, & Ryan, 2013).

7. CONCLUSIONS

The methods described here enable carrying out the genetic manipulations necessary to introduce genes of interest into heterologous expression hosts. This approach has been particularly successful for bisindoles because of the close relationships between the biosynthetic pathways and the relatively broad substrate selectivity for some of the enzymes. This work has allowed for manipulation of halide and glucose substituents, as well as bisindoles with modified core structures. With the recent identification of more bisindole natural products, including borregomycin, reductasporine, erdasporine, and lazaramide, along with their gene clusters, the future scope of this combinatorial synthesis method is expanding.

While in many cases combinatorial biosynthesis leads to novel and interesting bisindoles, this approach is limited by the inherent specificity of individual enzymes, and not all combinations of genes will lead to the expected products. A major future challenge in this area is in rational modification of enzymes in order to allow them to accept alternative substrates. Such protein engineering work would pave the way for predesigning combinations of genes in order to produce whichever microbial bisindole is desired.

REFERENCES

Asamizu, S., Hirano, S., Onaka, H., Koshino, H., Shiro, Y., & Nagano, S. (2012). Coupling reaction of indolepyruvic acid by StaD and its product: Implications for biosynthesis of indolocarbazole and violacein. *ChemBioChem, 13,* 2495–2500.

Bai, C., Zhang, Y., Zhao, X., Hu, Y., Xiang, S., Miao, J., et al. (2015). Exploiting a precise design of universal synthetic modular regulatory elements to unlock the microbial natural products in *Streptomyces*. *Proceedings of the National Academy of Sciences of the United States of America, 112,* 12181–12186.

Baltz, R. H. (2010). *Streptomyces* and *Saccharopolyspora* hosts for heterologous expression of secondary metabolite gene clusters. *Journal of Industrial Microbiology & Biotechnology, 37,* 759–772.

Bharate, S. B., Sawant, S. D., Singh, P. P., & Vishwakarma, R. A. (2013). Kinase inhibitors of marine origin. *Chemical Reviews, 113,* 6761–6815.

Bibb, M. J., Kelemen, G. H., Fernández-Abalos, J. M., & Sun, J. (1999). Green fluorescent protein as a reporter for spatial and temporal gene expression in *Streptomyces coelicolor* A3(2). *Microbiology, 145,* 2221–2227.

Chang, F.-Y., & Brady, S. F. (2011). Cloning and characterization of an environmental DNA-derived gene cluster that encodes the biosynthesis of the antitumor substance BE-54017. *Journal of the American Chemical Society, 133,* 9996–9999.

Chang, F.-Y., & Brady, S. F. (2013). Discovery of indolotryptoline antiproliferative agents by homology-guided metagenomic screening. *Proceedings of the National Academy of Sciences of the United States of America, 110,* 2478–2483.

Chang, F.-Y., & Brady, S. F. (2014). Characterization of an environmental DNA-derived gene cluster that encodes the bisindolylmaleimide methylarcyriarubin. *ChemBioChem, 15*, 815–821.

Chang, F.-Y., Kawashima, S. A., & Brady, S. F. (2014). Mutations in the proteolipid subunits of the vacuolar H^+-ATPase provide resistance to indolotryptoline natural products. *Biochemistry, 53*, 7123–7131.

Chang, F.-Y., Ternei, M. A., Calle, P. Y., & Brady, S. F. (2013). Discovery and synthetic refactoring of tryptophan dimer gene clusters from the environment. *Journal of the American Chemical Society, 135*, 17906–17912.

Chang, F.-Y., Ternei, M. A., Calle, P. Y., & Brady, S. F. (2015). Targeted metagenomics: Finding rare tryptophan dimer natural products in the environment. *Journal of the American Chemical Society, 137*, 6044–6052.

Cuenca-López, M. D., Serrano-Heras, G., Montero, J. C., Corrales-Sánchez, V., Gomez-Juarez, M., Gascón-Escribano, M. J., et al. (2015). Antitumor activity of the novel multi-kinase inhibitor EC-70124 in triple negative breast cancer. *Oncotarget, 6*, 27923–27937.

Deng, X., Duffy, S. P., Myrand-Lapierre, M.-E., Matthews, K., Santoso, A. T., Du, Y.-L., et al. (2015). Reduced deformability of parasitized red blood cells as a biomarker for anti-malarial drug efficacy. *Malaria Journal, 14*, 428.

Du, Y.-L., Ding, T., Patrick, B. O., & Ryan, K. S. (2013a). Xenocladoniamide F, minimal indolotryptoline from the cladoniamide pathway. *Tetrahedron Letters, 54*, 5635–5638.

Du, Y.-L., Ding, T., & Ryan, K. S. (2013b). Biosynthetic O-methylation protects cladoniamides from self-destruction. *Organic Letters, 15*, 2538–2541.

Du, Y.-L., & Ryan, K. S. (2015). Expansion of bisindole biosynthetic pathways by combinatorial construction. *ACS Synthetic Biology, 4*, 682–688.

Du, Y.-L., & Ryan, K. S. (2016). Catalytic repertoire of bacterial bisindole formation. *Current Opinion in Chemical Biology, 31*, 74–81.

Du, Y.-L., Williams, D. E., Patrick, B. O., Andersen, R. J., & Ryan, K. S. (2014). Reconstruction of cladoniamide biosynthesis reveals nonenzymatic routes to bisindole diversity. *ACS Chemical Biology, 9*, 2748–2754.

Eustáquio, A. S., Gust, B., Galm, U., Li, S.-M., Chater, K. F., & Heide, L. (2005). Heterologous expression of novobiocin and clorobiocin biosynthetic gene clusters. *Applied and Environmental Microbiology, 71*, 2452–2459.

Fernandez, P., Saint-Joanis, B., Barilone, N., Jackson, M., Gicquel, B., Cole, S. T., et al. (2006). The Ser/Thr protein kinase PknB is essential for sustaining mycobacterial growth. *Journal of Bacteriology, 188*, 7778–7784.

Fu, J., Bian, X., Hu, S., Wang, H., Huang, F., Seibert, P. M., et al. (2012). Full-length RecE enhances linear-linear homologous recombination and facilitates direct cloning for bioprospecting. *Nature Biotechnology, 30*, 440–446.

Goldman, P. J., Ryan, K. S., Hamill, M. J., Howard-Jones, A. R., Walsh, C. T., Elliott, S. J., et al. (2012). An unusual role for a mobile flavin in StaC-like indolocarbazole biosynthetic enzymes. *Chemistry & Biology, 19*, 855–865.

Gomez-Escribano, J. P., & Bibb, M. J. (2011). Engineering *Streptomyces coelicolor* for heterologous expression of secondary metabolite gene clusters. *Microbial Biotechnology, 4*, 207–215.

Gomez-Escribano, J. P., & Bibb, M. J. (2012). *Streptomyces coelicolor* as an expression host for heterologous gene clusters. *Methods in Enzymology, 517*, 279–300.

Gomez-Escribano, J. P., Song, L., Fox, D. J., Yeo, V., Bibb, M. J., & Challis, G. L. (2012). Structure and biosynthesis of the unusual polyketide alkaloid coelimycin P1, a metabolic product of the *cpk* gene cluster of *Streptomyces coelicolor* M145. *Chemical Science, 3*, 2716.

Gust, B., Challis, G. L., Fowler, K., Kieser, T., & Chater, K. F. (2003). PCR-targeted *Streptomyces* gene replacement identifies a protein domain needed for biosynthesis of the sesquiterpene soil odor geosmin. *Proceedings of the National Academy of Sciences of the United States of America, 100*, 1541–1546.

Howard-Jones, A. R., & Walsh, C. T. (2006). Staurosporine and rebeccamycin aglycones are assembled by the oxidative action of StaP, StaC, and RebC on chromopyrrolic acid. *Journal of the American Chemical Society, 128*, 12289–12298.

Howard-Jones, A. R., & Walsh, C. T. (2007). Nonenzymatic oxidative steps accompanying action of the cytochrome P450 enzymes StaP and RebP in the biosynthesis of staurosporine and rebeccamycin. *Journal of the American Chemical Society, 129*, 11016–11017.

Ishikawa, J., & Hotta, K. (1999). FramePlot: A new implementation of the frame analysis for predicting protein-coding regions in bacterial DNA with a high G+C content. *FEMS Microbiology Letters, 174*, 251–253.

Kieser, T., Bibb, M., Buttner, M., Chater, K., & Hopwood, D. (2000). *Practical streptomyces genetics*. Norwich: John Innes Foundation.

Kim, J. H., Feng, Z., Bauer, J. D., Kallifidas, D., Calle, P. Y., & Brady, S. F. (2010). Cloning large natural product gene clusters from the environment: Piecing environmental DNA gene clusters back together with TAR. *Biopolymers, 93*, 833–844.

Kimura, T., Kanagaki, S., Matsui, Y., Imoto, M., Watanabe, T., & Shibasaki, M. (2012). Synthesis and assignment of the absolute configuration of indenotryptoline bisindole alkaloid BE-54017. *Organic Letters, 14*, 4418–4421.

Kouprina, N., & Larionov, V. (2008). Selective isolation of genomic loci from complex genomes by transformation-associated recombination cloning in the yeast *Saccharomyces cerevisiae*. *Nature Protocols, 3*, 371–377.

Luo, Y., Zhang, L., Barton, K. W., & Zhao, H. (2015). Systematic identification of a panel of strong constitutive promoters from *Streptomyces albus*. *ACS Synthetic Biology, 4*, 1001–1010.

Makino, M., Sugimoto, H., Shiro, Y., Asamizu, S., Onaka, H., & Nagano, S. (2007). Crystal structures and catalytic mechanism of cytochrome P450 StaP that produces the indolocarbazole skeleton. *Proceedings of the National Academy of Sciences of the United States of America, 104*, 11591–11596.

Marschall, M., Stein-Gerlach, M., Freitag, M., Kupfer, R., van den Bogaard, M., & Stamminger, T. (2002). Direct targeting of human cytomegalovirus protein kinase pUL97 by kinase inhibitors is a novel principle for antiviral therapy. *The Journal of General Virology, 83*, 1013–1023.

Montiel, D., Kang, H.-S., Chang, F.-Y., Charlop-Powers, Z., & Brady, S. F. (2015). Yeast homologous recombination-based promoter engineering for the activation of silent natural product biosynthetic gene clusters. *Proceedings of the National Academy of Sciences of the United States of America, 112*, 8953–8958.

Nishizawa, T., Aldrich, C. C., & Sherman, D. H. (2005). Molecular analysis of the rebeccamycin L-amino acid oxidase from *Lechevalieria aerocolonigenes* ATCC 39243. *Journal of Bacteriology, 187*, 2084–2092.

Onaka, H., Taniguchi, S., Igarashi, Y., & Furumai, T. (2002). Cloning of the staurosporine biosynthetic gene cluster from *Streptomyces* sp. TP-A0274 and its heterologous expression in *Streptomyces lividans*. *The Journal of Antibiotics, 55*, 1063–1071.

Onaka, H., Taniguchi, S., Igarashi, Y., & Furumai, T. (2003). Characterization of the biosynthetic gene cluster of rebeccamycin from *Lechevalieria aerocolonigenes* ATCC 39243. *Bioscience, Biotechnology, and Biochemistry, 67*, 127–138.

Ryan, K. S. (2011). Biosynthetic gene cluster for the cladoniamides, bis-indoles with a rearranged scaffold. *PLoS One, 6*, e23694.

Ryan, K. S., & Drennan, C. L. (2009). Divergent pathways in the biosynthesis of bisindole natural products. *Chemistry & Biology, 16*, 351–364.

Ryan, K. S., Howard-Jones, A. R., Hamill, M. J., Elliott, S. J., Walsh, C. T., & Drennan, C. L. (2007). Crystallographic trapping in the rebeccamycin biosynthetic enzyme RebC. *Proceedings of the National Academy of Sciences of the United States of America, 104*, 15311–15316.

Salas, A. P., Zhu, L., Sánchez, C., Braña, A. F., Rohr, J., Méndez, C., et al. (2005). Deciphering the late steps in the biosynthesis of the anti-tumour indolocarbazole staurosporine: Sugar donor substrate flexibility of the StaG glycosyltransferase. *Molecular Microbiology, 58*, 17–27.

Sánchez, C., Butovich, I. A., Braña, A. F., Rohr, J., Méndez, C., & Salas, J. A. (2002). The biosynthetic gene cluster for the antitumor rebeccamycin. *Chemistry & Biology, 9*, 519–531.

Sánchez, C., Méndez, C., & Salas, J. A. (2006). Engineering biosynthetic pathways to generate antitumor indolocarbazole derivatives. *Journal of Industrial Microbiology and Biotechnology, 33*, 560–568.

Sánchez, C., Salas, A. P., Braña, A. F., Palomino, M., Pineda-Lucena, A., Carbajo, R. J., et al. (2009). Generation of potent and selective kinase inhibitors by combinatorial biosynthesis of glycosylated indolocarbazoles. *Chemical Communications*, 4118.

Sánchez, C., Zhu, L., Braña, A. F., Salas, A. P., Rohr, J., Méndez, C., et al. (2005). Combinatorial biosynthesis of antitumor indolocarbazole compounds. *Proceedings of the National Academy of Sciences of the United States of America, 102*, 461–466.

Schwandt, A., Mekhail, T., Halmos, B., O'Brien, T., Ma, P. C., Fu, P., et al. (2012). Phase-II trial of rebeccamycin analog, a dual topoisomerase-I and -II inhibitor, in relapsed "sensitive" small cell lung cancer. *Journal of Thoracic Oncology, 7*, 751–754.

Seghezzi, N., Amar, P., Koebmann, B., Jensen, P. R., & Virolle, M.-J. (2011). The construction of a library of synthetic promoters revealed some specific features of strong *Streptomyces* promoters. *Applied Microbiology and Biotechnology, 90*, 615–623.

Sherer, C., & Snape, T. J. (2015). Heterocyclic scaffolds as promising anticancer agents against tumours of the central nervous system: Exploring the scope of indole and carbazole derivatives. *European Journal of Medicinal Chemistry, 97*, 552–560.

Siegl, T., Tokovenko, B., Myronovskyi, M., & Luzhetskyy, A. (2013). Design, construction and characterisation of a synthetic promoter library for fine-tuned gene expression in actinomycetes. *Metabolic Engineering, 19*, 98–106.

Temme, K., Zhao, D., & Voigt, C. A. (2012). Refactoring the nitrogen fixation gene cluster from *Klebsiella oxytoca*. *Proceedings of the National Academy of Sciences of the United States of America, 109*, 7085–7090.

Tong, Y., Charusanti, P., Zhang, L., Weber, T., & Lee, S. Y. (2015). CRISPR-Cas9 based engineering of actinomycetal genomes. *ACS Synthetic Biology, 4*, 1020–1029.

Uguru, G. C., Mondhe, M., Goh, S., Hesketh, A., Bibb, M. J., Good, L., et al. (2013). Synthetic RNA silencing of actinorhodin biosynthesis in *Streptomyces coelicolor* A3(2). *PLoS One, 8*, e67509.

Williams, D. E., Davies, J., Patrick, B. O., Bottriell, H., Tarling, T., Roberge, M., et al. (2008). Cladoniamides A-G, tryptophan-derived alkaloids produced in culture by *Streptomyces uncialis*. *Organic Letters, 10*, 3501–3504.

Yamanaka, K., Reynolds, K. A., Kersten, R. D., Ryan, K. S., Gonzalez, D. J., Nizet, V., et al. (2014). Direct cloning and refactoring of a silent lipopeptide biosynthetic gene cluster yields the antibiotic taromycin A. *Proceedings of the National Academy of Sciences of the United States of America, 111*, 1957–1962.

Yeh, E., Garneau, S., & Walsh, C. T. (2005). Robust in vitro activity of RebF and RebH, a two-component reductase/halogenase, generating 7-chlorotryptophan during rebeccamycin biosynthesis. *Proceedings of the National Academy of Sciences of the United States of America, 102*, 3960–3965.

Zhang, C., Albermann, C., Fu, X., Peters, N. R., Chisholm, J. D., Zhang, G., et al. (2006). RebG- and RebM-catalyzed indolocarbazole diversification. *ChemBioChem, 7*, 795–804.

Zhang, W., Liu, Z., Li, S., Yang, T., Zhang, Q., Ma, L., et al. (2012). Spiroindimicins A–D: New bisindole alkaloids from a deep-sea-derived actinomycete. *Organic Letters, 14*, 3364–3367.

CHAPTER THREE

Enzymatic [4+2] Cycloadditions in the Biosynthesis of Spirotetramates and Spirotetronates

B. Pang*, G. Zhong*, Z. Tang*, W. Liu*,[†],[1]

*State Key Laboratory of Bioorganic and Natural Products Chemistry, Shanghai Institute of Organic Chemistry, Chinese Academy of Sciences, Shanghai, PR China
[†]Huzhou Center of Bio-Synthetic Innovation, Huzhou, PR China
[1]Corresponding author: e-mail address: wliu@mail.sioc.ac.cn

Contents

1. Introduction 40
2. Strategy 44
   2.1 Prediction of the Presence of [4+2] Cycloaddition Reactions in the PYR Biosynthetic Pathway 44
   2.2 Identification of the Candidates Coding for Cascade [4+2] Cycloadditions from the PYR Biosynthetic Gene Cluster 46
3. Methods 48
   3.1 In Vivo Validation of the Involvement of *pyrE3* and *pyrI4* in PYR Biosynthesis 48
   3.2 In Vitro Determination of the Functions of PyrE3 and PyrI4 for Pentacyclic Core Formation 50
   3.3 In Vivo and In Vitro Mechanistic Evaluation of the Generality of Pentacyclic Core Formation in the CHL Biosynthetic Pathway 54
   3.4 Examination of the Protein Natures of PyrE3 and ChlE3 57
4. Discussion and Perspectives 58
References 61

Abstract

The Diels–Alder reaction is a quintessential type of [4+2] cycloaddition that remains one of the most intriguing transformations in synthetic chemistry. This reaction has long been envisaged to participate in the biosynthesis of a number of cyclohexene-containing natural products, although the question of whether a bona fide Diels-Alderase exists remains unsolved. In nature, there are remarkably few enzymes known to have the activity of [4+2] cycloaddition. These enzymes are phylogenetically distinct and are often classified according to the specific chemical structures. The variation of protein ancestors and in many cases the instability/complexity of the substrates and products pose a significant challenge in identification of the [4+2] cycloaddition catalysts using general homology-based mining approaches. We here provide the detailed description of the multiple comparison-based strategy and methods for the characterization of two distinct types of dedicated [4+2] cyclases (eg, PyrE3 and PyrI4) in the biosynthesis of spirotetramates and spirotetronates, where they act in tandem for coordinated cross-bridging of a linear polyene intermediate into a enantiomerically pure pentacyclic core. The search of new protein scaffolds with the [4+2] cycloaddition activity could enrich the pool of the candidates for mechanistic examination of a true enzymatic Diels–Alder reaction. The protocols presented in this study would also be applicable to the study of other functionally similar but phylogenetically different proteins, eg, the spiroketal cyclases.

1. INTRODUCTION

The Diels–Alder reaction, which is considered one of the most important reactions for carbon–carbon bond formation, represents a quintessential type of [4+2] cycloaddition that occurs between a 1,3-diene and an alkene to form a cyclohexene through a single pericyclic transition state (Corey, 2002; Nicolaou, Snyder, Montagnon, & Vassilikogiannakis, 2002). Due to its central role in the theory and practice of organic chemistry, the development of relevant synthetic methodology, including the artificial design of biological Diels–Alder catalysts, has long been the focus of extensive efforts. However, few enzymes are known to perform [4+2] cycloadditions in nature (Kelly, 2008; Kim, Ruszczycky, & Liu, 2012). Because most of these enzymes possess multiple functions, lack catalytic efficiency, and/or participate in reactions that occur nonenzymatically, validation of the existence of a bona fide Diels-Alderase has proven challenging. This situation could change due to the recent emergence of monofunctional proteins, which effectively perform [4+2] cyclization reactions in a completely enzyme-dependent manner (Hashimoto et al., 2015; Tian et al., 2015), as exemplified by PyrE3 and PyrI4 in the biosynthetic pathway of cyclohexene-containing pyrroindomycins (PYRs).

PYRs were isolated from *Streptomyces rugosporus* during a screening for agents active against bacterial pathogens, such as methicillin-resistant *Staphylococcus aureus* and vancomycin-resistant *Enterococcus faecium* (Ding et al., 1994; Singh et al., 1994). These molecules represent the first spirotetramate products found in nature; they feature a pentacyclic core that contains two cyclohexene units present in both the dialkyldecalin system and the tetramate spiroconjugate portion. The process that synthesizes this enantiomerically pure pentacyclic core remained unclear until recently, an unprecedented enzymatic [4+2] cyclization cascade was discovered in the PYR biosynthetic pathway (Tian et al., 2015). Beginning with a linear polyene intermediate that bears two pairs of 1,3-diene and alkene groups, two dedicated proteins (ie, PyrE3 and PyrI4) act in tandem to catalyze the formation of the two cyclohexene units through [4+2] cycloadditions (Fig. 1). PyrE3 is a FAD-dependent enzyme responsible for *endo*-selective dialkyldecalin formation, and PyrI4, which shares no sequence similarity to any protein of known function, acts as an *exo*-selective spiroconjugate synthase. The two associated cyclization reactions occur in a regio- and stereoselective manner and proceed with catalytic efficiencies sufficient for a cross-bridging cascade. Remarkably, spirotetronate natural products that are structurally related to PYRs (eg, chlorothricin (CHL) and versipelostatin (VST)) appear to share this cross-bridging biosynthetic strategy (Hashimoto et al., 2015; Jia et al., 2006; Tian et al., 2015). The formation of a similar pentacyclic scaffold, which is common to these spirotetronate molecules, most likely relies on the two highly ordered [4+2] cycloadditions to first generate dialkyldecalin and then spiroconjugate despite differences in their linear polyene substrates, which vary in length,

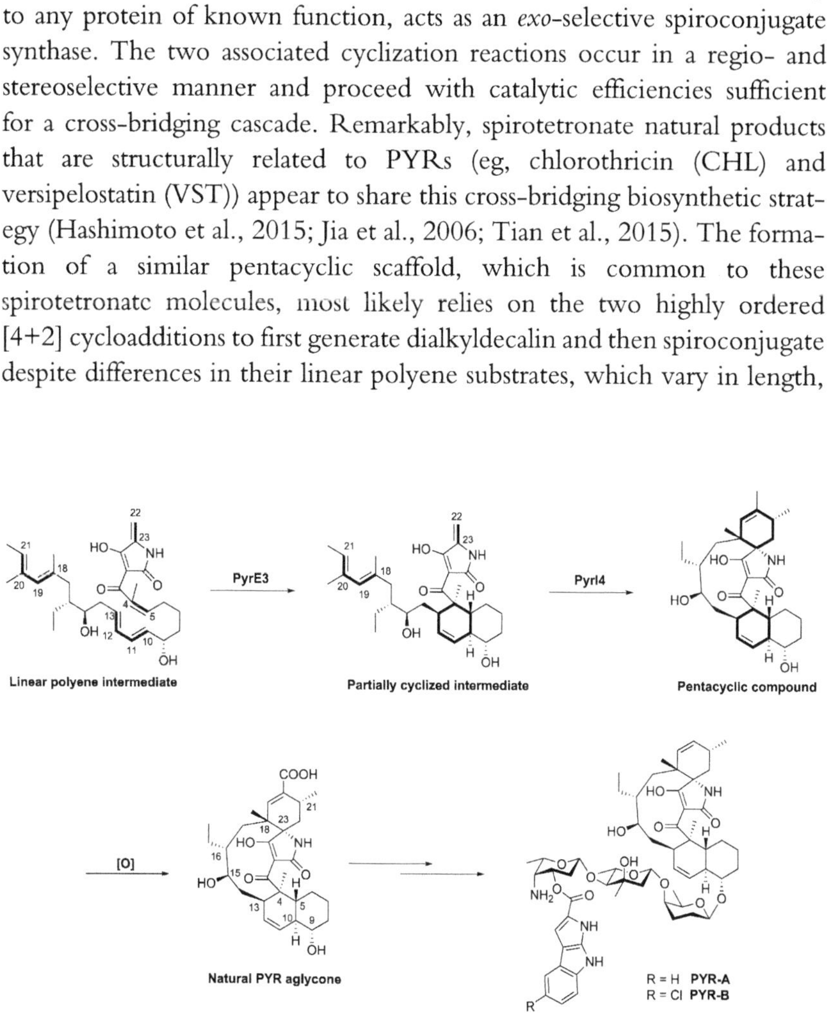

Fig. 1 The biosynthetic pathway of PYRs, in which the cross-bridging of the linear polyene intermediate to afford the pentacyclic core involves two tandem enzymatic [4+2] cycloaddition reactions. PyrE3-catalyzed dialkyldecalin formation and PyrI4-catalyzed spiroconjugate formation are shown in *bold*.

substitution pattern, and terminal heterocyclic functionality (containing tetronate rather than tetramate).

Notably, all types of the enzymes known to have [4+2] cycloaddition activity thus far are phylogenetically distinct and are classified according to the specific chemical structures of associated natural products (Fig. 2). For example, PyrE3-represented dialkyldecalin synthases are specific to the pathways of spirotetramates and spirotetronates and share no sequence homology with the lovastatin synthase LovB (Auclair et al., 2000; Ma et al., 2009), solanapyrone synthase Sol5 (Oikawa, Katayama, Suzuki, & Ichihara, 1995), or the newly identified Fsa2-like proteins in the biosynthesis of the fungal tetramic acids equisetin, pyrrolocin and Sch 210972 (Kakule et al., 2015; Kato et al., 2015; Sato et al., 2015), even though their products contain a similar 6,6′-bicyclic decalin system (Fig. 2). Each of these enzymes likely functions as a template through interactions with a specialized substrate, and the resulting proximity and polarization effects of the reactive groups (1,3-diene and alkene) would dominate the stereoselectivity of the product and accelerate the transformation rate. However, the variation of protein ancestors in evolution hampers the identification of potential candidates for [4+2] cycloaddition in the biosynthetic pathways of structurally unrelated cyclohexene-containing natural products using general homology-based mining approaches. Many of these cyclization reactions occur at late biosynthetic stages, and the instability and/or complexity of the substrates and products often pose considerable challenges to biochemical validation of the activities of the associated enzymes.

In this chapter, we highlight a strategy for the prediction and identification of PyrE3 and PyrI4, which are candidates for catalyzing [4+2] cycloaddition cascade in the biosynthetic pathway of spirotetramates PYRs. We provide detailed descriptions of the methods and protocols for the in vivo and in vitro characterization of both enzymes, and the methods and protocols of the in vivo and vitro examining the functional exchangeability of the homologs ChlE3 (to PyrE3) and ChlL (to PyrI4) from the biosynthetic pathway of CHL are described to validate the generality in formation of pentacyclic cores of structurally related spirotetronates. These strategy and methods are helpful for the discovery and characterization of new enzymes that are derived from unknown ancestors for Diels–Alder-like [4+2] cycloaddition activity. Furthermore, they can be applied to investigations of other functionally similar but phylogenetically different proteins as well, eg, the cyclases for spiroketalization (Sun et al., 2013; Takahashi et al., 2011).

Biosynthesis of Spirotetramates and Spirotetronates

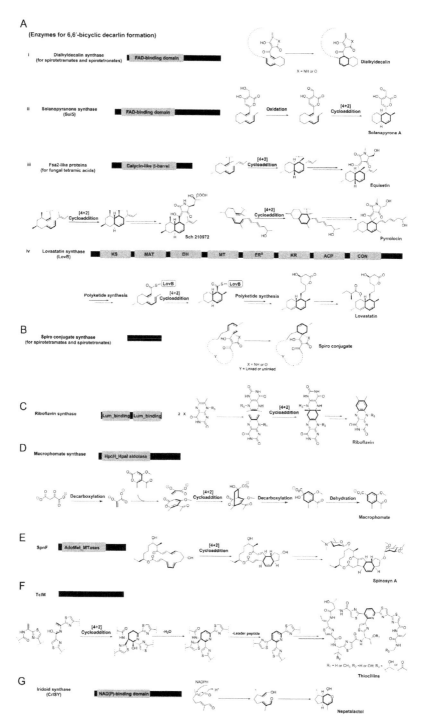

Fig. 2 See legend on next page.

2. STRATEGY
2.1 Prediction of the Presence of [4+2] Cycloaddition Reactions in the PYR Biosynthetic Pathway

In the biosynthetic pathway of PYRs, a modular polyketide synthase (PKS)/nonribosomal peptide synthetase (NRPS) hybrid system programs the assembly of the aglycone skeleton in a linear manner (Fig. 3) (Wu, Wu, Qu, & Liu, 2012). The PKSs are organized into typical modules to template short carboxylate polymerization and modification, assumedly following an assembly logic that is colinear with the functional domains of each module, which consists of the indispensable ketosynthase (KS), acyltransferase (AT), and acyl carrier protein (ACP), and the optional dehydratase (DH), enoylreductase (ER), and ketoreductase (KR). Given the presence of at least five modules sharing the KS-AT-DH-KR-ACP domain organization, the resulting ACP-tethered intermediate could be a polyene polyketide that has a minimum of five double bonds. Elongation of the carbon chain may be terminated by a NRPS module, which contains condensation (C), adenylation (A), and peptidyl carrier protein (PCP) domains for the incorporation of an amino acid. The full-length linear intermediate off-loads

Fig. 2 Known enzymes with the (putative) [4+2] cycloaddition activity and associated conversions in the pathways. The size of the enzymes and their domain organizations are compared here and shown in *black* and *gray*. For the reactions, the overall structures of the alkene- and 1,3-diene-containing precursors and the cyclohexene-containing products are shown in bold. The *dashed lines* indicate the variable side chains. (A) Enzymes for 6,6′-bicyclic decalin system formation. This is no significant sequence homology found between dialkyldecalin synthases (i) and solanapyrone synthase Sol5 (ii), even though they share the FAD-binding ability. The Fsa2-like proteins (iii) bear a calycin-like β-barrel scaffold. For the domains of LovB (iv), *KS*, ketosynthase; *MAT*, malonyl-CoA acyltransferase; *DH*, dehydratase; *MT*, methyltransferase; ER^0, inactive enoylreductase; *KR*, ketoreductase; *ACP*, acyl carrier protein, and *CON*, condensation protein. (B) Spiroconjugate synthase. The variable side chains can be coupled for intramolecular cyclization or uncoupled for intermolecular cyclization. (C) Riboflavin synthase, which contains two domains for binding of the lumazine substrate (Kim et al., 2010). (D) Macrophomate synthase, which phylogenetically belongs to the aldolase family (Ose et al., 2003). (E) SpnF, which shares sequence similarity with various S-adenosyl-L-methionine (AdoMet)-dependent methyltransferases (MTases) (Kim, Ruszczycky, Choi, Liu, & Liu, 2011). (F) TclM, the enzyme that is responsible for the formation of the six-membered aza-heterocycle in thiocillin biosynthesis (Wever et al., 2015). (G) Iridoid synthase, a homolog of progesterone 5β-reductase (P5βR) that catalyzes a NADPH-dependent, reduction-coupled oza-heterocyclization (Kries et al., 2016).

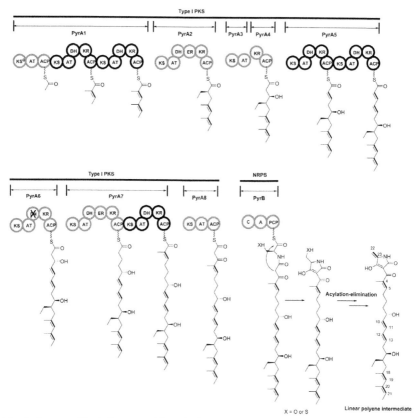

Fig. 3 Proposed logic for assembling the linear polyene intermediate in the PYR biosynthetic pathway. The modular type I PKS system (PyrA1–A8) consists of five functional modules sharing the KS-AT-DH-KR-ACP domain organization (shown in *bold*), corresponding to the five *trans*-double bonds formed in the linear intermediate. It should be noted that PyrA6 also has the same domain organization, in which, however, the DH domain (labeled by a *cross mark*) is likely redundant. These five double bonds, along with the exocyclic double bond (shown in *bold*) formed during the tailoring process, constitute two independent pairs of 1,3-diene and alkene groups for intramolecular [4+2] cycloaddition reactions.

from the assembly line likely through a Dickmann cyclization, producing a five-membered tetramate heterocycle. The following modifications may involve an acylation-elimination reaction (Fig. 3), leading to the generation of an additional double bond that is exocyclic to tetramate (Kanchanabanca et al., 2013). In general, the biosynthetic pathway of PYRs appears to be similar to those previously proposed for structurally related spirotetronates, including CHL, tetrocarcin A (TC-A), kijanimicin (KIJ), and lobophorins

(LOBs), for which the major difference involves the incorporation of a glycerate derivative rather than an amino acid as the three-carbon unit for tetronate formation (Fang et al., 2008; Jia et al., 2006; Li et al., 2013; Zhang et al., 2007).

Consequently, analysis of the template effects arising from the established catalytic logic on and off the assembly line suggests that a minimum of six double bonds formed during the biosynthetic processes of PYRs, as well as the spirotetronates CHL, TC-A, KIJ, and LOBs (Pang, Wang, & Liu, 2016). Taking their locations into account, these double bonds constitute two independent pairs of 1,3-diene and alkene groups, which are the structural hallmarks of the internal and terminal intramolecular [4+2] cycloaddition reactions, respectively, and correspond to the cyclohexene-containing dialkyldecalin and spiroconjugate portions of the characteristic pentacyclic cores. Therefore, this analysis supported an interesting hypothesis that had been the subject of considerable speculation: during the biosynthesis of spirotetramates and -tetronates, two sets of 1,3-diene and alkene groups are available in the cross-bridging process to establish a pentacyclic scaffold through two [4+2] cyclization reactions (Oikawa & Tokiwano, 2004).

2.2 Identification of the Candidates Coding for Cascade [4+2] Cycloadditions from the PYR Biosynthetic Gene Cluster

A search of potential [4+2] cyclases (ie, the dialkyldecalin synthase and the spiroconjugate synthase) was initiated using the biosynthetic machinery of PYRs as a model system. *S. rugosporus*, which is the PYR-producing strain, proved to be amenable to genetic manipulation, facilitating the in vivo correlation of target genes with their functions using a gene inactivation method (Wu et al., 2012). In organisms, biosynthetic genes relevant to a certain natural product are often clustered within a region on the genome for coordinated structural formation, regulation, and self-resistance. Taking advantage of this, the search focused on comparative analysis of the available biosynthetic gene clusters of spirotetramates and spirotetronates. Specifically, the genes for PYRs that were functionally unassigned or difficult to predict were identified, and then these genes served as targets for homolog searches from the clusters of spirotetronates CHL, TC-A, KIJ, and LOBs (Fig. 4) (Fang et al., 2008; Jia et al., 2006; Li et al., 2013; Wu et al., 2012; Zhang et al., 2007).

As a result, two candidates, *pyrE3* (1392 bp) and *pyrI4* (555 bp), were identified in the *pyr* cluster. The counterparts of both genes were found in the clusters of the aforementioned spirotetronates, with *chlE3*, *tcaE1*,

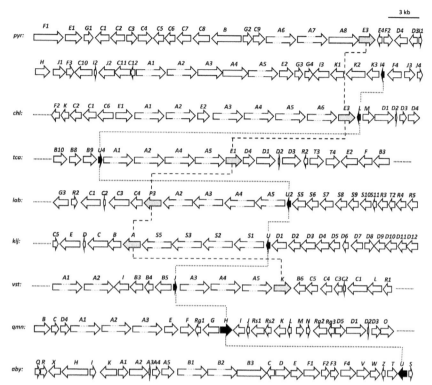

Fig. 4 Comparative analysis of the available biosynthetic gene clusters of spirotetramates and spirotetronates, including those for the spirotetramates PYRs (*pyr*) and the spirotetronates CHL (*chl*), TC-A (*tca*), LOBs (*lob*), KIJ (*kij*), VST (*vst*), QMNs (*qmn*), and ABY-C (*aby*). The genes responsible for the formation of the dialkyldecalin system and the spiroconjugate portion are labeled in *gray* and *black*, respectively.

lobP3, and *kijA* being homologous to *pyrE3* (46–52% identities) and *chlL*, *tcaU4*, *lobU2*, and *kijU* being homologous to *pyrI4* (26–48% identities) (Fig. 4). It is important to note that *kijU*, which was not initially identified from the reported sequence data (Zhang et al., 2007), was identified here by reexamination of the flanking regions between the *kijS1* and *kijD1* genes in the *kij* cluster. A similar result was reported in the study of homolog encoding VstJ in VST biosynthesis (Hashimoto et al., 2015). These genes have been proposed to play a role in the construction of certain structural features that are shared between spirotetramates and spirotetronates (ie, presumably the dialkyldecalin and the spiroconjugate moieties).

To further distinguish the potential functions of *pyrE3* and *pyrI4*, the sequence comparison was extended to the gene clusters encoding the spirotetronate metabolites abyssomicin C (ABY-C) and quartromicins (QMNs), which share the spiroconjugate portion but lack the dialkyldecalin system (Gottardi et al., 2011; He et al., 2012). The *pyrE3* homolog was absent in both the *aby* and *qmn* clusters, which do have *pyrI4* homologs (Fig. 4). In the *aby* cluster, reexamination of the flanking regions between the *abyI* and *abyK* genes revealed *abyU* (429 bp), which was not initially identified but codes for an unknown protein that has been deposited into GenBank (with the accession number WP_013733060.1). Interestingly, the *qmnH* gene in the *qmn* cluster is 1131 bp long and contains two homologs (coding for QmnH-N and QmnH-C, respectively), which individually share 24% identity to *pyrI4* and 45% identity to each other. Consequently, taking the structural relevance into account, PyrI4 is most likely involved in the formation of the more common spiroconjugate portion, in contrast to PyrE3, which may be responsible for the construction of the less common dialkyldecalin system.

3. METHODS

3.1 In Vivo Validation of the Involvement of *pyrE3* and *pyrI4* in PYR Biosynthesis

3.1.1 *Inactivation of* pyrE3 *or* pyrI4 *and Associated Homologous Complementation in* S. rugosporus

The inactivation of *pyrE3* or *pyrI4* in *S. rugosporus* was performed by in-frame deletion to avoid potential polar effects on the expression of adjacent biosynthetic genes. For PCR amplification, the genomic DNA of the PYR-producing *S. rugosporus* wild-type strain served as the template. For complementation *in trans*, the expression of *pyrE3* or *pyrI4* was under the control of the constitutive promoter *PermE** in each mutant strain (Chen et al., 2008; Kieser, Bibb, Buttner, Chater, & Hopwood, 2000; Reeves, English, Lampel, Post, & Vanden Boom, 1999).

1. Mutant construction and selection. Clone the two DNA fragments, which contain the upstream and downstream regions of each target gene, respectively, by PCR amplification into the *Eco*RI–*Hin*dIII site of pKC1139 (Kieser et al., 2000), and construct the recombinant plasmid in which the target gene is mutated by in-frame deletion of its internal coding region. Introduce this pKC1139 derivative from *Escherichia coli* ET12567 (pUZ8002) (MacNeil et al., 1992) into *S. rugosporus* wild-type strain through intergeneric conjugation.

Culture the conjugates at 37°C and identify apramycin-resistant colonies as the integrating mutants, in which a single-crossover homologous recombination occurs. Culture these mutants around five rounds in the absence of apramycin until apramycin-sensitive colonies are obtained. Examine the genotype of the obtained apramycin-sensitive strains by PCR amplification and sequence to confirm the genotype of the double-crossover mutants (for $\Delta pyrE3$, a 831-bp in-frame coding region was deleted; and for $\Delta pyrI4$, a 396-bp in-frame coding region was deleted).

2. Examination of the chemical profiles. Spread each in-frame deletion mutant strain on agar plates and incubate at 30°C for sporulation and growth. Inoculate the spore-containing agar piece into seed medium and incubate at 28°C and 220 r.p.m. for 36 h. Transfer 5 mL of the seed culture broth into 100 mL of the fermentation medium and incubate for 7 days at 28°C and 220 r.p.m.

   Extract 50 mL of each fermentation broth three times with equal volumes of ethyl acetate (EtOAc). Evaporate the organic solvent under reduced pressure and then dissolve the residue in methanol. Analyze the methanol solution by high-performance liquid chromatography (HPLC) or LC-electrospray ionization mass spectrometer (ESI-MS) on an Agilent Zorbax column (SB-C18, 5 μm, 4.6 × 250 mm, Agilent Technologies Inc., USA) using gradient elution with mobile phase A (H_2O supplemented with 0.1% HCOOH) and mobile phase B (CH_3CN supplemented with 0.1% HCOOH) at a flow rate of 1 mL/min: 0–5 min, 10–30% phase B; 5–10 min, 30–65% phase B; 10–20 min, 65% phase B; 20–25 min, 65–100% phase B; 25–30 min, 100% phase B; and 30–35 min, 100–10% phase B (λ at 335 nm).

3. Homologous complementation of *pyrE3* or *pyrI4* in the corresponding *S. rugosporus* mutant strain. Amplify the target gene-containing fragment by PCR. Digest the PCR product by *Eco*RI and *Eco*RV, and then coligate with an *Xba*I–*Eco*RI fragment that contains the constitutive promoter *PermE** arising from pLL6214 (Li et al., 2008). Insert the ligation product into the *Xba*I–*Eco*RV site of pSET152 to afford the recombinant plasmid in which the expression of the target gene is under the control of the *PermE** promoter.

   Introduce each pSET152 derivative from *E. coli* ET12567 (pUZ8002) (MacNeil et al., 1992) into the corresponding *S. rugosporus* ($\Delta pyrE3$ or $\Delta pyrI4$) mutant strain through intergeneric conjugation, and identify recombinant apramycin-resistant strains. Culture these recombinant strains and analyze the product profiles according to the protocols described earlier.

This procedure established the relevance of PyrE3 and PyrI4 to PYR biosynthesis.

3.1.2 Characterization of the Intermediate Isolated from the ΔpyrE3 S. rugosporus Mutant Strain

1. Extract 120 L of the culture broth of the *ΔpyrE3 S. rugosporus* mutant strain three times with equal volumes of EtOAc.
2. Fractionate the concentrated EtOAc extract by octadecylsilyl column chromatography (LiChroprep RP-18, 40–63 μm, Merck KGaA, Germany) using gradient elution with methanol in H_2O (30–100%).
3. Further fractionate the concentrated, target compound-containing fraction by semipreparative HPLC on an Agilent Zorbax column (SB-C18, 5 μm, 9.4 × 250 mm, Agilent Technologies Inc., USA) using isocratic elution with 75% CH_3CN supplemented with 0.1% HCOOH at a flow rate of 4 mL/min (λ at 220 nm).
4. Freeze-dry to obtain the target compound (13.4 mg) as a white amorphous powder.
5. Characterize the purified target compound by high-resolution (HR)-ESI-MS and 1D- and 2D-nuclear magnetic resonance (NMR).

This procedure revealed the target compound is a linear polyene intermediate, which simultaneously contains two unsaturated systems, with an internal pair ($\Delta^{4,5}$-alkene and $\Delta^{10,11}$, $\Delta^{12,13}$-diene) corresponding to the dialkyldecalin and a terminal pair ($\Delta^{22,23}$-methylene and $\Delta^{18,19}$, $\Delta^{20,21}$-diene) corresponding to the tetramate spiroconjugate (Figs. 1 and 3).

3.2 In Vitro Determination of the Functions of PyrE3 and PyrI4 for Pentacyclic Core Formation

3.2.1 Expression and Purification of PyrE3 and PyrI4 from E. coli

1. Amplify the target protein-encoding fragments from the genomic DNA of *S. rugosporus* by PCR, and clone each PCR product into the *Nde*I–*Hind*III site of pET-28a(+) (Novagen, Merck KGaA, Germany).
2. Transform each resulting pET-28a(+) derivative into *E. coli* BL21(DE3), and select kanamycin-resistant transformants on LB agar.
3. Incubate the culture of *E. coli* transformants in LB medium containing 50 μg/mL kanamycin at 37°C and 250 r.p.m. until the cell density reach 0.6–0.8 at OD_{600}. Induce protein expression by the addition of

isopropyl-β-D-thiogalactopyranoside (IPTG) to a final concentration of 0.2 mM and further incubate for 40 h at 16°C.

4. Harvest the cells by centrifuging at 5000 r.p.m. for 20 min at 4°C, and then resuspend the cells in lysis buffer (50 mM Tris–HCl, 300 mM NaCl, 10% glycerol, pH 7.0).
5. Disrupt the cells using a FB-110X low temperature ultra-pressure continuous flow cell disrupter (Shanghai Litu Mechanical Equipment Engineering Co., Ltd., China). Collect the soluble fraction and purify each protein using a HisTrap FF column (GE Healthcare, USA) according to the manufacturer's protocol.
6. Verifying the protein purity by SDS-PAGE. Concentrate each target protein using an Amicon Ultra-15 Centrifugal Filter (3 kDa, Merck Millipore, Germany). Desalt the protein using a PD-10 Desalting Column (GE Healthcare, USA) into 50 mM Tris–HCl (pH 7.5) containing 100 mM NaCl and 10% (w/w) glycerol.
7. Determine the protein concentration using a Bradford assay with bovine serum albumin as the standard.
8. Flash-freeze each purified protein in liquid nitrogen and store at −80°C. This procedure generated purified PyrE3 and PyrI4 proteins as N-terminal His$_6$-fusion proteins.

3.2.2 Assay of the Activities of PyrE3, PyrI4, and Their Combination Along with Characterization of the Products

The in vitro assays of the activities of purified proteins were performed in Tris–HCl buffer. To terminate the reaction, an equal volume of methanol was added to each reaction mixture. Each reaction mixture was centrifuged, and the supernatant was subjected to HPLC or LC-ESI-MS analysis, which was performed on an Agilent Zorbax column (SB-C18, 5 μm, 4.6 × 250 mm, Agilent Technologies Inc., USA) using gradient elution with mobile phase A (H$_2$O supplemented with 0.1% HCOOH) and mobile phase B (CH$_3$CN supplemented with 0.1% HCOOH) at a flow rate of 1 mL/min: 0–5 min, 65% phase B; 5–15 min, 65–100% phase B; 15–20 min, 100% phase B; 20–25 min, 100–65% phase B; and 25–30 min, 65% phase B (λ at 280 nm).

1. Assay of the activity of both PyrE3 and PyrI4 along with characterization of their pentacyclic product and the PYR aglycone. Incubate the purified PyrE3 and PyrI4 (40 μM for each) with the linear polyene intermediate (200 μM) isolated from the Δ$pyrE3$ S. rugosporus mutant strain in 50 mM Tris–HCl buffer (pH 7.0) at 30°C. After 15 min of incubation, terminate the reaction and analyze the resulting product.

To characterize the pentacyclic product, perform the reaction at 30°C in a 20 mL mixture containing 3.5 mg of the linear polyene intermediate, 5% methanol, 50 μM PyrE3, and 50 μM PyrI4 in 50 mM Tris–HCl buffer (pH 7.0). After completion of the reaction (monitored by HPLC), extract the reaction mixture three times with equal volumes of EtOAc. Concentrate the organic extract under reduced pressure and then subject the crude residue to semipreparative HPLC on an Agilent Zorbax column (SB-C18, 5 μm, 4.6 × 250 mm, Agilent Technologies Inc., USA) using isocratic elution with 45% CH_3CN supplemented with 0.1% HCOOH at a flow rate of 4 mL/min (λ at 220 nm) to yield the purified target product (2.0 mg, 57% yield). Characterize the compound by HR-ESI-MS and 1D- and 2D-NMR.

To characterize the PYR aglycone, extract 10 L of the culture broth of the *S. rugosporus* wild-type strain three times with equal volumes of EtOAc. Fractionate the concentrated EtOAc extract by octadecylsilyl column chromatography (LiChroprep RP-18, 40–63 μm, Merck KGaA, Germany) using gradient elution with methanol in H_2O (10–100%). Subject the PYR-containing fractions to acid hydrolysis with 0.5 M CF_3COOH in 10% methanol (v/v) at 60°C for 1 h. Separate the hydrolysate by semipreparative HPLC on an Agilent Zorbax column (SB-C18, 5 μm, 4.6 × 250 mm, Agilent Technologies Inc., USA) using isocratic elution with 45% CH_3CN supplemented with 0.1% HCOOH at a flow rate of 4 mL/min (λ at 220 nm) to yield the purified PYR aglycone (5.7 mg). Characterize the purified compound by HR-ESI-MS, 1D- and 2D-NMR, and single-crystal X-ray diffraction for the determination of the chemical structure and relative configuration, as well as with gas-phase and solution density functional theory (DFT) conformational analysis and time-dependent DFT (TDDFT)-electronic circular dichroism (ECD) calculations for the establishment of the absolute configuration.

2. Assay of the activity of PyrE3 along with characterization of its partially cyclized product. Incubate PyrE3 (40 μM) with the linear polyene intermediate (200 μM) and 10% methanol (v/v) in 50 mM Tris–HCl buffer (pH 7.5) at 30°C. After 15 min of incubation, terminate the reaction and analyze the resulting product.

To characterize the partially cyclized product, perform reaction at 30°C in a 40 mL mixture containing 4.5 mg of the linear polyene intermediate, 5% methanol, and 50 μM PyrE3 in 50 mM Tris–HCl buffer (pH 7.0) at 30°C. After incubation for 30 min, extract the reaction mixture three times with equal volumes of EtOAc. Concentrate the organic extract under reduced pressure and then subject the crude residue to semipreparative HPLC on an Agilent Zorbax column (SB-C18, 5 µm, 4.6 × 250 mm, Agilent Technologies Inc., USA) using gradient elution with mobile phase A (H_2O supplemented with 0.1% HCOOH) and phase B (CH_3CN supplemented with 0.1% HCOOH) at a flow rate of 4 mL/min: 0–5 min, 65% phase B; 5–15 min, 65–100% phase B; 15–20 min, 100% phase B, to yield the purified target product (3.1 mg, 66% yield). Characterize the purified compound by HR-ESI-MS and 1D- and 2D-NMR.

To evaluate the pH dependence of the PyrE3-catalyzed conversion, conduct the reactions at 30°C for 1.0 min in 50 mM MES (pH 6.0 or 6.5) or 50 mM Tris–HCl (pH 7.0, 7.5, 8.0, 8.5, or 9.0) buffer in the presence of 0.33 µM PyrE3 and 200 µM linear polyene intermediate. Terminate each reaction and analyze the substrate consumption and product production.

For kinetic analysis, conduct a time-course experiment to determine the initial rate conditions in a 20-µL reaction mixture containing 0.33 µM PyrE3, 200 µM linear polyene intermediate, 10% methanol (v/v), and 50 mM Tris–HCl buffer (pH 7.5). Initiate the reactions by addition of PyrE3 followed by incubation at 30°C and termination at 0, 0.5, 1.0, 2.0, 5.0, or 10.0 min. Analysis the samples by HPLC, and calculate the product yield according to the standard concentration curve of the purified partially cyclized product. Determine the period of time that the production is linear with respect to time, and fit the production data to a linear equation to obtain the initial velocity. To determine the kinetic parameters for the partially cyclized product, conduct the PyrE3-catalyzed reactions at 30°C with varying concentrations of the substrate (2.5, 5, 10, 25, 50, 100, 200, 400, 600, or 800 µM). Perform the assays in triplicate, and fit the resulting initial velocities to the Michaelis–Menten equation using OriginPro 8 (Originlab Co., USA) to extract the K_m and k_{cat} parameters.

3. Assay of the activity of PyrI4. Incubate purified PyrI4 (40 µM) with the partially cyclized substrate (the product of PyrE3, 200 µM) and 10% methanol (v/v) in 50 mM Tris–HCl buffer (pH 7.5) at 30°C. After 15 min of incubation, terminate the reaction and analyze the resulting product. To exclude metal ion dependence, conduct the reaction after the addition of 1 mM ethylene diamine tetraacetic acid.

Determine the pH dependence as described for PyrE3, except that the concentration of PyrI4 is 0.3 µM.

For kinetic analysis, conduct a time-course experiment to determine the initial rate conditions in a 20 μL reaction mixture containing 200 μM partially cyclized substrate, 0.3 μM PyrI4, 10% methanol (v/v) and 50 mM Tris–HCl buffer (pH 7.5). Initiate the reactions by addition of PyrI4 followed by incubation at 30°C and termination at 0, 0.5, 1.0, 2.0, 5.0, or 10.0 min. Analysis samples and calculate the initial velocity as described above. To determine the kinetic parameters for the pentacyclic compound, conduct PyrI4-catalyzed reactions at 30°C with varying concentrations of the substrate (0, 25, 50, 100, 200, 400, 600, or 800 μM). Perform the assays in triplicate, and extract the K_m and k_{cat} parameters according to the methods described above.

The results from this procedure revealed that PyrE3 exclusively converts the linear polyene intermediate isolated from the Δ$pyrE3$ S. *rugosporus* mutant strain into a partially cyclized intermediate that contains the intact dialkyldecalin system and retains the free terminal pair, consisting of $\Delta^{22,23}$-methylene and $\Delta^{18,19},\Delta^{20,21}$-diene. Then, PyrI4 converts the partially cyclized intermediate into a highly rigidified pentacyclic compound that is identical to the natural PYR aglycone in stereochemistry (Fig. 1).

3.3 In Vivo and In Vitro Mechanistic Evaluation of the Generality of Pentacyclic Core Formation in the CHL Biosynthetic Pathway

3.3.1 Inactivation of chlE3 or chlL in S. antibioticus

The inactivation of *chlE3* was conducted by in-frame deletion, using the methods described in Section 3.1. *chlL* was inactivated by gene replacement.

1. Construction and selection of the Δ*chlL* mutant. Clone the two DNA fragments, which contain the upstream and downstream regions of *chlL* from the genomic DNA of the CHL-producing *S. antibioticus* wild-type strain by PCR amplification and coligate with an EcoRI–BglII fragment from pUO9090 (Jia et al., 2006) that contains the apramycin-resistance gene. Insert the ligation product into the XbaI–HindIII site of pTL1001, which is a pKC1139-derived *E. coli-Streptomyces* shuttle vector with thiostrepton resistance (Jia et al., 2006). The recombinant plasmid in which *chlL* is mutated by replacement with an apramycin-resistance gene. Introduce this recombinant plasmid from *E. coli* ET12567 (pUZ8002) (MacNeil et al., 1992) into *S. antibioticus* through intergeneric conjugation.

   Culture the conjugates at 37°C and identify apramycin-resistant colonies as the integrating mutants, in which a single-crossover homologous recombination occurs. Culture these integrating mutants and select apramycin-resistant and thiostrepton-sensitive colonies. Validation of the Δ*chlL* mutant, in which the 102-bp coding region of *chlL* was replaced by the 1.5 kb apramycin-resistance gene, using the methods described in Section 3.1

2. Examination of the chemical profiles. Spread each mutant strain (Δ *chlE3* and Δ *chlL*) onto 25 mL agar plates and incubate at 30°C for 5 days. Chop the agar from each plate and extract three times with 50 mL of methanol followed by sonication to improve the extraction efficiency. Evaporate the organic solvent under reduced pressure and then dissolve the residue in methanol. Analyze the methanol solution by HPLC or LC-ESI-MS on an Agilent Zorbax column (SB-C18, 5 μm, 4.6 × 250 mm, Agilent Technologies Inc., USA) using gradient elution with mobile phase A (H_2O supplemented with 0.1% HCOOH) and mobile phase B (CH_3CN supplemented with 0.1% HCOOH) at a flow rate of 1 mL/min: 0–18 min, 30–100% phase B; 18–25 min, 100% phase B; 25–27 min, 100–30% phase B; 27–30 min, 30% phase B (λ at 220 nm).

This procedure established the relevance of ChlE3 and ChlL to CHL biosynthesis.

3.3.2 Exchanges Between pyrE3 and chlE3 as well as pyrI4 and chlL by Heterologous Complementation

For heterologous complementation *in trans*, the expression of each target gene was under the control of the constitutive promoter *PermE** in the corresponding mutant strain.

1. Heterologous complementation of *pyrE3* in the *S. antibioticus* Δ*chlE3* mutant strain. Introduce the pSET152 derivative for homologous complementation of *pyrE3* (see Section 3.1) from *E. coli* ET12567 (pUZ8002) (MacNeil et al., 1992) into the *S. antibioticus* Δ*chlE3* mutant strain through intergeneric conjugation, and identify apramycin-resistant recombinant strains.

2. Heterologous complementation of *pyrI4* in the *S. antibioticus* Δ*chlL* mutant strain. Amplify the *pyrI4*-containing fragment by PCR using the genomic DNA of the *S. rugosporus* wild-type strain as the template. Digest the PCR product by *Xba*I and *Hind*III and then coligate with an *Xba*I–*Eco*RI fragment that contains the constitutive *PermE** promoter arising from pLL6214 (Li et al., 2008). Insert the ligation product into the *Eco*RI–*Hind*III site of pTGV2 (Jia et al., 2006) to generate the recombinant plasmid, in which the expression of *pyrI4* is under the

control of the promoter *PermE**. Introduce this pTGV2 derivative from *E. coli* ET12567 (pUZ8002) (MacNeil et al., 1992) into the *S. antibioticus* Δ*chlL* mutant strain through intergeneric conjugation, and identify thiostrepton-resistant recombinant strains.
3. Heterologous complementation of *chlE3* or *chlL* in the corresponding *S. rugosporus* mutant strain. Amplify the target gene-containing fragment by PCR using the genomic DNA of wild-type *S. antibioticus* as the template. Construct each pSET152-derived recombinant plasmid, in which the target gene is controlled by the *PermE** promoter, using the methods described in Section 3.1. Introduce each pSET152 derivative into the corresponding *S. rugosporus* (Δ*pyrE3* or Δ*pyrI4*) mutant strain through intergeneric conjugation and identify apramycin-resistant recombinant strains.
4. Culture these recombinant strains and analyze the product profile according to the protocols described earlier.

Each heterologous complementation partially restored the production of PYRs or CHL, indicating that the *chlE3* and *chlL* genes are functionally exchangeable with *pyrE3* and *pyrI4*, respectively.

3.3.3 Expression and Purification of ChlE3 and ChlL from E. coli

The expression and purification of ChlE3 and ChlL as N-terminal His_6-fusion proteins from *E. coli* BL21(DE3) were performed according to the protocols described in Section 3.2.

3.3.4 Assay of the Activities of ChlE3, ChlL, and Their Combination

In vitro assays of the purified proteins were carried out in Tris–HCl buffer. The reaction termination and sample analysis methods were described in Section 3.2.
1. Assay of the activity of both ChlE3 and ChlL. Incubate purified ChlE3 and ChlL (40 μM for each) with the linear polyene intermediate (200 μM), isolated from the Δ*pyrE3 S. rugosporus* mutant strain, in 50 mM Tris–HCl buffer (pH 7.0) at 30°C. After 15 min of incubation, terminate the reaction and analyze the resulting product.
2. Assay of the activity of ChlE3. Incubate the purified ChlE3 (40 μM) with the linear polyene intermediate (200 μM) and 10% methanol (v/v) in 50 mM Tris–HCl buffer (pH 7.5) at 30°C. After 15 min of incubation, terminate the reaction and analyze the resulting product.

Determine the pH dependence using the methods described in Section 3.2, except that the concentration of ChlE3 is 5 μM and that the reactions are performed for 2 min.

For the kinetic analysis, conduct a time-course experiment to determine the initial rate conditions as described for PyrE3, except that the concentration of ChlE3 is 5 μM. To determine the kinetic parameters for the partially cyclized product, conduct ChlE3-catalyzed reactions at 30°C with varying concentrations of the substrate (5, 10, 25, 50, 100, 200, 400, or 800 μM). Perform the assays in triplicate, and extract the K_m and k_{cat} parameters according to the methods described earlier.

3. Assay of the activity of ChlL. Incubate purified ChlL (40 μM) with the partially cyclized substrate (200 μM) and 10% methanol (v/v) in 50 mM Tris–HCl buffer (pH 7.0) at 30°C. After 15 min of incubation, terminate the reaction and analyze the resulting product.

Determine pH dependence using the methods described in Section 3.2, except the concentration of ChlL is 20 μM and that the reactions are performed for 5 min.

For the kinetic analysis, conduct a time-course experiment to determine the initial rate conditions as described for PyrI4, except that the pH is 7.0 and that 20 μM ChlL is used. To determine the kinetic parameters for the pentacyclic product, conduct the ChlL-catalyzed reactions at 30°C with varying concentrations of the substrate (25, 50, 100, 200, 400, or 800 μM). Perform the assays in triplicate, and extract the K_m and k_{cat} parameters as described earlier.

The results from this procedure revealed that, although kinetically less efficient than PyrE3 and PyrI4, both ChlE3 and ChlL are active in a highly ordered process: ChlE3 catalyzes the conversion of the linear polyene intermediate isolated from the Δ$pyrE3$ S. *rugosporus* mutant strain into the partially cyclized intermediate, which is then transformed by ChlL into the pentacyclic compound.

3.4 Examination of the Protein Natures of PyrE3 and ChlE3

1. Perform computational modeling of PyrE3 and ChlE3 using the protein structure prediction platform I-TASSER (http://zhanglab.ccmb.med.umich.edu/I-TASSER) (Roy, Kucukural, & Zhang, 2010). Select the potential residues that bind FAD based on computational predictions.
2. For site-specific mutation of PyrE3 or ChlE3, conduct rolling-cycle PCR amplification, followed by subsequent DpnI digestion according

to the Mut Express II standard procedure (Vazyme Biotech Co., Ltd., China). Confirm each mutation by sequencing. Express the mutant proteins PyrE3^{E32A}, PyrE3^{D275}, ChlE3^{E32A}, and ChlE3^{D280A} in *E. coli* BL21 (DE3) and purify to homogeneity. Assay the activities of these mutant proteins according to the methods described above for native proteins.

3. For spectroscopic analysis, record the UV spectra of the recombinant proteins PyrE3 and ChlE3 and their mutants (1 mg/mL) on a Unico 4802 UV/Vis spectrophotometer (UNICO Instruments Co., Ltd., China). For mass spectrometric analysis of the cofactors, release the possible noncovalent cofactor from the PyrE3 and ChlE3 recombinant proteins and their mutants by heating at 100°C for 5 min, and subject the supernatant to HR-ESI-MS analysis. For protein conformational analysis, record the ECD spectra of PyrE3 and ChlE3 and their mutants (with a 1-nm bandwidth resolution and in 1-nm steps) in a 1-cm path length quartz cuvette at 25°C using an AVIV 400 CD spectrophotometer (AVIV Biochemical, Inc., Lakewood, NJ) at a concentration of 0.2 mg/mL in 50 mM NaH$_2$PO$_4$ (pH 7.6) buffer. For scanning, the wavelength is varied from 190 to 300 nm.

The results from this procedure indicated that PyrE3 and ChlE3 are flavoproteins containing the flavin cofactor FAD in the oxidized form and that mutations at the FAD-binding site resulted in production of the PyrE3^{E32A}, PyrE3^{D275}, ChlE3^{E32A}, and ChlE3^{D280A} proteins that are inactive and markedly changed in protein structure.

4. DISCUSSION AND PERSPECTIVES

The coupling of the PyrE3-represented dialkyldecalin synthases and the PyrI4-represented spiroconjugate synthases plays a central role in the formation of the rigid pentacyclic core shared by spirotetramates and spirotetronates, via the catalysis of two tandem [4+2] cyclization reactions for coordinated cross-bridging of a linear polyene intermediate. Because these two types of unusual cyclases are phylogenetically distinct from those previously known to exhibit this activity, identification of them from the associated pathways depends on a key idea of comparative analyses for the commonality and specificity in biosynthesis. These analyses rationally revealed the necessity of two enzymatic [4+2] cycloadditions according to the template effects of assembly lines and the functional prediction of the enzymes associated with the two reactions by taking into account the structural relevance. The following characterization combined various in vivo and in vitro approaches for validation of the necessity of *pyrE3* and *pyrI4* for the biosynthesis of spirotetramate PYRs, structural elucidation of the linear PYR intermediate, enzymatic construction of the pentacyclic core of PYRs, dissection of PyrE3 and PyrI4 as the dialkyldecalin synthase and spiroconjugate synthase, respectively, and finally, evaluation of the generality of the spirotetronate CHL in the formation of the pentacyclic core.

The multiple comparison-based strategy and methods presented here provide an example that is useful for the identification and characterization of unusual enzymes whose activities are dependent on the chemical structure, such as the phylogenetically distinct spiroketal cyclases. These cyclases (eg, RevJ and AveC in the biosynthesis of reveromycin A, which is an antitumor agent, and avermectin agrochemicals, respectively), catalyze the dehydrative ketalization of dihydroxy ketone to form a spiroketal unit (Fig. 5) (Sun et al., 2013; Takahashi et al., 2011). They share a number of correspondences to the enzymes for [4+2] cycloaddition during catalytic processes, including generality in the control of the stereochemistry and enhancement of the transformation rate, as well as the specificity in the evolution of the activity of each enzyme using a variable protein fold that likely occurs via mutual interaction with a specialized chemical structure. For example, in the biosynthetic pathway of avermectins, the identification of AveC, which is representative of a new type of spiroketal cyclases that have no sequence similarity with any protein of known function, relied on a rational query of its role by comparative analysis of its counterparts in the pathways of structurally related metabolites (ie, milbemycins, meilingmycins, and nemadectin). Similarly, further characterization of AveC was conducted by simplification of the chemical profile for gene inactivation-based product tracing, characterization of the extremely unstable intermediates and products, and comparative in vitro biotransformations and in vivo complementations for final validation (Sun et al., 2013). Apparently, different protein folds can be utilized to match the substrates whose overall structures are unrelated for stereoselective accommodation of the reactive groups (eg, 1,3-diene and alkene groups for Diels–Alder-like [4+2] cyclizations and the dihydroxy ketone motif for spiroketalization) and creation of a number of phylogenetically distinct but functionally identical catalysts. Understanding this nature should facilitate elucidating the underlying principles of the application and catalyst design of associated enzymatic reactions in organic chemistry and synthetic biology.

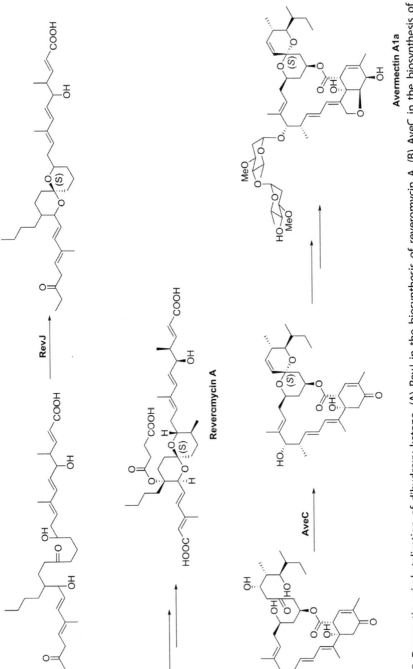

Fig. 5 Enzymatic spiroketalization of dihydroxy ketone. (A) RevJ in the biosynthesis of reveromycin A. (B) AveC in the biosynthesis of avermectin A1a.

REFERENCES

Auclair, K., Sutherland, A., Kennedy, J., Witter, D. J., Van den Heever, J. P., Hutchinson, C. R., et al. (2000). Lovastatin nonaketide synthase catalyzes an intramolecular Diels–Alder reaction of a substrate analogue. *Journal of the American Chemical Society, 122,* 11519–11520.

Chen, Y., Deng, W., Wu, J., Qian, J., Chu, J., Zhuang, Y., et al. (2008). Genetic modulation of the overexpression of tailoring genes *eryK* and *eryG* leading to the improvement of erythromycin A purity and production in *Saccharopolyspora erythraea* fermentation. *Applied and Environmental Microbiology, 74,* 1820–1828.

Corey, E. J. (2002). Catalytic enantioselective Diels–Alder reactions: Methods, mechanistic fundamentals, pathways, and applications. *Angewandte Chemie International Edition, 41,* 1650–1667.

Ding, W., Williams, D. R., Northcote, P., Siegel, M. M., Tsao, R., Ashcroft, J., et al. (1994). Pyrroindomycins, novel antibiotics produced by *Streptomyces rugosporus* sp. LL-42D005. I. Isolation and structure determination. *The Journal of Antibiotics, 47,* 1250–1257.

Fang, J., Zhang, Y., Huang, L., Jia, X., Zhang, Q., Zhang, X., et al. (2008). Cloning and characterization of the tetrocarcin A gene cluster from *Micromonospora chalcea* NRRL 11289 reveals a highly conserved strategy for tetronate biosynthesis in spirotetronate antibiotics. *Journal of Bacteriology, 190,* 6014–6025.

Gottardi, E. M., Krawczyk, J. M., von Suchodoletz, H., Schadt, S., Mühlenweg, A., Uguru, G. C., et al. (2011). Abyssomicin biosynthesis: Formation of an unusual polyketide, antibiotic-feeding studies and genetic analysis. *Chembiochem, 12,* 1401–1410.

Hashimoto, T., Hashimoto, J., Teruya, K., Hirano, T., Shin-ya, K., Ikeda, H., et al. (2015). Biosynthesis of versipelostatin: Identification of an enzyme-catalyzed [4+2]-cycloaddition required for macrocyclization of spirotetronate-containing polyketides. *Journal of the American Chemical Society, 137,* 572–575.

He, H., Pan, H., Wu, L., Zhang, B., Chai, H.-B., Liu, W., et al. (2012). Quartromicin biosynthesis: Two alternative polyketide chains produced by one polyketide synthase assembly line. *Chemistry & Biology, 19,* 1313–1323.

Jia, X., Tian, Z., Shao, L., Qu, X., Zhao, Q., Tang, J., et al. (2006). Genetic characterization of the chlorothricin gene cluster as a model for spirotetronate antibiotic biosynthesis. *Chemistry & Biology, 13,* 575–585.

Kakule, T. B., Jadulco, R. C., Koch, M., Janso, J. E., Barrows, L. R., & Schmidt, E. W. (2015). Native promoter strategy for high-yielding synthesis and engineering of fungal secondary metabolites. *ACS Synthetic Biology, 4,* 625–633.

Kanchanabanca, C., Tao, W., Hong, H., Liu, Y., Hahn, F., Samborskyy, M., et al. (2013). Unusual acetylation–elimination in the formation of tetronate antibiotics. *Angewandte Chemie International Edition, 52,* 5785–5788.

Kato, N., Nogawa, T., Hirota, H., Jang, J.-H., Takahashi, S., Ahn, J. S., et al. (2015). A new enzyme involved in the control of the stereochemistry in the decalin formation during equisetin biosynthesis. *Biochemical and Biophysical Research Communications, 460,* 210–215.

Kelly, W. L. (2008). Intramolecular cyclizations of polyketide biosynthesis: Mining for a "Diels-Alderase"? *Organic & Biomolecular Chemistry, 6,* 4483–4493.

Kieser, T., Bibb, M. J., Buttner, M. J., Chater, K. F., & Hopwood, D. A. (2000). *Practical Streptomyces genetics.* Norwich, UK: John Innes Foundation.

Kim, R.-R., Illarionov, B., Joshi, M., Cushman, M., Lee, C. Y., Eisenreich, W., et al. (2010). Mechanistic insights on riboflavin synthase inspired by selective binding of the 6,7-dimethyl-8-ribityllumazine exomethylene anion. *Journal of the American Chemical Society, 132,* 2983–2990.

Kim, H. J., Ruszczycky, M. W., Choi, S.-H., Liu, Y.-N., & Liu, H.-W. (2011). Enzyme-catalysed [4+2] cycloaddition is a key step in the biosynthesis of spinosyn A. *Nature, 473,* 109–112.

Kim, H. J., Ruszczycky, M. W., & Liu, H.-W. (2012). Current developments and challenges in the search for a naturally selected Diels-Alderase. *Current Opinion in Chemical Biology, 16*, 124–131.

Kries, H., Caputi, L., Stevenson, C. E. M., Kamileen, M. O., Sherden, N. H., Geu-Flores, F., et al. (2016). Structural determinants of reductive terpene cyclization in iridoid biosynthesis. *Nature Chemical Biology, 12*, 6–8. http://dx.doi.org/10.1038/nchembio.1955. Advance online publication.

Li, L., Deng, W., Song, J., Ding, W., Zhao, Q., Peng, C., et al. (2008). Characterization of the saframycin A gene cluster from *Streptomyces lavendulae* NRRL 11002 revealing a non-ribosomal peptide synthetase system for assembling the unusual tetrapeptidyl skeleton in an iterative manner. *Journal of Bacteriology, 190*, 251–263.

Li, S., Xiao, J., Zhu, Y., Zhang, G., Yang, C., Zhang, H., et al. (2013). Dissecting glycosylation steps in lobophorin biosynthesis implies an iterative glycosyltransferase. *Organic Letters, 15*, 1374–1377.

Ma, S. M., Li, J. W.-H., Choi, J. W., Zhou, H., Lee, K. K. M., Moorthie, V. A., et al. (2009). Complete reconstitution of a highly reducing iterative polyketide synthase. *Science, 326*, 589–592.

MacNeil, D. J., Gewain, K. M., Ruby, C. L., Dezeny, G., Gibbons, P. H., & MacNeil, T. (1992). Analysis of *Streptomyces avermitilis* genes required for avermectin biosynthesis utilizing a novel integration vector. *Gene, 111*, 61–68.

Nicolaou, K. C., Snyder, S. A., Montagnon, T., & Vassilikogiannakis, G. (2002). The Diels–Alder reaction in total synthesis. *Angewandte Chemie International Edition, 41*, 1668–1698.

Oikawa, H., Katayama, K., Suzuki, Y., & Ichihara, A. (1995). Enzymatic activity catalysing *exo*-selective Diels-Alder reaction in solanapyrone biosynthesis. *Journal of the Chemical Society, Chemical Communications, 13*, 1321–1322.

Oikawa, H., & Tokiwano, T. (2004). Enzymatic catalysis of the Diels–Alder reaction in the biosynthesis of natural products. *Natural Product Reports, 21*, 321–352.

Ose, T., Watanabe, K., Mie, T., Honma, M., Watanabe, H., Yao, M., et al. (2003). Insight into a natural Diels–Alder reaction from the structure of macrophomate synthase. *Nature, 422*, 185–189.

Pang, B., Wang, M., & Liu, W. (2016). Cyclization of polyketides and non-ribosomal peptides on and off their assembly lines. *Natural Product Reports, 33*, 162–173. http://dx.doi.org/10.1039/C5NP00095E. Advance online publication.

Reeves, A. R., English, R. S., Lampel, J. S., Post, D. A., & Vanden Boom, T. J. (1999). Transcriptional organization of the erythromycin biosynthetic gene cluster of *Saccharopolyspora erythraea*. *Journal of Bacteriology, 181*, 7098–7106.

Roy, A., Kucukural, A., & Zhang, Y. (2010). I-TASSER: A unified platform for automated protein structure and function prediction. *Nature Protocols, 5*, 725–738.

Sato, M., Yagishita, F., Mino, T., Uchiyama, N., Patel, A., Chooi, Y.-H., et al. (2015). Involvement of lipocalin-like CghA in decalin-forming stereoselective intramolecular [4+2] cycloaddition. *Chembiochem, 16*, 2294–2298.

Singh, M. P., Petersen, P. J., Jacobus, N. V., Mroczenski-Wildey, M. J., Maiese, W. M., Greenstein, M., et al. (1994). Pyrroindomycins, novel antibiotics produced by *Streptomyces rugosporus* LL-42D005. II. Biological activities. *The Journal of Antibiotics, 47*, 1258–1265.

Sun, P., Zhao, Q., Yu, F., Zhang, H., Wu, Z., Wang, Y., et al. (2013). Spiroketal formation and modification in avermectin biosynthesis involves a dual activity of AveC. *Journal of the American Chemical Society, 135*, 1540–1548.

Takahashi, S., Toyoda, A., Sekiyama, Y., Takagi, H., Nogawa, T., Uramoto, M., et al. (2011). Reveromycin A biosynthesis uses RevG and RevJ for stereospecific spiroacetal formation. *Nature Chemical Biology, 7*, 461–468.

Tian, Z., Sun, P., Yan, Y., Wu, Z., Zheng, Q., Zhou, S., et al. (2015). An enzymatic [4+2] cyclization cascade creates the pentacyclic core of pyrroindomycins. *Nature Chemical Biology, 11,* 259–265.

Wever, W. J., Bogart, J. W., Baccile, J. A., Chan, A. N., Schroeder, F. C., & Bowers, A. A. (2015). Chemoenzymatic synthesis of thiazolyl peptide natural products featuring an enzyme-catalyzed formal [4+2] cycloaddition. *Journal of the American Chemical Society, 137,* 3494–3497.

Wu, Q., Wu, Z., Qu, X., & Liu, W. (2012). Insights into pyrroindomycin biosynthesis reveal a uniform paradigm for tetramate/tetronate formation. *Journal of the American Chemical Society, 134,* 17342–17345.

Zhang, H., White-Phillip, J. A., Melançon, C. E., Kwon, H.-J., Yu, W.-L., & Liu, H.-W. (2007). Elucidation of the kijanimicin gene cluster: Insights into the biosynthesis of spirotetronate antibiotics and nitrosugars. *Journal of the American Chemical Society, 129,* 14670–14683.

CHAPTER FOUR

Application and Modification of Flavin-Dependent Halogenases

K.-H. van Pée[1], D. Milbredt, E.P. Patallo, V. Weichold, M. Gajewi
Allgemeine Biochemie, TU Dresden, Dresden, Germany
[1]Corresponding author: e-mail address: karl-heinz.vanpee@chemie.tu-dresden.de

Contents

1. Introduction	66
2. Inactivation of Halogenases Under Reaction Conditions	67
2.1 Improvement of Halogenase Stability by Error-Prone PCR	67
2.2 Stabilization of Halogenases by Formation of Cross-Linked Enzyme Aggregates (CLEAs)	70
3. The Biocatalytic Scope of Flavin-Dependent Tryptophan Halogenases	73
3.1 Substrate Specificity of Tryptophan Halogenases	73
3.2 Modification of Biosynthetic Pathways Using Tryptophan Halogenases	78
3.3 Chemical Substitution of Enzymatically Introduced Halogen Atoms	81
3.4 Modification of the Regioselectivity of Tryptophan Halogenases	82
4. Synthesis of PCP-Bound Halogenase Substrates	83
4.1 Enzymatic Synthesis of Pyrrolyl-S-PCPs	84
4.2 Chemoenzymatic Synthesis of Pyrrolyl-S-PCPs	86
4.3 Transfer of Pyrrolyl-S-CoA Thioesters to Carrier Proteins and Halogenation of the Substrate	87
4.4 Release of Halogenated Pyrrole-2-Carboxylic Acid from PCPs	87
5. Conclusions	88
Acknowledgments	89
References	89

Abstract

The application of flavin-dependent halogenases is hampered by their lack of stability under reaction conditions. However, first attempts to improve halogenase stability by error-prone PCR have resulted in mutants with higher temperature stability. To facilitate the screening for mutants with higher activity, a high-throughput assay was developed. Formation of cross-linked enzyme aggregates (CLEAs) of halogenases has increased halogenase lifetime by a factor of about 10, and CLEAs have been used to produce halogenated tryptophan in gram scale. Analyses of the substrate specificity of tryptophan halogenases have shown that they accept a much broader range of substrates than previously thought. The introduction of tryptophan halogenase genes into bacteria and plants led to the in vivo formation of peptides containing halogenated tryptophan or novel tryptophan-derived alkaloids, respectively. The halogen atoms in these compounds could be chemically exchanged against other substituents by cross-coupling reactions leading to novel compounds. Site-directed mutageneses have been used to modify the substrate specificity and the regioselectivity of flavin-dependent tryptophan halogenases. Since many flavin-dependent halogenases only accept protein-bound substrates, enzymatic and chemoenzymatic syntheses for protein-tethered substrates were developed, and the synthesized substrates were used in enzymatic halogenation reactions.

1. INTRODUCTION

Flavin-dependent halogenases are a promising group of halogenating enzymes. They catalyze halogenation reactions with high regioselectivity and without the formation of by-products except water and thus provide the opportunity for industrial-scale production of chlorinated pharmaceuticals and organic compounds. However, the low stability of halogenases under reaction conditions, leading to fast inactivation, hampers their application in industry. Thus, it is of great importance to increase their stability under reaction conditions. While their high substrate specificity is of advantage in their natural environment, it is a disadvantage for their extensive application and therefore needs to be broadened. Furthermore, all so far known flavin-dependent halogenases have a rather low reactivity with k_{cat}-values around 1–7 min^{-1} which also needs considerable improvement. Flavin-dependent halogenases are a two-component system consisting of a flavin reductase providing the actual halogenase with FADH$_2$ produced from NADH and FAD. For industrial application of the two-component system, effective regeneration of the reducing agents is also required. The best-investigated flavin-dependent halogenases are the tryptophan 7-halogenases PrnA from pyrrolnitrin biosynthesis (Keller et al., 2000) and RebH from rebeccamycin biosynthesis (Sanchez et al., 2002), the tryptophan 5-halogenase PyrH from pyrroindomycin B biosynthesis (Zehner et al., 2005), and the tryptophan 6-halogenase Thal/ThdH from thienodolin biosynthesis (Milbredt, Patallo, & van Pée, 2014; Seibold et al., 2006) which all accept L- and D-tryptophan as their natural substrates. Additionally, work on halogenases catalyzing the halogenation of peptidyl carrier protein (PCP)-bound pyrrole carboxylic acid or PCP-tethered peptides showed how complex natural substrates of halogenases can be produced by enzymatic or chemical synthesis. The last 10 years have seen considerable progress toward the understanding and improvement of the properties of halogenases which will be described later.

2. INACTIVATION OF HALOGENASES UNDER REACTION CONDITIONS

2.1 Improvement of Halogenase Stability by Error-Prone PCR

Flavin-dependent halogenases were found to completely lose their activity already 2–3 h after the start of the reaction (Fig. 1). One possible reason for this instability could be a simple temperature sensitivity, but there is also the possibility of an altered protein structure due to reaction with the halogenating intermediate or other side products formed at the active site. The halogenating agent produced by the enzymes is hypohalous acid which is directed to the substrate through a 10-Å long tunnel (Dong et al., 2005). This hypohalous acid might not only react with the substrate but also with amino acids located along the tunnel. For example, the reaction of amino acids with hypochlorous acid can lead to formation of chloramines, 3-chlorotyrosines or 3-hydroxytryptophans (Stadtman & Levine, 2006). On the other hand, the reaction of $FADH_2$ with oxygen can lead to the formation of hydrogen peroxide or even superoxide anion radical. These could in turn react with the halogenase and damage the protein structure and

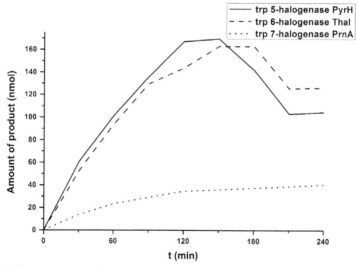

Fig. 1 Time course of conversion of tryptophan to 5-chlorotryptophan, 6-chlorotryptophan, and 7-chlorotryptophan by PyrH, Thal/ThdH, and PrnA, respectively. The amount of tryptophan at the start of the reaction was 400 nmol for PyrH and Thal/ThdH and 50 nmol for PrnA. The reason for the observed decrease in formed product by PyrH and Thal/ThdH after 2.5–3 h is not known.

thereby result in its loss of activity. Product inhibition was also considered as a possible reason for the loss of activity during incubation under reaction conditions. Up to now, there are no studies dealing with the mechanisms of inactivation of flavin-dependent halogenases. Nevertheless, the last few years have seen investigations dealing with ways to increase the stability of flavin-dependent halogenases.

Muffler, Kuetchou Ngnigha, and Ulber (2010) characterized the activity of the tryptophan-5-halogenase PyrH from *Streptomyces rugosporus* LL-42D005 which is involved in the biosynthesis of the antibiotic pyrroindomycin B (Zehner et al., 2005). Checking the deactivation constants of PyrH, they recognized the instability of PyrH at 30°C. After 3 h at this temperature, the enzyme showed nearly no activity, anymore. PyrH was longer active at lower temperatures (after 6 h of incubation, 70% activity at 18°C and 90% activity at 0°C were left). For this reason, they recommended a lower temperature for the enzyme activity assay than the normally used 30°C. However, this results in a decrease of the activity of the halogenases, since most of them have their temperature optimum around 30°C, eg, PrnA at 30°C (Flecks et al., 2008) and Thal/ThdH at 32°C (Ernyei, 2008). The addition of compounds known to stabilize enzymes such as glycerol, glucose, or sucrose had no stabilizing effect under reaction conditions (Fig. 2).

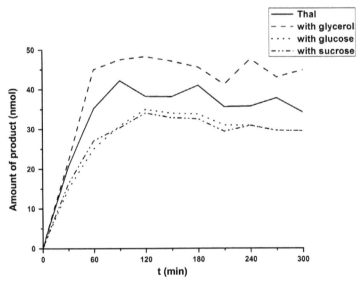

Fig. 2 Time course of conversion of tryptophan to 6-chlorotryptophan by Thal/ThdH in comparison with reactions containing glycerol (2.2%), glucose (250 mg/mL), and sucrose (250 mg/mL), respectively. The amount of tryptophan at the start of reaction was 60 nmol.

Investigations into the linearity of the reaction of the three tryptophan halogenases PrnA, PyrH, and Thal/ThdH showed that the reaction velocities of these enzymes are only linear for a short period of time of about 30 min and then start to decline. This is of high importance for the determination of kinetic data (Fig. 1). Therefore, kinetic data should always be taken in the first minutes of the reaction to be sure to stay in the linear region. However, since activity measurements have to be done by HPLC analysis, rather high amounts of the enzymes are required to make sure that the product peaks are high enough for quantification.

The first problem to solve when working with flavin-dependent halogenases is the issue of overexpression of the halogenase gene. Until now, nobody has achieved to show halogenating activity in a wild-type strain producing a halogenase. Only after high overexpression of the halogenase gene and production of soluble halogenase, halogenating activity can be detected. Overexpression of halogenase genes in soluble form in *E. coli* seems to be quite problematic and forced the group of Jared Lewis to construct a maltose-binding protein halogenase fusion protein to obtain enough of the tryptophan 7-halogenase RebH and the flavin reductase RebF from rebeccamycin biosynthesis in *Lechevalieria aerocolonigenes* for their work (Payne, Andorfer, & Lewis, 2013; Sanchez et al., 2002). In our hands, expression of halogenase genes from different sources in *Pseudomonas* strains works excellent with extremely high overexpression (Keller et al., 2000; Seibold et al., 2006; Zehner et al., 2005), although sometimes the clones completely lose their ability to express the genes. In these cases, fresh transformation of *Pseudomonas* with the plasmids isolated from *E. coli*, not from *Pseudomonas*, will solve the problem.

During their further work with RebH, Poor, Andorfer, and Lewis (2014) also encountered the problem of instability. During the enzyme reaction in a preparative scale, an extensive precipitation of RebH occurred. In their opinion, an increase of thermostability could lead to a longer lifetime of RebH, an increase of stress tolerance and might also enable higher reaction temperatures. Due to the lack of a high-throughput assay, every single *E. coli* clone obtained by error-prone PCR had to be grown separately, lysed, and was then incubated at elevated temperatures, followed by incubation under reaction conditions for 16 and up to 48 h, and analyzed for overall product formation by HPLC. The mutations of variants showing higher product formation after this treatment were combined using overlap extension PCR. This procedure was repeated twice, and the most promising variants of RebH were characterized in detail.

Table 1 Kinetic Data for the Halogenation of 2-Methyltryptamine by the Wild-Type Tryptophan 7-Halogenase RebH and Its 3-LSR Variant at Different Temperatures (Poor et al., 2014)

Enzyme	T (°C)	K_m (μM)	k_{cat} (s^{-1})	k_{cat}/K_m (s^{-1} μM^{-1})
RebH wild type	21	16.8 ± 3.8	0.060 ± 0.0057	3.6 × 10^{-3}
3-LSR variant	21	40.2 ± 3.7	0.013 ± 0.00036	3.3 × 10^{-4}
RebH wild type	40	280.1 ± 18.4	0.25 ± 0.0095	8.8 × 10^{-4}
3-LSR variant	40	202.5 ± 12.7	0.15 ± 0.0093	7.9 × 10^{-4}

The two RebH variants 3-LR and 3-LSR showed the best conversion and were further characterized. The mutant RebH variant 3-LR showed a 5°C higher optimal temperature and 100% more product than wild-type RebH at its optimal temperature. Additionally, the yield of product for the conversion of several other substrates was higher, compared to the wild-type enzyme. An examination of the reaction profiles showed an about threefold longer lifetime of 3-LSR than the wild-type enzyme. Furthermore, steady-state kinetic data were collected at 21°C and 40°C for wild-type RebH and the mutant 3-LSR. Both enzymes show a similar k_{cat}/K_m at 40°C, whereas the wild type has a significantly higher k_{cat}/K_m at 21°C. This is due to a significantly lower catalytic efficiency of the 3-LSR mutant at 21°C (Table 1; Poor et al., 2014). Thus, the higher product yield of the RebH variant is due to a longer lifetime, compensating a lower catalytic activity. This shows that directed evolution is one way to create better halogenases but seems not to be the most efficient.

2.2 Stabilization of Halogenases by Formation of Cross-Linked Enzyme Aggregates (CLEAs)

Characterization of kinetic parameters such as K_m, k_{cat} and total turnover numbers of different tryptophan halogenases showed that in vitro enzymatic halogenation is an inefficient reaction (Table 2) which is in a large part due to the instability of the halogenases (Dong et al., 2005; Frese, Guzowska, Voß, & Sewald, 2014; Seibold et al., 2006; Yeh, Garneau, & Walsh, 2005; Zehner et al., 2005). Thus, the biotechnological application of these enzymes is still restricted. Enzyme immobilization has been proven as an efficient method to deal with enzyme stability problems. In contrast to soluble enzymes, immobilized enzymes are more stable, easy to handle, and facilitate efficient recovery and reuse. Cross-linked enzyme aggregates (CLEAs) are biocatalysts prepared via physical enzyme aggregation followed

Table 2 k_{cat} and K_m Values of Different Tryptophan Halogenases for the Chlorination of L-Tryptophan

Halogenase	k_{cat} (min^{-1})	K_m (μM)	References
PrnA	6.8	50	Lang et al. (2011)
Thal/ThdH	2.8	110	Seibold et al. (2006)
PyrH	0.5	150	Zehner et al. (2005)
RebH	1.4	2	Yeh et al. (2005)

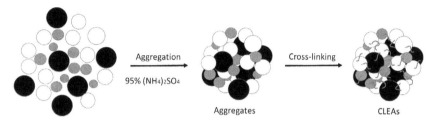

Fig. 3 Scheme for the preparation of CLEAs containing the tryptophan 7-halogenase RebH, the flavin reductase PrnF, and the alcohol dehydrogenase ADH (Frese & Sewald, 2014).

by treatment with a cross-linking agent, most commonly glutaraldehyde (Sheldon, 2011) to covalently link the proteins. Preparation of CLEAs includes two simple steps: precipitation of native enzymes with ammonium sulfate or polyethylene glycol and a cross-linking reaction of free amino groups of lysine residues with neighboring enzyme molecules by addition of glutaraldehyde.

Recently, Frese and Sewald (2014) described the formation of CLEAs for halogenases. These CLEAs contained the tryptophan 7-halogenase RebH from *L. aerocolonigenes* and enzymes required for cofactor regeneration, a flavin reductase from *Pseudomonas fluorescens* BL915 (PrnF; K.-H. van Pée, unpublished) and an alcohol dehydrogenase from *Rhodococcus* sp. (ADH; Fig. 3). Using these CLEAs, halogenation of the natural substrate L-tryptophan as well as D-tryptophan and L-5-hydroxytryptophan was achieved in a gram scale, showing that the long-term stability and the preparative application of tryptophan halogenases can be achieved.

2.2.1 Preparation of CLEAs
1. Inoculate 1.5 L LB medium with an overnight culture of *E. coli* BL21 (DE3) pGro7 containing the plasmid pET28aRebH and the appropriate

antibiotics kanamycin and chloramphenicol. Incubate at 37°C and 150 rpm shaking until OD_{600} reaches 0.4–0.6.
2. Induce the expression of the halogenase gene by addition of 100 μM IPTG and 2 mg/mL arabinose and incubate for further 20 h at 25°C.
3. Harvest cells by centrifugation (4000 × g, 30 min, 4°C).
4. Resuspend cells in 25 mL of 100 mM Na_2HPO_4, pH 7.4, lyse cells twice by French Press, and centrifuge (10,000 × g, 30 min, 4°C).
5. Add 2.5 U/mL flavin reductase PrnF and 1 U/mL alcohol dehydrogenase to the cleared lysate and mix thoroughly.
6. Add 16.2 g of ammonium sulfate (~95% saturation) at 4°C in a tube rotator for 1 h.
7. Add glutaraldehyde to a final concentration of 0.5% (w/v) and incubate for additional 2 h at 4°C in a tube rotator.
8. Centrifuge the resulting CLEAs (10,000 × g, 30 min, 4°C) and wash three times with 50 mL of 100 mM Na_2HPO_4, pH 7.4.
9. Resuspend the solid biocatalyst in 50 mL of 100 mM Na_2HPO_4, pH 7.4, and store overnight at 4°C prior to use.

2.2.2 Halogenation Using CLEAs

CLEAs obtained by the procedure described earlier can be used for batchwise halogenation of tryptophan (Frese & Sewald, 2014). For this purpose, 30 mL of 15 mM Na_2HPO_4, pH 7.4, containing 3 mM tryptophan, 100 μM NAD, 10 μM FAD, 5% (v/v) isopropanol, and 30 mM NaCl or NaBr (for chlorination or bromination, respectively) are used to resuspend CLEAs in a 50-mL Erlenmeyer flask and incubated under slow stirring at 25°C. After 24 h, the reaction solution is centrifuged for 20 min at 10,000 × g, and the supernatant is analyzed by RP-HPLC.

CLEAs can be reused after a washing step with 20 mL of 10 mM Na_2HPO_4 buffer, pH 7.4. It has been demonstrated that CLEAs containing RebH can be recycled at least 10 times with no significant loss of activity.

In our hands, the life time of PyrH under reaction conditions was improved up to about 20 h compared to about 2 h for the free enzyme (Fig. 4). However, after about 20 h, the enzyme became inactive. Obviously, the halogenase inside the CLEAs was stabilized by a yet unknown mechanism.

2.2.3 Gram-Scale Halogenation Reaction Using CLEAs

A serious disadvantage of halogenation reactions using tryptophan halogenases is their low overall turnover number that strongly reduces their

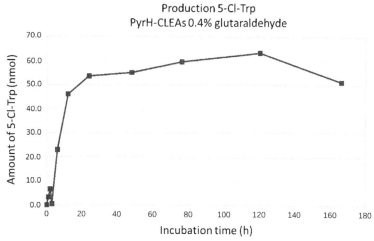

Fig. 4 Plot of the conversion of tryptophan to 5-chlorotryptophan by CLEAs containing the tryptophan 5-halogenase PyrH and the flavin reductase PrnF against time.

industrial application. The CLEA methodology is suitable to overcome stability and enzyme recovery problems. Payne et al. (2013) used 10 L *E. coli* culture overexpressing *rebH* to obtain around 100 mg chlorinated product. Frese and Sewald (2014) achieved gram-scale reaction using CLEAs containing the overproduced halogenase RebH from 6 L *E. coli* culture. The reaction mixture consisted of 15 mM Na$_2$HPO$_4$, pH 7.4, 1 mM tryptophan, 100 µM NAD, 1 µM FAD, 5% (v/v) isopropanol, and 30 mM NaBr in a final volume of 5 L at 25°C and 150 rpm. After 8 days, complete conversion of L-tryptophan was observed, yielding 1.8 g of L-Br-tryptophan.

The conversion of tryptophan to 5-chlorotryptophan by CLEAs of the tryptophan 5-halogenase PyrH and the flavin reductase PrnF is shown in Fig. 4.

3. THE BIOCATALYTIC SCOPE OF FLAVIN-DEPENDENT TRYPTOPHAN HALOGENASES

3.1 Substrate Specificity of Tryptophan Halogenases

The first findings about the substrate specificity of flavin-dependent tryptophan halogenases were published in 2001 for the tryptophan 7-halogenase PrnA from *P. fluorescens* BL915 (Hölzer, Burd, Reißig, & van Pée, 2001). It was shown that this enzyme is able to chlorinate a number of tryptophan, indole, and phenylpyrrole derivatives, but only with its natural substrate tryptophan regioselective chlorination in the position 7 of the indole ring was achieved. With all the compounds different from tryptophan, halogenation proceeded with a relaxed regioselectivity and chlorination mostly occurred at the electronically most activated 2-position of the indole ring. In some cases, even a mixture of different chlorinated products or dichlorination was observed. These findings led to the conclusion that only the natural substrate tryptophan can be bound correctly in the active site of the enzyme PrnA to allow regioselective chlorination and that this might also be the case for other flavin-dependent tryptophan halogenases from other secondary metabolite biosynthetic pathways. During the following years, a multitude of new halogenase genes, among them the genes of several flavin-dependent tryptophan halogenases, was isolated from secondary metabolite gene clusters. In a few cases, the genes were heterologously expressed, the enzymes characterized in regard to their natural substrates and the kinetic parameters of the reaction were determined (Weichold, Milbredt, & van Pée, 2016). In some cases also the 3D structures of the enzymes were solved (Bitto et al., 2008; Dong et al., 2005; Yeh, Blasiak, Koglin, Drennan, & Walsh, 2007; Zhu et al., 2009). No further investigations were published on the substrate specificity of tryptophan halogenases until more than one decade later. The next data on this issue were published by Payne et al. (2013) who described the chlorination and also the bromination of indole, several tryptophan derivatives and also less structurally related compounds like tryptoline and the two differently monosubstituted naphthalenes, 2-aminonaphthalene and 1-hydroxynaphthalene. In 2014, Frese et al. (2014) described the conversion of several fluoro-, methyl-, amino-, and hydroxy-substituted tryptophan derivatives by RebH. Extending the biocatalytic scope of tryptophan halogenases even further, Shepherd et al. (2015) showed that the tryptophan 5-halogenase PyrH and the tryptophan 7-halogenase PrnA are not only able to convert tryptophan derivatives to their respective halogenated analogs but also accept nonindolic substrates such as the aniline derivatives kynurenine and anthranilamide. All the mentioned groups working on the issue of the conversion of compounds others than tryptophan by tryptophan halogenases described different experiments to obtain new halogenated compounds. The protocols differ in expression hosts and fusion tags for protein purification. The reaction conditions during the large scale conversions as well as the methods used for quenching of the reactions and product purification are different which makes it quite difficult to compare the existing protocols. A short overview summarizing the reaction conditions applied for the large-scale conversion of different compounds is given in Table 3.

Table 3 Overview of the Conditions Used for Large Scale Conversions of Different Substrates by Tryptophan Halogenases

Hölzer et al. (2001)	Payne et al. (2013)	Frese et al. (2014)	Shepherd et al. (2015)
Halogenase(s)			
PrmA (produced in *Pseudomonas fluorescens* BL915)	His-RebH (produced in *E. coli* BL21 (DE3))	His-RebH (produced in *E. coli* BL21 (DE3))	His-PrmA and different mutants His-PyrH (produced in *E. coli* Arctic Express)
Substrates			
Indole derivatives, phenylpyrrole derivatives	Indole, tryptophan derivatives, tryptoline, substituted naphthalenes	Hydroxy-, amino-, fluoro-, methyl-substituted tryptophans	Kynurenine, anthranilic acid, anthranilamide
Assay mixtures for the enzymatic halogenation reactions for large scale bioconversions			
4.45 mU PrmA 550 µU Fre 0.1–0.6 mM substrate 50 mM NADH 10 µM FAD 100 mM NaCl → total volume of 64 mL for tryptophan and indole derivatives and 200 mL for phenylpyrrole derivatives → incubation for 16 h at 30°C	25 µM His-RebH 2.5 µM MBP-RebF 0.5 mM substrate 100 µM NAD 100 µM FAD 10 mM NaCl 50 U/mL glucose dehydrogenase 20 mM glucose → in crystallization dishes (125 × 65 or 100 × 50 mm)	His-RebH in cell lysate of 1.5 L culture 2.5 U/mL His-PrnF 2–6 mM substrate 1 mM NADH 10 µM FAD 30 mM NaCl 1 U/mL alcohol dehydrogenase 5% isopropanol (v/v) 20 mM DTT	25 µM His-halogenase 2.5 µM His-Fre 3 mM substrate 10 µM NADH 10 µM FAD 100 mM MgCl$_2$ 12.5 µM glucose dehydrogenase 2 200 mM glucose

Continued

Table 3 Overview of the Conditions Used for Large Scale Conversions of Different Substrates by Tryptophan Halogenases—cont'd

Hölzer et al. (2001)	Payne et al. (2013)	Frese et al. (2014)	Shepherd et al. (2015)
	→ covered with perforated aluminum foil → stirring of reaction mixture with magnetic bar at 60 rpm	→ total volume of 30 mL in Erlenmeyer flask (50 mL) → incubation for 16–24 h at 25°C in orbital shaker at 100 rpm	→ total volume of 10 mL → incubation at 30°C in orbital shaker

Quenching method and product purification

Hölzer et al. (2001)	Payne et al. (2013)	Frese et al. (2014)	Shepherd et al. (2015)
→ extraction with *tert*-butyl methyl ether → evaporation to dryness under vacuum → purification by HPLC	→ addition of 5 M HCl until pH < 2 → addition of NaCl to saturation → filtration of precipitated protein → extraction with $CHCl_2$ → strong cation exchange chromatography → purification by chromatography	→ dialysis of reaction mixture against 5 × 500 mL Millipore water + 0.1% TFA → evaporation of solvent under reduced pressure → preparative HPLC	→ incubation at 95°C for 5 min → centrifugation for 10 min at 4°C and 12,000 × g → semipreparative HPLC

3.1.1 A High-Throughput Assay for the Detection of Halogenase Activity

It has been shown that the substrate specificity of tryptophan halogenases is not as high as originally assumed and that it is even possible to create variants of the native enzymes with modified substrate specificity. Unfortunately, the construction of such improved halogenases is severely hampered by the issue of mutant screening. Every single clone of a mutant library needs to be grown, followed by cell lysis and incubation of the cell-free extracts with potential substrates and finally the enzyme assays need to be analyzed for possible conversion products by HPLC or other methods. This makes the screening of mutant libraries for halogenases accepting new substrates a really tough challenge. To overcome this problem, it is necessary to develop high-throughput techniques using simple detection methods. A publication by Hoshford, Shepherd, Micklefield, and Wong (2014) describes such a high-throughput assay for arylamine halogenation in aqueous solutions. This colorimetric assay is based on a peroxidase (HRP)-catalyzed in situ oxidation of 4-methyl-catechol and the following addition of chlorinated or nonhalogenated arylamines (benzoquinone–amine-coupling reaction) which can be analyzed spectrophotometrically due to a shift of the absorbance maxima of the halogenated compounds in comparison to their nonhalogenated analogs (Fig. 5). Unfortunately, this reaction requires the presence of an arylamine functionality in the molecule which makes this assay suitable for only a limited number of potential halogenase substrates.

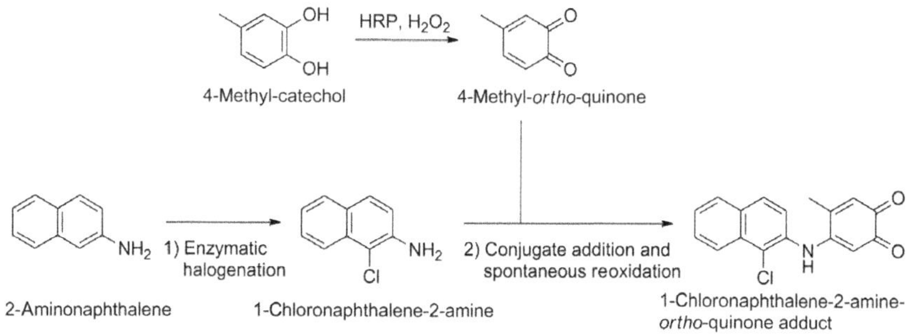

Fig. 5 Reaction sequence for RebH halogenation of 2-naphthylamine followed by in situ oxidation of 4-methyl-catechol and formation of the chlorinated 2-naphthylamine-*ortho*-quinone adduct.

Quantification of the halogenated arylamines can be done according to the method described by Hoshford et al. (2014) for the halogenation of 2-naphthylamine by the tryptophan 7-halogenase RebH:
1. Halogenation reaction
   - The halogenation reaction is carried out in 50 mM K$_2$HPO$_4$ (pH 7.4) with 5% isopropanol (v/v).
   - Add 2-naphthylamine (0.6 mM), NAD (100 μM), FAD (100 μM), NaCl (10 mM), Fre (2.5 μM), glucose dehydrogenase 2 (5 μM), and glucose (20 mM) to make a final volume of 300 μL.
   - Start the reaction by adding RebH to a final concentration of 25 μM.
   - Incubate the reaction at 21°C under constant shaking.
   - Stop the reaction by heating at 95°C for 5 min and remove precipitated protein by centrifugation.
2. Conjugate addition and quantification of halogenated arylamine
   - To 125 μL of the quenched halogenation reaction, add 4-methylcatechol to a final concentration of 0.5 mM (stock solution of 200 mM in 50 mM K$_2$HPO$_4$, pH 7.4) and hydrogen peroxide to a final concentration of 3 mM.
   - To start the reaction, add 1 μL horseradish peroxidase (0.1 mg/mL stock solution in 50 mM K$_2$HPO$_4$, pH 7.4)
   - Incubate under shaking for 5 min.
   - Measure the absorption of the 2-naphthylamine adduct at 516 nm in a spectrophotometer.
   - Quantify product formation by fitting to a calibration curve prepared from known concentrations of 2-naphthylamine and the chlorinated reaction product in reaction buffer containing all the components of the halogenation reaction except the RebH solution.

3.2 Modification of Biosynthetic Pathways Using Tryptophan Halogenases

All the before-mentioned publications dealt with the substrate specificity of mostly native tryptophan halogenase enzymes in in vitro reactions, but during the last few years, a lot of effort was placed into the construction of enzyme variants with modified regioselectivity or substrate specificity obtained by error-prone or site-directed mutagenesis and the use of native halogenases and respective variants in metabolic engineering approaches. Introducing genes coding for tryptophan halogenases with different

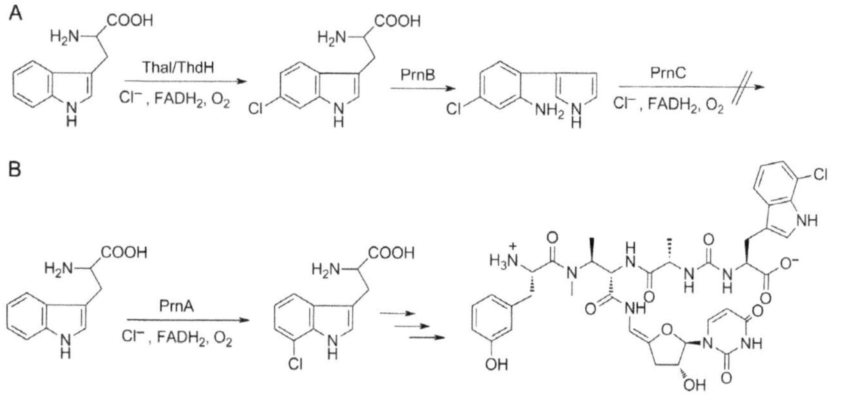

Fig. 6 Effect of the introduction of (A) the tryptophan 6-halogenase gene *thal/thdH* into the pyrrolnitrin producer *Pseudomonas chlororaphis* on pyrrolnitrin biosynthesis (Seibold et al., 2006) and (B) the tryptophan 7-halogenase gene *prnA* into the pacidamycin producer *Streptomyces coeruleorubidus* on pacidamycin biosynthesis (Roy, Grüschow, Cairns, & Goss, 2010).

regioselectivities into bacteria or plants producing halogenated secondary metabolites resulted in the production of variants of those halogenated compounds. Seibold et al. (2006) introduced the tryptophan 6-halogenases gene (*thal/thdH*) from the biosynthesis of thienodolin in *Streptomyces albogriseolus* (Milbredt et al., 2014) into the pyrrolnitrin producer *Pseudomonas chlororaphis* ACN. Although this strain still contained the tryptophan 7-halogenase gene *prnA*, overexpression of the tryptophan 6-halogenase gene *thal/thdH* was so high that all available tryptophan was converted to 6-chlorotryptophan thus no 7-chlorotryptophan was available for pyrrolnitrin biosynthesis. However, 6-chlorotryptophan formed by the reaction of Thal/ThdH was accepted as a substrate by the second enzyme in the pyrrolnitrin biosynthetic pathway, PrnB, resulting in the formation of the phenylpyrrole derivative 3-(2'-amino-4'-chlorophenyl)pyrrole. But the next enzyme, the monodechloroaminopyrrolnitrin 3-halogenase (PrnC; Hohaus et al., 1997; Kirner et al., 1998) did not accept this compound as a substrate which was thus not further metabolized and accumulated. This example shows the importance of the substrate specificity of the downstream enzymes in the biosynthetic pathway (Fig. 6A). The gene for the tryptophan 6-halogenase Thal/ThdH was introduced into *P. chlororaphis* by biparental conjugation as described later.

3.2.1 Introduction of thal/thdH into P. chlororaphis and Isolation of 3-(2'-Amino-4'-Chlorophenyl)Pyrrole

The donor strain *E. coli* S17-1 containing the plasmid pCIBhisthal (Seibold et al., 2006) and the recipient strain *P. chlororaphis* ACN were precultured in 5 mL of HNB medium at 37°C and 30°C, respectively, overnight. 100 µL of each culture were pipetted onto the center of an agar plate containing *Pseudomonas* minimal medium (PMM; Kirner et al., 1996) and the antibiotic tetracycline (20 µg/mL). After incubation at 30°C for 8 h the bacteria were scratched from the plate, diluted, and plated onto PMM agar plates containing tetracyclin for selection of *Pseudomonas* clones containing the plasmid pCIBhisthal, since under these conditions, only recombinant *Pseudomonas* clones were able to grow. The obtained clones could be used for overproduction of the tryptophan 6-halogenase Thal/ThdH or for the isolation of 3-(2'-amino-4'-chlorophenyl)pyrrole. For this purpose, *P. chlororaphis* ACN containing pCIBhisthal was grown in 6 x 1 L of HNB medium at 30°C for 96 h. After removal of the cells by centrifugation, the medium was extracted twice with 1.5 L of methyl-*tert*-butyl ether. The organic extracts were combined and evaporated to dryness in vacuo. The residue was dissolved in 2 mL of methanol and loaded onto a LH-20 column and eluted with methanol. Elution of metabolites was monitored at 254 nm. Fractions showing an UV spectrum with a maximum at 303 nm were combined and purified further by preparative HPLC using an Eurosphere-100 RP-18 column with water/methanol (35:65) as the eluent. The structure of the purified product was elucidated by HPLC-MS and ^{1}H-NMR.

3.2.2 In Vivo Modification of Biosynthetic Pathways Using Halogenase Genes

Roy et al. (2010) used the introduction of the gene of the tryptophan 7-halogenase PrnA from pyrrolnitrin biosynthesis (Kirner et al., 1998) into the pacidamycin producer *Streptomyces coeruleorubidus* for the modification of the peptide antibiotic pacidamycin. Due to the activity of PrnA, tryptophan was chlorinated to 7-chlorotryptophan and used as a substrate by the non-ribosomal peptide synthetase involved in pacidamycin biosynthesis resulting in the production of chloropacidamycin (Fig. 6B). Runguphan, Qu, and O'Connor (2010) were able to introduce the gene of the tryptophan 7-halogenase RebH into the medicinal plant *Catharanthus roseus*. They also introduced the flavin reductase gene *rebF* into the plant which is not necessary in the bacterial systems, because there are several flavin reductases in the bacteria which can provide the halogenase with reduced flavin. It is not clear, whether transformation of plants with a flavin reductase gene is absolutely necessary. Runguphan et al. (2010) obtained halogenated tryptophan-derived alkaloids, some of which were identified as halogenated analogs of known alkaloids produced by the medicinal plant, but due to the substrate specificity of downstream enzymes present in the plant, also new chlorinated tryptophan-based compounds were formed. It was observed that the indole alkaloid formation from 7-chlorotryptophan in *C. roseus* is limited by the activity of the tryptophan decarboxylase converting tryptophan to tryptamine. Based on the crystal structure of RebH and thus the knowledge of the exact coordination of its natural substrate tryptophan in the active site of the enzyme, a tyrosine residue interacting with the carboxylic acid group of tryptophan was exchanged by a sterically more demanding tryptophan residue. This mutation resulted in the preferential halogenation of tryptamine compared to tryptophan. This clearly demonstrates that it is possible to change the substrate specificity of flavin-dependent tryptophan halogenases by site-specific mutagenesis.

3.3 Chemical Substitution of Enzymatically Introduced Halogen Atoms

Halogen atoms can have a profound influence on the biological activity of compounds and in addition they can serve as a useful handle for further chemical derivatization. Bromine and chlorine atoms can be substituted to afford aryl and heteroaryl analogs. Roy, Goss, Wagner, and Winn (2008) developed a method for Suzuki-Miyaura cross-coupling of halogenated tryptophans obtained by enzymatic synthesis from the corresponding haloindoles and serine by a simple one-pot reaction (Goss & Newill, 2006). They used the water-soluble phosphine ligand *tris*(4,6-dimethyl-3-sulfonatophenyl) phosphine trisodium salt (TXPTS) and various arylboronic acids and achieved yields of up to 90%. Since the reaction also worked with small peptides, they adapted the reaction to chlorinated pacidamycin. They achieved the modification of not only purified chloropacidamycin but also the modification of chloropacidamycin in crude extracts (Roy et al., 2010).

Runguphan and O'Connor (2013) described modification of chlorinated alkaloids obtained by in vivo chlorination of the precursor tryptophan by the tryptophan 7-halogenase RebH in crude extracts by Pd-catalyzed Suzuki-Miyaura cross-coupling. Chlorinated alkaloids obtained via the exogenous addition of 6-chlorotrytophan to cultures of the plant *C. roseus* could also be modified by cross-coupling.

3.4 Modification of the Regioselectivity of Tryptophan Halogenases

The regioselectivity of the tryptophan 7-halogenase PrnA could be modified by site-directed mutagenesis (Lang et al., 2011). Based on the three-dimensional structure of PrnA and the binding of the substrate tryptophan seen in this structure, amino acid residues involved in the binding of the substrate via ionic interactions with the amino and carboxylic acid group, and/or with the indole ring were exchanged. The exchange of a phenylalanine residue shielding the pyrrole ring from attack by the halogenating agent against a much smaller alanine residue led to a modification of the regioselectivity. Obviously, this mutation allowed the substrate a higher degree of flexibility in the active site and resulted in the possibility of an attack of the substrate also in the 5-position in addition to halogenation at the 7-position. The procedure for the construction of the PrnAF103A variant with modified regioselectivity and the use of this mutant for the formation of 7-chloro- and 5-chlorotryptophan is described below.

1. The primers used for the construction of the His-tagged PrnAF103A variant by overlap extension PCR were:

    primer a: 5′-GACTCTAGAGG**GGATCC**CATGAACAAGCC GATCAAGAAT-3′ (sense)

    primer b: 5′-GTTCGGCACGTTGCCGGCCAAATGGTAGAA GT-3′ (antisense)

    primer c: 5′-ACTTCTACCATTTGGCCGGCAACGTGCC GAAC-3′ (sense)

    primer d: 5′-ATC**AAGCTT**CTACAGGCTTTCCTGCGCTGC GAGCTT-3′ (antisense).

    Mutagenic sites are underlined, and restriction sites are in bold. The template used was pUC-*prnA* (Dong et al., 2005).

2. The fusion fragment obtained by overlap extension PCR was ligated into pBluescript II SK (+) and introduced into *E. coli* TG1 by electroporation.

3. For expression of the mutated halogenase gene, the gene was ligated into the *E. coli-Pseudomonas* shuttle vector pCIBhis (Wynands & van Pée, 2004) and introduced into *P. fluorescens* BL915 ΔORF 1–4 (Hammer, Hill, Lam, van Pée, & Ligon, 1997) by triparental conjugation (Ditta, Stanfield, Corbin, & Helinski, 1980).

4. Purification of His-tagged enzyme was performed by Ni-chelating affinity chromatography using a sepharose FF column.

5. Halogenating activity was analyzed using the following reaction mixture: 50 μL enzyme-containing protein solution, 10 mU flavin reductase, 1 μM FAD, 2.4 mM NADH, 12.5 mM MgCl$_2$, 100 U catalase, and 0.25 mM L-tryptophan in a total volume of 200 μL in 10 mM potassium phosphate buffer, pH 7.2. After incubation at 30°C for 30 min, the reaction was stopped by incubation at 95°C for 5 min. Denatured proteins were removed by centrifugation, and the solution was analyzed by HPLC. One unit of halogenating activity is defined as the formation of 1 μmol product per min.
6. For identification of the reaction products formed by the PrnAF103A variant, a large scale preparation was performed. The assay mixture consisted of 850 μL enzyme solution in 10 mM potassium phosphate buffer, pH 7.2, 50 mU flavin reductase, 10 μM FAD, 2.4 mM NADH, 12.5 mM NaBr, and 0.25 mM L-tryptophan in a total volume of 1000 μL.
7. After incubation at 30°C for 4 h, the reaction was stopped by boiling in a water bath for 5 min. Denatured protein was removed by centrifugation, and the supernatant was loaded onto a solid phase extraction column, equilibrated with methanol and water. After washing with 10% methanol, the brominated products were eluted with 100% methanol. The eluates of 10 large scale preparations were concentrated in vacuo.
8. The two halogenated products were separated and purified by HPLC using a RP18 column with methanol:water (40:60), 0.1% TFA (v/v) as the eluent.
9. The reaction products were identified by ^{1}H-NMR and ESI-MS.

4. SYNTHESIS OF PCP-BOUND HALOGENASE SUBSTRATES

The so far characterized tryptophan halogenases all accept a free substrate. However, most of the flavin-dependent halogenases and also most of the nonhaem iron-dependent halogenases require a substrate bound to a peptidyl carrier or acyl carrier protein (Weichold et al., 2016). The preparation of these substrates is highly challenging. Attempts to circumvent the synthetic problems by using the much easier to synthesize SNAC-derivatives of the substrates with the small N-acetylcysteamine moiety mimicking the phosphopantetheinyl arm of holo-carrier proteins in the case of thioesterases and NRPS condensation domains (Ehmann, Trauger, Stachelhaus, & Walsh, 2000; Yeh, Kohli, Bruner, & Walsh, 2004) failed, probably due to the substrate specificity of the halogenases and a proposed conformational change upon interaction with the protein-coupled substrate (Pang, Garneau-Tsodikova, & Tsodikov, 2015). Instead, chemoenzymatically or purely enzymatically synthesized protein-bound substrates were used (Agarwal et al., 2014; Dorrestein, Yeh, Garneau-Tsodikova, Kelleher, & Walsh, 2005; Lin, Van Lanen, & Shen, 2007; Schmartz, Zerbe, Abou-Hadeed, & Robinson, 2014).

So far, there are only a few examples for synthesis and use of protein-bound substrates to show halogenase activity. One method has previously been described by Li et al. (2009) for the synthesis of a vancomycin precursor hexapeptide. Only the hexapeptide precursor with a β-hydroxytyrosine residue at the C-terminus and not the corresponding dipeptide with a β-hydroxytyrosine residue at the C-terminus was accepted by the $FADH_2$-dependent halogenase VhaA (Schmartz et al., 2014).

Synthesis of a PCP-bound β-tyrosine and halogenation by SgcC3 from C-1027 biosynthesis was shown by Lin, Huang, and Shen (2012).

For some nonhaem iron, α-ketoglutarate- and O_2-dependent halogenases, activity has been shown by providing them with enzymatically derived carrier protein-tethered amino acids. Exemplarily, carrier protein-tethered L-threonine is used by SyrB2 from syringomycin E biosynthesis and peptide-bound L-allo-isoleucine is accepted by CmaB from coronatine biosynthesis yielding the 4-chloro-derivatives (Vaillancourt, Yeh, Vosburg, O'Connor, & Walsh, 2005; Vaillancourt, Yin, & Walsh, 2005). Trichlorination in position 5 of peptide-bound leucine can be achieved by BarB1 and BarB2 from barbamide biosynthesis (Galonic, Vaillancourt, & Walsh, 2006).

For the formation of PCP-tethered pyrrole-2-carboxylic acid substrates, two different main strategies have been used. The first one relies on a fully enzymatic catalysis based on the natural production of pyrrole residues from proline (Fig. 7) (Dorrestein et al., 2005). However, the enzymatic synthesis of PCP-bound pyrrole-2-carboxylic acid requires the heterologous expression of a number of genes, and the purification of the corresponding proteins and enzymes can therefore be rather complicated. The second strategy uses organic synthesis to create a coenzyme A-coupled pyrrolyl residue that is then enzymatically transferred to an apo-PCP by a nonspecific phosphopantetheinyl transferase (Garneau-Tsodikova, Stapon, Kahne, & Walsh, 2006).

4.1 Enzymatic Synthesis of Pyrrolyl-S-PCPs

For enzymatic synthesis of pyrrolyl-S-PCPs, the PCPs are required in their holo-form, but expression of PCP genes in *E. coli* yields predominantly the

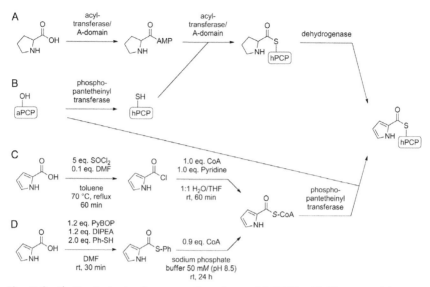

Fig. 7 Synthetic strategies for generation of pyrrolyl-S-PCPs. (A) The natural biosynthetic pathway that can also be reproduced in vitro. (B) Completion of apo-PCP (aPCP) by transferring a phosphopantetheine structure to it forms holo-PCP (hPCP) in vivo and in vitro. (C) and (D) Synthesis of pyrrolyl-S-CoA via pyrrole-2-carboxylic acid chloride or S-phenylpyrrole-2-carbothioate which can be transferred to apo-PCP by a nonspecific phosphopantetheinyl transferase (for conditions of the enzymatic in vitro reactions see text).

apo-PCPs (Gocht & Marahiel, 1994). Thus, it is necessary to attach a phosphopantetheinyl moiety which can be done by treatment with a phosphopantetheinyl transferase and coenzyme A or coexpression with the transferase gene. For example, Thomas, Burkart, and Walsh (2002) obtained more than 95% holo-PCP by coexpressing *sfp*, the gene of an unspecific phosphopantetheinyl transferase from *Bacillus subtilis*, together with the gene of the PCP, compared to only 25% without coexpression.

Enzymatic synthesis of PCP-pyrrole-2-carbothioates was demonstrated for pyoluteorin biosynthesis by Dorrestein et al. (2005) using previously published protocols (Thomas et al., 2002).

To generate holo-PCP, apo-PCP (52 μM) was treated with Sfp (1.1 μM) and coenzyme A (295 μM) supplied with $MgCl_2$ (7.1 μM) in 300 μL reaction volume and was incubated for 1 h.

For adenylation of proline and transfer to the PCP, the enzyme CouN4 from coumermycin biosynthesis was used, and for oxidation to form the pyrrolyl-residue ClON3 from clorobiocin biosynthesis was employed.

The previously provided reaction mixture was used without further purification and supplied with CouN4 (84 µg/mL), L-proline (4.2 mM), ATP (3.3 mM) followed by ClonN3 (2.7 µM), and FAD (150 µM) in a total volume of 720 µL in 50 mM Tris/HCl (pH 7.4) buffer containing 1 mM TCEP.

Incubation at room temperature for 3 h led to the formation of pyrrolyl-S-PCP which could be detected by LC/MS/MS and could be used as the substrate for the halogenase PltA from pyoluteorin biosynthesis.

Another protocol for the one-step preparation of pyrrolyl-S-PCP with proteins from the biosynthesis of nargenicin A_1 was described by Maharjan et al. (2012).

4.2 Chemoenzymatic Synthesis of Pyrrolyl-S-PCPs

4.2.1 Synthesis of Pyrrolyl-S-CoA via Acid Chloride

Garneau-Tsodikova et al. (2006) described a method for activation of pyrrole carboxylic acid via the acid chloride. Pyrrole-2-carboxylic acid (100 mg) was stirred in dried toluene (5 mL) with a catalytic amount of DMF (7 µL, 0.1 eq.) which is meant to form the actually chlorinating (chloromethylene)dimethyliminium later on (Fritz & Oehl, 1971; Jugie, Martin, & Smith, 1975). Then thionyl chloride (328 µL, 5 eq.) was added, and the mixture was heated at 70°C for 1 h using a reflux condenser. Afterwards, the reaction mixture was concentrated by a rotary evaporator, 5 mL of dry toluene were added and concentrated again. The residue was then dried under vacuum for at least 30 min. It can also be stored at −20°C under dry conditions for several days.

To coenzyme A trilithium salt (20 mg, 1 eq.) in a 1:1 mixture of THF/ H_2O with pyridine (1.9 µL, 1 eq.), pyrrole-2-carbonyl chloride (3.3 mg, 1 eq.) was added. The mixture was stirred for 1 h at room temperature and then purified by semipreparative HPLC in 50 µL portions diluted by H_2O to 1 mL. For this purpose a 60 min gradient from H_2O with 0.1% TFA to 50% MeCN/H_2O with 0.1% TFA was used on an Agilent Zorbax 5 µm 300SB-C18 semiprep HPLC column with a flow rate of 5 mL/min and detection of CoA thioesters at 310 nm.

4.2.2 Synthesis of Pyrrolyl-S-CoA via S-Phenyl Thioates

Agarwal et al. (2014) used a traditional peptide-coupling reaction as described by Li et al. (2009) to obtain S-phenyl-pyrrole-2-carbothioate that could be transferred to coenzyme A via a transesterification reaction. To a solution of pyrrole-2-carboxylic acid (20 mg) in DMF (500 µL) PyBOP

(112 mg, 1.2 eq.), DIPEA (38 μL, 1.2 eq.) and thiophenol (37 μL, 2 eq.) were added. The mixture was stirred at room temperature for 30 min, quenched by saturated brine, and extracted with ethyl acetate. After drying over $MgSO_4$, the organic phases were concentrated in vacuo and chromatographed on a silica column using a 40:1 mixture of hexane/ethyl acetate which yielded 27% of S-phenyl-pyrrole-2-carbothioate.

4 mg of the S-phenyl-pyrrole-2-carbothioate were dissolved in 200 μL THF and added to 300 μL of sodium phosphate buffer (50 mM, pH 8.5) containing coenzyme A sodium salt (14 mg, 0.9 eq.). The mixture was stirred for 24 h at room temperature and subsequently chromatographed on a reverse-phase C18 semipreparative HPLC column using a gradient from 5% to 100% MeCN with 0.1% TFA against H_2O with 0.1% TFA over 30 min. Lyophilization of product-containing fractions yielded 7% (2% overall) of the pyrrolyl–CoA thioester.

4.3 Transfer of Pyrrolyl-S-CoA Thioesters to Carrier Proteins and Halogenation of the Substrate

A procedure for the transfer of synthesized CoA thioesters to apo-PCP was previously described by Li et al. (2009) for a hexapeptidyl-CoA thioester which works also well for the pyrrolyl-CoA thioester. All components were dissolved in 50 mM Tris/HCl (pH 7.5) reaction buffer and added as follows. The purified apo-PCP (120 μM) was supplied with the pyrrolyl-CoA thioester (100 μM), the unspecific phosphopantetheinyl transferase Sfp (5 μM) and $MgCl_2$ (50 mM). The mixture was incubated for 30 min at 37°C and then purified by HPLC.

Agarwal et al. (2014) did not purify the product but directly used the reaction mixture for enzymatic halogenation. To 100 μL of the PCP-pyrrole-2-carbothioate-containing mixture 5 mM NADPH, 100 μM FAD, 100 mM KBr for bromination, 20 μM halogenase, and 6 μM SsuE flavin reductase were added in 50 mM Tris buffer (pH 8.0). The reaction was maintained by periodic addition of additional NADPH (5 mM every hour).

So far, no kinetic data for the halogenation of PCP-tethered pyrrole-2-carboxylic acid could be obtained due to the very low activity of these reactions.

4.4 Release of Halogenated Pyrrole-2-Carboxylic Acid from PCPs

Although PCP-coupled substrates and products can be analyzed by MALDI-TOF (Garneau-Tsodikova et al., 2006; Thomas et al., 2002) or

LC/MS/MS after proteolytic digestion (Agarwal et al., 2014; Dorrestein et al., 2006), in many cases a release from the PCP is desired.

Thomas et al. (2002) described a method to release the products from the PCP hydrolytically by treatment with KOH. For this purpose the proteins from a sample (500 µL) were precipitated by addition of trichloroacetic acid up to 5–10% (w/v) and pelleted by centrifugation. The pellet was then washed twice with H_2O and resuspended in 200 µL of 0.1 M KOH. Incubation at 65°C for 15 min, addition of TFA for neutralization, and removal of precipitated proteins by centrifugation or filtration lead to halogenated pyrrole-2-carboxylic acid. However, in our hands release by KOH treatment did not work satisfactorily; we could hardly detect any released pyrrol-2-carboxylic acid.

Another strategy inspired by Li et al. (2009) was employed by Schmartz et al. (2014) who treated the vancomycin precursor hexapeptide-PCP conjugate with 1/10 volume of an aqueous hydrazine solution (25% v/v) at 30°C for 30 min to lyse the thioester bond freeing the peptide hydrazide. However, in our hands this did not lead to detectable amounts of released pyrrole carbonyl hydrazide.

Dorrestein et al. (2005) used the type II thioesterase TycF from tyrocidine biosynthesis (Mootz & Marahiel, 1997) for the enzymatic release of pyrrole-2-carboxylic acid from the PCP according to a protocol from Yeh et al. (2004), by incubating the PCP-pyrrole-2-carbothioate in 50 mM Hepes pH 7.5 with 5 mM $MgCl_2$ and 1 mM TCEP with 10 µM TycF to release the substrate.

5. CONCLUSIONS

The lack of stability of flavin-dependent halogenases under reaction conditions poses a serious problem for their use in industrial applications. To improve enzyme stability, the construction of mutants is required. However, for screening of mutants, the methods available for high-throughput screening have to be improved considerably, especially in their general applicability. Elucidation of the reason for the loss of activity should allow the modification of the enzymes in such a way that they cannot be inactivated anymore. The substrate scope of tryptophan halogenases has been broadened substantially. Error-prone PCR and site-directed mutagenesis in combination with high-throughput assays should allow the improvement of halogenating activity toward substrates, now only halogenated with a very low activity and should thus tremendously increase the suitability of halogenases for industrial application. The in vivo use of halogenases will lead to the production of novel compounds with new or improved biological activities and the possibility to exchange the halogen substituents via cross-coupling reactions increases the possibility for the formation of novel compounds tremendously. The requirement of protein-bound substrates hampers the use of many halogenases in in vitro reactions enormously, since the amount of substrate obtained by these syntheses is rather low as well as the activities of the enzymes. Here, the in vivo use seems to be a good alternative. The future for enzymatic halogenation reactions looks very promising, especially with a steadily increasing number of potent groups working on this subject.

ACKNOWLEDGMENTS
Thanks to Sarah O'Connor for critically reading this manuscript. We would also like to express our thanks to ERA-IB, the Sächsische Staatsministerium für Wissenschaft und Kunst (SMWK), the Bundesministerium für Bildung und Forschung (BMBF), and the Deutsche Forschungsgemeinschaft (DFG) for supporting the research on enzymatic halogenation performed in the authors' laboratory.

REFERENCES
Agarwal, V., El Gamal, A. A., Yamanaka, K., Poth, D., Kersten, R. D., Schorn, M., et al. (2014). Biosynthesis of polybrominated aromatic organic compounds by marine bacteria. *Nature Chemical Biology*, *10*, 640–647.
Bitto, E., Huang, Y., Bingman, C. A., Singh, S., Thorson, J. S., & Phillips, G. N., Jr. (2008). The structure of flavin-dependent tryptophan 7-halogenase RebH. *Proteins*, *70*, 289–293.
Ditta, G., Stanfield, S., Corbin, D., & Helinski, D. R. (1980). Broad host range DNA cloning system for Gram-negative bacteria: Construction of a gene bank of *Rhizobium meliloti*. *Proceedings of the National Academy of Sciences of the United States of America*, *77*, 7347–7351.
Dong, C., Flecks, S., Unversucht, S., Haupt, C., van Pée, K.-H., & Naismith, J. H. (2005). Tryptophan 7-halogenase (PrnA) structure suggests a mechanism for regioselective chlorination. *Science*, *309*, 2216–2219.
Dorrestein, P. C., Bumpus, S. B., Calderone, C. T., Garneau-Tsodikova, S., Aron, Z. D., Straight, P. D., et al. (2006). Facile detection of acyl- and peptidyl-intermediates on thiotemplate carrier domains via phosphopantetheinyl elimination reactions during tandem mass spectrometry. *Biochemistry*, *45*, 12756–12766.
Dorrestein, P. C., Yeh, E., Garneau-Tsodikova, S., Kelleher, N. L., & Walsh, C. T. (2005). Dichlorination of a pyrrolyl-S-carrier protein by $FADH_2$-dependent halogenase PltA during pyoluteorin biosynthesis. *Proceedings of the National Academy of Sciences of the United States of America*, *102*, 13843–13848.
Ehmann, D. E., Trauger, J. W., Stachelhaus, T., & Walsh, C. T. (2000). Aminoacyl-SNACs as small-molecule substrates for the condensation domains of nonribosomal peptide synthetases. *Chemistry & Biology*, *7*, 765–772.
Ernyei, A. J. (2008). *Untersuchungen zur Substratspezifität der Tryptophan-5- und der Tryptophan-6-Halogenase und Optimierung der Reaktionen*. Doctoral thesis, Fachrichtung Chemie und Lebensmittelchemie, Technische Universität Dresden.

Flecks, S., Patallo, E. P., Zhu, X., Ernyei, A. J., Seifert, G., Schneider, A., et al. (2008). New insights into the mechanism of enzymatic chlorination of tryptophan. *Angewandte Chemie International Edition, 47*, 9533–9536.

Frese, M., Guzowska, P. H., Voß, H., & Sewald, N. (2014). Regioselective enzymatic halogenation of substituted tryptophan derivatives using the FAD-dependent halogenase RebH. *ChemCatChem, 6*, 1270–1276.

Frese, M., & Sewald, N. (2014). Enzymatic halogenation of tryptophan on a gram scale. *Angewandte Chemie International Edition, 54*, 298–301.

Fritz, H., & Oehl, R. (1971). Halogen-Sauerstoff-Austausch zwischen Formamidchloriden und Formamiden, nachgewiesen bei der Einfuhrung deuterierter Formylgruppen nach Vilsmeier. *Liebigs Annalen der Chemie, 749*, 159–167.

Galonic, D. P., Vaillancourt, F. H., & Walsh, C. T. (2006). Halogenation of unactivated carbon centers in natural product biosynthesis: Trichlorination of leucine during barbamide biosynthesis. *Journal of the American Chemical Society, 128*, 3900–3901.

Garneau-Tsodikova, S., Stapon, A., Kahne, D., & Walsh, C. T. (2006). Installation of the pyrrolyl-2-carboxyl pharmacophore by CouN1 and CouN7 in the late biosynthetic steps of the aminocoumarin antibiotics clorobiocin and coumermycin A1. *Biochemistry, 45*, 8568–8578.

Gocht, M., & Marahiel, M. A. (1994). Analysis of core sequences in the D-Phe activating domain of the multifunctional peptide synthetase TycA by site-directed mutagenesis. *Journal of Bacteriology, 176*, 2654–2662.

Goss, R. J. M., & Newill, P. L. A. (2006). A convenient enzymatic synthesis of L-halotryptophans. *Chemical Communications, 2006*, 4924–4925.

Hammer, P. E., Hill, D. S., Lam, S. T., van Pée, K.-H., & Ligon, J. M. (1997). Four genes from *Pseudomonas fluorescens* that encode the biosynthesis of pyrrolnitrin. *Applied and Environmental Microbiology, 63*, 2147–2154.

Hohaus, K., Altmann, A., Burd, W., Fischer, I., Hammer, P. E., Hill, D. S., et al. (1997). NADH-dependent halogenases are more likely to be involved in halometabolite biosynthesis than haloperoxidases. *Angewandte Chemie International Edition in English, 36*, 2012–2013.

Hölzer, M., Burd, W., Reißig, H.-U., & van Pée, K.-H. (2001). Substrate specificity and regioselectivity of tryptophan 7-halogenase from *Pseudomonas fluorescens* BL915. *Advanced Synthesis and Catalysis, 343*, 591–595.

Hoshford, J., Shepherd, S. A., Micklefield, J., & Wong, L. S. (2014). A high-throughput assay for arylamine halogenation based on a peroxidase-mediated-quinone amine coupling with applications in the screening of enzymatic halogenations. *Chemistry - A European Journal, 20*, 16759–16763.

Jugie, G., Martin, G., & Smith, J. A. S. (1975). Nuclear magnetic resonance investigations of carbonium ion intermediates. Part III. A chlorine-35 quadrupole resonance study of several (R-chloromethylene)dimethylammonium salts (Vilsmeier-Haack and Viehe reagents). *Journal of the Chemical Society. Perkin Transactions II, 516*, 925–927.

Keller, S., Wage, T., Hohaus, K., Hölzer, M., Eichhorn, E., & van Pée, K.-H. (2000). Purification and partial characterization of tryptophan 7-halogenase (PrnA) from *Pseudomonas fluorescens*. *Angewandte Chemie International Edition, 39*, 2300–2302.

Kirner, S., Hammer, P. E., Hill, D. S., Altmann, A., Fischer, I., Weislo, L. J., et al. (1998). Functions encoded by pyrrolnitrin biosynthetic genes from *Pseudomonas fluorescens*. *Journal of Bacteriology, 180*, 1939–1943.

Kirner, S., Krauss, S., Sury, G., Lam, S. T., Ligon, J. M., & van Pée, K.-H. (1996). The non-haem chloroperoxidase from *Pseudomonas fluorescens* and its relationship to pyrrolnitrin biosynthesis. *Microbiology, 142*, 2129–2135.

Lang, A., Polnick, S., Nicke, T., William, P., Patallo, E. P., Naismith, J. H., et al. (2011). Changing the regioselectivity of the tryptophan 7-halogenase PrnA by site-directed mutagenesis. *Angewandte Chemie International Edition, 50*, 2951–2953.

Li, D. B., Woithe, K., Geib, N., Abou-Hadeed, K., Zerbe, K., & Robinson, J. A. (2009). In vitro studies of phenol coupling enzymes involved in vancomycin biosynthesis. *Methods in Enzymology, 458*, 487–509.

Lin, S., Huang, T., & Shen, B. (2012). Tailoring enzymes acting on carrier protein-tethered substrates in natural product biosynthesis. *Methods in Enzymology, 516*, 321–343.

Lin, S., Van Lanen, S. G., & Shen, B. (2007). Regiospecific chlorination of (S)-β-tyrosyl-S-carrier protein catalyzed by SgcC3 in the biosynthesis of the enediyne antitumor antibiotic C-1027. *Journal of the American Chemical Society, 129*, 12432–12438.

Maharjan, S., Aryal, N., Bhattarai, S., Koju, D., Lamichhane, J., & Sohng, J. K. (2012). Biosynthesis of the nargenicin A1 pyrrole moiety from *Nocardia* sp. CS682. *Applied Microbiology and Biotechnology, 93*, 687–696.

Milbredt, D., Patallo, E. P., & van Pée, K.-H. (2014). A tryptophan 6-halogenase and an amidotransferase are involved in thienodolin biosynthesis. *ChemBioChem, 15*, 1011–1020.

Mootz, H. D., & Marahiel, M. A. (1997). The tyrocidine biosynthesis operon of *Bacillus brevis*: Complete nucleotide sequence and biochemical characterization of functional internal adenylation domains. *Journal of Bacteriology, 179*, 6843–6850.

Muffler, K., Kuetchou Ngnigha, A. R., & Ulber, R. (2010). Bestimmung kinetischer Parameter der FADH$_2$-abhängigen Tryptophan-5-Halogenase aus *Streptomyces rugosporus*. *Chemie Ingenieur Technik, 82*, 121–127.

Pang, A. H., Garneau-Tsodikova, S., & Tsodikov, O. V. (2015). Crystal structure of halogenase PltA from the pyoluteorin biosynthetic pathway. *Journal of Structural Biology, 192*, 349–357.

Payne, J. T., Andorfer, M. C., & Lewis, J. C. (2013). Regioselective arene halogenation using the FAD-dependent halogenase RebH. *Angewandte Chemie International Edition, 52*, 5271–5274.

Poor, C. B., Andorfer, M. C., & Lewis, J. C. (2014). Improving the stability and catalyst lifetime of the halogenase RebH by directed evolution. *ChemBioChem, 15*, 1286–1289.

Roy, A. R., Goss, R. J. M., Wagner, G. K., & Winn, M. (2008). Development of fluorescent aryltryptophans by Pd mediated cross-coupling of unprotected halotryptophans in water. *Chemical Communications*, 4831–4833.

Roy, A. D., Grüschow, S., Cairns, N., & Goss, R. J. M. (2010). Gene expression enabling synthetic diversification of natural products: Chemogenetic generation of pacidamycin analogs. *Journal of the American Chemical Society, 132*, 12243–12245.

Runguphan, W., & O'Connor, S. E. (2013). Diversification of monoterpene indole alkaloid analogs through cross-coupling. *Organic Letters, 15*, 2850–2853.

Runguphan, W., Qu, X., & O'Connor, S. E. (2010). Integrating carbon-halogen bond formation into medicinal plant metabolism. *Nature, 468*, 461–464.

Sanchez, C., Butovich, I. A., Brana, A. F., Rohr, J., Mendez, C., & Salas, J. A. (2002). The biosynthetic gene cluster for the antitumor rebeccamycin. Characterization and generation of indolocarbazole derivatives. *Chemistry & Biology, 9*, 519–531.

Schmartz, P. C., Zerbe, K., Abou-Hadeed, K., & Robinson, J. A. (2014). Bis-chlorination of a hexapeptide-PCP conjugate by the halogenase involved in vancomycin biosynthesis. *Organic & Biomolecular Chemistry, 12*, 5574–5577.

Seibold, C., Schnerr, H., Rumpf, J., Kunzendorf, A., Hatscher, C., Wage, T., et al. (2006). A flavin-dependent tryptophan 6-halogenase and its use in modification of pyrrolnitrin biosynthesis. *Biocatalysis and Biotransformation, 24*, 401–408.

Sheldon, R. A. (2011). Characteristic features and biotechnological applications of cross-linked enzyme aggregates (CLEAs). *Applied Microbiology and Biotechnology, 92*, 467–477.

Shepherd, S. A., Karthikeyan, C., Latham, J., Struck, A.-W., Thompson, M. L., Menon, B. R. K., et al. (2015). Extending the biocatalytic scope of regiocomplementary flavin-dependent halogenase enzymes. *Chemical Science, 6*, 3454–3460.

Stadtman, E. R., & Levine, R. L. (2006). Chemical modification of proteins by reactive oxygen species. In I. Dalle-Donne, A. Scaloni, & D. A. Butterfield (Eds.), *Redox proteomics: From protein modifications to cellular dysfunction and disease* (pp. 3–23). Hoboken: John Wiley & Sons, Inc.

Thomas, M. G., Burkart, M. D., & Walsh, C. T. (2002). Conversion of L-proline to pyrrolyl-2-carboxyl-S-PCP during undecylprodigiosin and pyoluteorin biosynthesis. *Chemistry & Biology, 9,* 171–184.

Vaillancourt, F. H., Yeh, E., Vosburg, D. A., O'Connor, S. E., & Walsh, C. T. (2005). Cryptic chlorination by a non-haem iron enzyme during cyclopropyl amino acid biosynthesis. *Nature, 436,* 1191–1194.

Vaillancourt, F. H., Yin, J., & Walsh, C. T. (2005). SyrB2 in syringomycin E biosynthesis is a nonheme Fe^{II} α-ketoglutarate- and O$_2$-dependent halogenase. *Proceedings of the National Academy of Sciences of the United States of America, 102,* 10111–10116.

Weichold, V., Milbredt, D., & van Pée, K.-H. (2016). Specific enzymatic halogenation has come of age. *Angewandte Chemie International Edition.* http://dx.doi.org/10.1002/anie.201509573.

Wynands, I., & van Pée, K.-H. (2004). A novel halogenase gene from the pentachloropseudilin producer *Actinoplanes* sp. ATCC 33002 and detection of *in vitro* halogenase activity. *FEMS Microbiology Letters, 237,* 363–367.

Yeh, E., Blasiak, L. C., Koglin, A., Drennan, C. L., & Walsh, C. T. (2007). Chlorination by a long-lived intermediate in the mechanism of flavin-dependent halogenases. *Biochemistry, 46,* 1284–1292.

Yeh, E., Garneau, S., & Walsh, C. T. (2005). Robust *in vitro* activity of RebF and RebH, a two-component reductase/halogenase, generating 7-chlorotryptophan during rebeccamycin biosynthesis. *Proceedings of the National Academy of Sciences of the United States of America, 102,* 3960–3965.

Yeh, E., Kohli, R. M., Bruner, S. D., & Walsh, C. T. (2004). Type II thioesterase restores activity of a NRPS module stalled with an aminoacyl-S-enzyme that cannot be elongated. *ChemBioChem, 5,* 1290–1293.

Zehner, S., Kotzsch, A., Bister, B., Süssmuth, R. D., Méndez, C., Salas, J. A., et al. (2005). A regioselective tryptophan 5-halogenase is involved in pyrroindomycin biosynthesis in *Streptomyces rugosporus* LL-42D005. *Chemistry & Biology, 12,* 445–452.

Zhu, X., De Laurentis, W., Leang, K., Herrmann, J., Ihlefeld, J., van Pée, K.-H., et al. (2009). Structural insights into regioselectivity in the enzymatic chlorination of tryptophan. *Journal of Molecular Biology, 391,* 74–85.

CHAPTER FIVE

Engineering Flavin-Dependent Halogenases

J.T. Payne*,†,1, M.C. Andorfer*, J.C. Lewis*,1
*University of Chicago, Chicago, IL, United States
†Stanford University, Stanford, CA, United States
[1]Corresponding authors: e-mail address: jtpayne@stanford.edu; jaredlewis@uchicago.edu

Contents

1. Introduction	93
2. Improving the Stability of FDHs	98
2.1 Improving FDH Stability via Directed Evolution	98
2.2 Alternative Screening Methods for FDH Evolution	105
2.3 Improving Enzyme Stability Through Immobilization	108
3. Altering the Regioselectivity of FDHs	109
3.1 Altered Regioselectivity via Iterative Mutagenesis and Screening	110
4. Expanding the Substrate Scope of FDHs	114
4.1 Expanding FDH Substrate Scope via Targeted Mutations	114
4.2 Expanding FDH Substrate Scope via Random Mutations	115
5. Conclusions	122
References	122

Abstract

In the two decades since the discovery of the first flavin-dependent halogenase (FDH), great strides have been made in elucidating the diversity of enzymes in this class as well as their structures and functions. More recently, efforts to engineer these enzymes for synthetic applications have yielded their first successes. FDH variants with improved stability, expanded substrate scope, and altered regioselectivity have been developed. Here, we review these efforts and provide representative protocols that have been employed for FDH engineering. Random and structure-guided mutagenesis approaches and screening methods, including HPLC, mass spectrometry, and spectrophotometric methods, are discussed. The protocols described herein should be generalizable to many FDHs and a wide variety of engineering goals.

1. INTRODUCTION

Halogenation is an important process in the production of pharmaceuticals and agrochemicals; an estimated one-quarter of all such compounds possess a halogen substituent (Herrera-Rodriguez, Khan, Robins, & Meyer, 2011). The importance of halogenation is in part due to the profound effect that halogen substituents can have on the biological activity of a compound (Hernandes, Cavalcanti, Moreira, de Azevedo Junior, & Leite, 2010; Sun, Keefer, & Scott, 2011), and examples of these effects can be seen in antibiotics (Bunders et al., 2011; Harris, Kannan, Kopecka, & Harris, 1985), anticancer compounds (Williams et al., 2005), and psychoactive compounds (Smith et al., 2008). Despite the importance of halogenation, current halogenation methods often rely on harsh reaction conditions, produce mixtures of products, and are unable to selectively halogenate compounds at electronically disfavored positions (Smith & El-Hiti, 2004).

In contrast to halogenation methods employing chemical reagents, which have been used for over a century, the importance and power of biological halogenation has only more recently come to light. Through the 1960s, halogenated natural products were thought to be infrequently occurring "chance products of nature" (Petty, 1961), but the number of known halogenated natural products has since increased dramatically to nearly 5000 (Gribble, 2010). Along with this increased inventory of structures has come an increased understanding of the enzymes responsible for their biosynthesis.

The first characterized halogenases were haloperoxidases (van Pée, 2013), and the first of these came from investigations of the *Caldariomyces fumago* caldariomycin biosynthetic pathway (Fig. 1A) (Morris & Hager, 1966). This enzyme was suspected to halogenate an electron-rich enol substrate to afford the product caldariomycin, although its role in caldariomycin biosynthesis was never definitively shown (van Pée & Patallo, 2006). Despite this, monochlorodimedone, an analog of the precursor to caldariomycin, has been used as a reliable spectrophotometric assay for haloperoxidase activity (Fig. 1A) (van Pée, 1990). This activity is believed to involve oxidation of halide anion (X^-, $X=I$, Br, Cl) to the corresponding hypohalous acid (HOX), the active halogenating species, which is released from the enzyme active site such that halogenation occurs freely in solution (Manoj, 2006). As such, haloperoxidases tend to halogenate electron-rich substrates with the same regioselectivity as conventional halogenation reagents (ie, that resulting from the chemoselectivity of HOX) (Wagner & König, 2012). Only relatively recently have haloperoxidases capable of enantioselective halogenation been discovered, indicating that in at least some cases halogenation occurs within the enzyme (Bernhardt, Okino, Winter, Miyanaga, & Moore, 2011; Carter-Franklin & Butler, 2004; Diethelm, Teufel, Kaysser, & Moore, 2014).

Fig. 1 (A) Reaction catalyzed by the first discovered haloperoxidase and the spectrophotometric assay using monochlorodimedone that was subsequently developed and (B) regioselective halogenations catalyzed by different FDHs on the same substrate, L-tryptophan, affording three unique products (Keller et al., 2000; Seibold et al., 2006; Yeh, Garneau, & Walsh, 2005; Zehner et al., 2005). Panel (A): Adapted from van Pée, K. -H. (1990). Bacterial haloperoxidases and their role in secondary metabolism. Biotechnology Advances, 8, 185–205.

Prior to the discovery of these enantioselective haloperoxidases, characterization of natural products with halogen substitution at electronically disfavored positions led to speculation that other classes of halogenases must be responsible for their biosynthesis since free HOX would not lead to such selectivity (van Pée, 2013). Indeed, flavin-dependent halogenases (FDHs) were discovered in the late 1990s (Dairi, Nakano, Aisaka,

Katsumata, & Hasegawa, 1995), and one of these, PrnA, was found to be responsible for the selective indole 7-chlorination of L-tryptophan in the course of pyrrolnitrin biosynthesis (Fig. 1B). This finding clearly showed how an FDH could catalyze halogenation at an electronically disfavored site on a substrate (Kirner et al., 1998). Later structural analysis of PrnA revealed that reduced FAD is bound in one area of the enzyme, which then reacts with O_2 to generate a flavin peroxide intermediate (Vaillancourt, Yeh, Vosburg, Garneau-Tsodikova, & Walsh, 2006). This flavin peroxide then reacts with halide to generate a HOX, which is believed to travel to a second enzyme active site in which the L-tryptophan substrate is bound. The HOX is then either hydrogen bonded to (Flecks et al., 2008) or reacts with (Yeh, Blasiak, Koglin, Drennan, & Walsh, 2007) a specific lysine residue, such that it is proximal to only a single C–H bond of the bound substrate. The FDH is thus able to accomplish C–H halogenation selectively at a single position of the substrate, even if that position is significantly disfavored electronically relative to other positions on the substrate (Lewis, Coelho, & Arnold, 2011).

In subsequent years, numerous other FDHs have been discovered, some of which have been found to selectively halogenate L-tryptophan at either the 5-, 6-, or 7-position, demonstrating the significant power of these enzymes for regioselective halogenation (Fig. 1B) (Keller et al., 2000; Seibold et al., 2006; Yeh et al., 2005; Zehner et al., 2005). Other FDHs have been found to halogenate additional substrate classes, including phenols, pyrroles, and even unactivated hydrocarbons (Wagner & König, 2012), but in many of these cases the substrate is bound to a carrier protein (van Pée, 2013), thus complicating synthetic applications. As of Dec. 2015, BLAST searches of these characterized FDHs return hundreds of additional putative halogenases.

To date, tryptophan halogenases are the best-characterized FDHs and have thus been most extensively explored for biotechnological applications. For example, three L-tryptophan FDHs were integrated into heterologously expressed biosynthetic pathways for indolocarbazole compounds to generate numerous chlorinated and brominated natural product analogs (Sánchez et al., 2005). Two of these FDHs were incorporated into the medicinal plant *Catharanthus roseus* to produce novel chlorinated alkaloids *in planta* (Runguphan, Qu, & O'Connor, 2010). Similarly, L-tryptophan FDHs were integrated into the pacidamycin biosynthetic pathway in order to produce chlorinated pacidamycin derivatives (Roy, Grüschow, Cairns, & Goss, 2010). All of these examples relied on wild-type FDHs halogenating their native substrate, L-tryptophan. Despite this limitation, the novel chlorinated products from two of the aforementioned metabolic engineering efforts were also employed for subsequent transition metal catalyzed cross-coupling reactions to create further novel products (Roy et al., 2010; Runguphan & O'Connor, 2013). This chemoenzymatic methodology has been further expanded upon to allow access to an even greater diversity of products (Durak, Payne, & Lewis, 2016).

A number of studies have also focused on exploring the activity of wild-type FDHs on nonnative substrates. One of the earliest such explorations was performed using the L-tryptophan 7-halogenase PrnA (Hölzer, Burd, Reißig, & van Pée, 2001). The authors reported that numerous nonnative indoles, including 5- and N-methyltryptamine, as well as 3- and 5-methylindole, were halogenated by PrnA, while others, including indole itself, gramine, and indole-3-acetic acid were not. More significantly, the authors reported that the nonnative substrates that were accepted were invariably halogenated at the electronically favored 2-position, rather than at the relatively electronically disfavored 7-position as seen with the native substrate. This was nonetheless an encouraging first demonstration that nonnative substrates could be halogenated by an FDH and led to later work showing that PrnA can halogenate kynurenine and a range of nonnative anthranilic acids and anthranilamides (Shepherd et al., 2015).

In more recent years, several reports have focused on the substrate scope of another L-tryptophan 7-halogenase, RebH (Yeh et al., 2005). In 2011, Glenn et al. reported that RebH could functionalize the nonnative substrate tryptamine selectively at the indole 7-position (Glenn, Nims, & O'Connor, 2011). This work was a follow-up to the authors' earlier work in which wild-type RebH and an L-tryptophan 5-halogenase, PyrH, were integrated into the biosynthetic pathway of several catharanthine alkaloids to afford halogenated derivatives of these alkaloids (Runguphan et al., 2010). Selective tryptamine, rather than L-tryptophan, halogenation helped to overcome a bottleneck in these biosynthetic pathways. In 2013, Payne et al. reported that wild-type RebH halogenated a significantly wider range of nonnative substrates, including tryptamine derivatives, the tricyclic tryptoline, indole, and substituted naphthalenes (Payne, Andorfer, & Lewis, 2013). Notably, many of these substrates were halogenated selectively at the indole 6- and/or 7-positions, although these substrates were still all comparable to or smaller than the native substrate, L-tryptophan. Frese et al. later reported that a range of L-tryptophan derivatives with methyl, amino, fluoro, or hydroxyl substituents at the indole 5- or 6-positions could also be selectively halogenated away from the preferred indole 2-position (Frese, Guzowska, Voß, & Sewald, 2014). One additional report showed that a fungal FDH, Rdc2, which natively halogenates monocillin II, also accepted two hydroxyisoquinoline substrates (Zeng, Lutle, Gage, Johnson, & Zhan, 2013). As was seen with the L-tryptophan FDHs, only substrates comparable in size or smaller than the native substrate are accepted.

The activities and/or conversions reported for PrnA, RebH, and Rdc2 on nonnative substrates tend to be significantly lower than those observed on the native substrate. With RebH, for example, the total turnover number observed with tryptoline was nearly 50-fold lower than that observed with L-tryptophan (Payne et al., 2013). In order to obtain FDHs useful for biotechnological applications, including metabolic engineering and in vitro biocatalysis, engineering efforts have thus been performed to increase the activities observed on nonnative substrates and to further expand FDH substrate scope. While most efforts at FDH protein engineering are relatively recent, significant progress has been made in improving the stability, expanding the substrate scope, and altering the regioselectivity of these enzymes (vide infra). These efforts have furnished protocols that, while initially applied to an individual FDH to accomplish a specific task, can be generalized to a wide range of FDHs and potential engineering efforts. The recent work that has been done on engineering FDHs and the protocols employed are thus discussed in this work, with care taken to describe how the protocols may be tailored toward novel efforts. Protocols for activity assays and purification of FDHs were described in a recent work (van Pée, 2012) and are thus not described herein.

2. IMPROVING THE STABILITY OF FDHs
2.1 Improving FDH Stability via Directed Evolution

The stability of an enzyme is an extremely important factor in determining its synthetic utility. More stable enzymes tend to have longer lifetimes and are easier to store and handle (Liao, 1993; Polizzi, Bommarius, Broering, & Chaparro-Riggers, 2007; Wu & Arnold, 2013), and it has been demonstrated that the thermostability of an enzyme can improve its evolvability (Bloom, Labthavikul, Otey, & Arnold, 2006). As mutations are introduced to the parent enzyme to expand its substrate scope or increase its activity, these mutations will tend to be destabilizing. The more thermostable the parent enzyme is, the greater the mutational load it will likely tolerate over the course of subsequent evolution. The use of directed evolution to optimize enzymes for use in industrial processes is increasingly common, and more thermostable enzymes will thus make better candidates for subsequent tailoring to industrial applications (as well as for their ability to tolerate higher reaction temperatures) (Liao, 1993; Wu & Arnold, 2013; Zhao & Arnold, 1999). Furthermore, it has been suggested that higher thermostability of an enzyme facilitates protein crystallography efforts (Derewenda, 2010), which are an invaluable resource for structural details that can guide targeted mutagenesis efforts. The stability of enzymes toward organic solvents is also important, as organic solvents are frequently employed when reaction components are not fully soluble in aqueous media (Liao, 1993).

Some of the earliest examples of directed evolution were focused on improving the thermostability (Liao, 1993) and organic solvent tolerance (You & Arnold, 1994) of enzymes. In the case of halogenases, both the organic solvent and pH tolerance of haloperoxidases have been improved via directed evolution (Hasan et al., 2006; Yamada, Higo, Yoshikawa, China, & Ogino, 2014). In both of these examples of haloperoxidase engineering, colorimetric assays were employed for high-throughput screening; a similar screen for FDHs is discussed in Section 2.3. The first example of FDH directed evolution was focused on improving the thermostability of RebH (Poor, Andorfer, & Lewis, 2014). In three rounds of mutagenesis and UPLC screening, eight nonsynonymous mutations were introduced (Scheme 1) that resulted in a nearly 20°C increase in the T_m of the protein relative to wild-type RebH. It was also demonstrated that the resultant RebH variants demonstrated increased activity at higher reaction temperatures and longer enzyme lifetime in reactions with several nonnative substrates. One of these variants served as the starting point for a future directed evolution effort to expand the substrate scope of RebH, and a crystal structure was obtained of another. This work demonstrated the application of directed evolution to improve the thermostability of an FDH to facilitate biocatalysis, crystallization, and further engineering efforts, and the procedure employed should be generalizable to improve the thermostability of any other FDH.

Scheme 1 Mutations incorporated via error-prone PCR into RebH to improve its thermostability.

2.1.1 Procedure for Evolving FDH Thermostability

While the procedure described here should be generalizable to any FDH, the specific examples given in parentheses for each step indicate the specific conditions used by Poor et al. (2014).

1. Select a parent FDH (eg, RebH) and obtain either plasmid or genomic DNA.
2. Generate a mutant library by error-prone PCR: with RebH, 1-2 nonsynonymous mutations were obtained on an average using *Taq* polymerase from New England Biolabs with the manufacturer's recommended conditions and 100–150 μM $MnCl_2$. A 50 μL reaction volume was used for PCR with the following procedure: 95°C 30 s (95°C 30 s, 55°C 30 s, 72°C 90 s) for 20 cycles, 72°C 10 min. Primers should be constructed to generate restriction digest sites (eg, *Nde*I and *Hind*III sites) to allow for integration into the expression plasmid in step 7.
    a. *Note*: a range of $MnCl_2$ concentrations should be tested prior to library screening in order to determine the proper concentration to elicit the desired mutation rate. Sequencing 24 colonies from step 10 was used to determine the average mutation rate.
    b. *Note*: $MnCl_2$ supplementation need not be used to generate the mutant library. For example, random libraries may be generated through the use of the Agilent GeneMorph random mutagenesis kit following the manufacturer's protocol. Site-directed libraries may be generated using sequence-overlap extension (SOE) to integrate point mutations, NNK codons, NDT codons, etc., or via codon mutagenesis methods. Employing one of these alternative methods may alter steps 2 through 8.
3. Using a QIAgen Gel Extraction kit, isolate the mutant library DNA by agarose gel purification and extraction.
4. Digest the mutant library DNA using the restriction digests appropriate for the sites generated in step 2. Also digest the desired expression plasmid (eg, pET28) using the same restriction digests.
5. Isolate the digested mutant library DNA and digested expression plasmid DNA by agarose gel purification and extraction.
6. Treat the digested expression plasmid DNA with Antarctic Phosphatase to minimize self-ligation in the following step. Heat inactivation of the Antarctic Phosphatase without subsequent purification provides sufficient ligation efficiency.
7. Ligate the digested mutant library DNA with the digested expression plasmid DNA using T4 DNA Ligase. For RebH, 10 μL ligation

reaction volumes, with an insert:vector molar ratio of 6–10:1, and 10–100 ng/μL total DNA concentration have been found to give the most reliable results.
8. Using a Zymo Clean & Concentrate kit, desalt the ligation reaction prior to transformation, eluting with 8 μL of molecular biology grade water.
9. Transform 4 μL of the desalted ligation reaction into BL21 (DE3) *Escherichia coli* containing the pGro7 chaperone plasmid via electroporation.
   a. *Note*: coexpression with the pGro7 chaperone plasmid gives an approximate sevenfold increase in the titer of soluble RebH obtained in crude lysate (Payne et al., 2013), and similar or even greater increases have been seen with other FDHs (data unpublished). In order to obtain electrocompetent cells containing pGro7, the commercially obtained pGro7 plasmid was itself transformed into BL21 (DE3) *E. coli*, selected for on agar plates containing chloramphenicol, and then used to prepare electrocompetent cells following the procedure described in Sambrook and Russell.
10. After allowing the electroporated cells to recover for 45–60 min in SOC media, plate the cells onto LB/agar plates containing chloramphenicol (for the pGro7 chaperone plasmid) and the appropriate antibiotic for the FDH expression plasmid (eg, kanamycin for pET28). Plating 100–200 μL of the transformation mixture has consistently provided 100–300 colonies per plate. Allow the colonies to grow for 16 h until they are large enough to be easily picked by hand or using a colony picker.
11. Pick (either using a colony picker or by hand) the desired number of colonies (~1000 were screened per round of evolution by Poor et al. (2014)) and array them into 1 mL 96-well plates containing 300 μL LB with the appropriate antibiotics. Leave 2–4 wells in each plate uninoculated to monitor for potential contamination and an additional 2–8 wells in each plate should be inoculated solely with cells containing the parent plasmid (and pGro7 if employed) to determine the parent activity. Significant plate-to-plate variation has been observed, so it is essential that wells containing only parent are used in *each* plate.
12. Incubate the 1-mL plates overnight at 37°C and 250 rpm.
13. Using 50–100 μL of the culture from the 1-mL plates (noting that the wells left uninoculated have not grown, indicating no contamination), inoculate 2 mL 96-well plates containing 1 mL TB with the appropriate antibiotics. Save the remaining culture in the 1-mL plates at 4°C.

14. Incubate the 2-mL plates at 37°C and 250 rpm until they reach an OD_{600} of 0.9–1.0, then induce enzyme expression. For pET28/RebH coexpressed with pGro7, final concentrations of 10 μM IPTG and 0.2 mg/mL L-arabinose were employed.
15. Incubate the induced 2 mL plates at 30°C and 250 rpm for 20 h, then spin the plates down at 3600 rpm for 15 min at 4°C and decant the supernatant.
16. Resuspend the cell pellets by gently vortexing in 200 μL HEPES buffer (25 mM, pH 7.4), then spin the plates down again 3600 rpm for 15 min at 4°C and decant the supernatant. This wash step removes residual L-tryptophan and chloride from the growth media that could otherwise interfere with library screening.
17. Resuspend the cell pellets by gently vortexing in 100 μL HEPES buffer (25 mM, pH 7.4) containing 0.75 mg/mL lysozyme. Incubate at 37°C and 250 rpm for 30–45 min, then flash freeze in liquid nitrogen and thaw in a 37°C water bath. Add 10 μL of HEPES buffer (25 mM, pH 7.4) containing 1 mg/mL DNase I and then incubate at 37°C and 250 rpm for 15 min. Spin down the plates at 3600 rpm for 15 min at 4°C and transfer 50 μL of the supernatant to a microtiter plate.
18. In order to screen for increasing thermostability, the crude lysate is at this point subjected to a heat pretreatment step prior to subsequent reaction screening. The exact temperature and duration of the heat pretreatment can be determined empirically by taking samples of crude lysate and pretreating at different temperatures until only ~20% activity of the parent is observed. For the evolution of thermostability of RebH, this resulted in a 2 h, 42°C pretreatment for the first round, a 2 h, 51°C pretreatment for the second round, and a 3 h, 54°C pretreatment for the third round. The microtiter plates containing crude lysate were sealed using AeraSeals from Research Products International, incubated in a water bath at the temperature and for the duration described, and then immediately cooled in an ice-water bath.
19. Add solutions of substrate (eg, L-tryptophan for RebH, 10 mM in HEPES buffer (25 mM, pH 7.4), 0.5 mM final concentration), halide (eg, NaCl, 1.5 M in HEPES buffer (25 mM, pH 7.4), 10 mM final concentration), FAD (10 mM in HEPES buffer (25 mM, pH 7.4), 100 μM final concentration), NAD (10 mM in HEPES buffer (25 mM, pH 7.4), 100 μM final concentration), FAD reductase (eg, MBP-RebF, 50–100 μM in HEPES storage buffer (25 mM, pH 7.4, 10% glycerol), 2.5 μM final concentration—other reductases have been employed for

other FDHs), and glucose dehydrogenase (eg, GDH-105 from Codexis, 1800 U/mL in HEPES buffer (25 mM, pH 7.4), 9 U/mL final concentration) to the pretreated crude lysate. Add additional HEPES buffer (25 mM, pH 7.4) such that the final reaction volume is 75 µL and then add a solution of glucose (1 M in HEPES buffer (25 mM, pH 7.4), 20 mM final concentration) to initiate the reaction. Reseal the microtiter plates with AeraSeals and then shake overnight at 600 rpm at room temperature.

20. Quench the reactions with 75 µL of methanol to precipitate out the proteins, add an internal standard to control for differences in handling (phenol at a final concentration of 0.5 mM is commonly employed, but a different standard may be required depending on the substrate and analysis method used), and spin the microtiter plates down at 3600 rpm for 15 min at 4°C.

21. Filter the supernatant using a 0.45-µm filter plate placed directly on a 96-well microtiter plate prior to UPLC analysis. 96-well microtiter plates can then be rearrayed into 384-well plates to increase analysis throughput.

22. Seal the plates with aluminum foil using a plate sealer to minimize evaporation and prevent particulate contamination, then analyze the reactions (eg, by UPLC) for halogenated product production. For RebH, reactions were analyzed for 7-chlorotryptophan production using an Agilent 1200 UPLC with an Agilent Eclipse Plus C18 2.1 × 50 mm column, 1.8 µM particle size; solvent A = H_2O/0.1% TFA, solvent B = CH_3CN; 0–0.5 min, B = 16%; 0.5–1.5 min, B = 16–80%.

23. Analyze the reaction data to determine the relative improvement in halogenated product production relative to the parent wells added to each plate in step 11. The wells that show the greatest degree of improvement should be grown on a larger scale and purified as described in steps 25–34 (using the saved primary culture in the 1-mL plates in step 13), plasmid isolated and sequenced to confirm the presence of nonsynonymous mutations, and halogenase variants analyzed in vitro as described in step 35 (for FPLC purification of either non-His-tagged or His-tagged halogenases, see van Pée (2012)).

24. If multiple improved variants are found displaying different nonsynonymous mutations, it is recommended that the beneficial mutations be analyzed both individually and in combination. Mutations can be recombined via SOE PCR; for RebH, PCR conditions were as follows: 98°C 30 s (98°C 10 s, 72°C 50 s) for 35 cycles, 72°C 10 min. Once

plasmids are obtained containing the combined mutations, the halogenase variants should be immediately assayed in vitro as described in steps 25–36. DNA shuffling can also be used to combine mutations; various methods exist with individual advantages and disadvantages, which have been reviewed elsewhere (Fox & Huisman, 2008).

25. Use the saved primary culture in 1 mL plates from step 13 to generate fresh primary cultures by inoculating LB with the appropriate antibiotics and growing overnight at 37°C and 250 rpm.
26. Add 500 μL of the fresh primary culture to 50 mL TB with the appropriate antibiotics in a 250-mL Erlenmeyer flask covered with aluminum foil then incubate at 37°C and 250 rpm.
27. Once the cultures reach an OD_{600} of 0.6–0.8, induce enzyme expression. For pET28/RebH coexpressed with pGro7, final concentrations of 100 μM IPTG and 2.0 mg/mL L-arabinose were employed (both are 10-fold higher than the concentrations used for 96-well plates).
28. Incubate the induced 50 mL cultures at 30°C and 250 rpm for 20 h, then spin the cultures down at 3600 rpm for 15 min at 4°C and decant the supernatant.
29. Resuspend the cell pellets in 10 mL cold HEPES buffer (20 mM, pH 7.4, 150 mM NaCl).
30. Lyse the resuspended pellets by sonication (Qsonica S-4000 with a 0.5″ horn; 8 × 30 s with 45 s rests, 20% duty cycle delivering 40–50 W), keeping the lysing cultures cold with a circulating ice-water bath.
31. Clarify the lysed cultures by centrifugation at 15,000 rpm for 45 min at 4°C.
32. Purify the clarified lysate. For FPLC purification of either non-His-tagged or His-tagged halogenases, see van Pée (2012). His-tagged halogenases can also be purified by Ni-NTA affinity chromatography using a peristaltic pump or centrifugation following the resin manufacturer's recommended protocol. Verify the presence of purified halogenase by SDS-PAGE.
33. Combine the pure protein fractions and exchange into HEPES storage buffer (20 mM, pH 7.4, 150 mM NaCl, 10% glycerol) using dialysis or a spin filter (eg, Amicon 30 kDa 15 mL filters, spun at $4000 \times g$ for 15 min at 4°C).
34. Determine the protein concentration using a BCA assay or using A_{280} and extinction coefficients calculated based on amino acid composition (Protein Calculator v3.3, http://www.scripps.edu/~cdputnam/protcalc.html).

35. Determine the residual activity of each halogenase variant using the reaction conditions described in step 19, substituting purified halogenase from step 33 (25 µM final concentration) for the crude lysate and performing the reactions in 1.5-mL microcentrifuge tubes. Quenching and analysis steps are the same as steps 20–23 described earlier for crude lysate reactions.
36. To determine the T_m of each halogenase variant, monitor the thermal denaturation at 222 nm by circular dichroism spectroscopy in 2°C increments from 20°C to 90°C with 2 min equilibration at each temperature. The halogenase variant should be at a concentration of 20 µM in HEPES storage buffer (20 mM, pH 7.4, 150 mM NaCl, 10% glycerol). For RebH, an AVIV 202 CD Spectrometer with Peltier temperature controller was used. The midpoint of the denaturation curve was determined with SigmaPlot after fitting to a four-parameter sigmoid.

The steps described earlier may be repeated as many times as necessary until a sufficiently thermostabilized FDH is obtained.

2.1.2 Procedure for Evolving FDH Organic Solvent Tolerance

Poor et al. also observed that several of the thermostabilized variants of RebH displayed significantly increased tolerance to organic solvents (Lewis, Poor, & Payne, 2014). In particular, one of the intermediate variants generated containing two nonsynonymous mutations displayed 2.5-fold higher activity in the presence of 30% DMSO than wild-type RebH did. While RebH was not explicitly evolved for increased organic solvent tolerance, the procedure described in Section 2.1.1 can be easily modified to do so. Step 18 is simply modified as follows:

18. In order to screen for increasing organic solvent tolerance, a quantity of the desired organic solvent is added to the crude lysate prior to subsequent reaction screening. The exact quantity of organic solvent to add can be determined empirically by taking samples of crude lysate and adding different volumes until only ~20% activity of the parent is observed. Adjust the amount of HEPES buffer added in step 19 to accommodate the organic solvent added in this step while still maintaining a 75 µL final reaction volume.

2.2 Alternative Screening Methods for FDH Evolution

The procedure described in Section 2.1.1 to evolve RebH for enhanced thermostability involved UPLC analysis to analyze the halogenation reactions (step 22) (Poor et al., 2014). Chromatographic separations can be used to screen enzymes for essentially any reaction even (or especially) when multiple products are formed, the products have only minor structural differences (eg, regioisomers), or impurities are present. It is important to note that enzymes are increasingly used to address reactions that suffer from precisely these complications, making chromatographic separation an important tool for directed evolution. The cost of this generality, however, is screening time. The UPLC method developed to monitor chlorination of L-tryptophan, for example, was optimized to 90 s per sample, with an additional ∼50 s per sample of pre- and post-run time required. This results in approximately 2 days of UPLC analysis time for a ∼1000 member mutant library. While this is a manageable amount of time if a dedicated UPLC is available and can be reduced further using SFC, it does place a limit on the number of enzyme variants that can be examined.

In some cases, colorimetric or fluorimetric assays could potentially be used to increase the size of the mutant libraries that can be evaluated and to decrease screening times. For example, haloperoxidase engineering efforts have utilized the monochlorodimedone assay described in Fig. 1A (Yamada et al., 2014). Hosford et al. recently reported the development of a colorimetric method for the analysis of arylamine halogenation (Scheme 2) (Hosford, Shepherd, Micklefield, & Wong, 2014). This method requires only UV/vis analysis to provide a quantitative measure of arylamine halogenation, and the authors demonstrated its consistency with the results obtained from HPLC analysis of the RebH catalyzed chlorination of

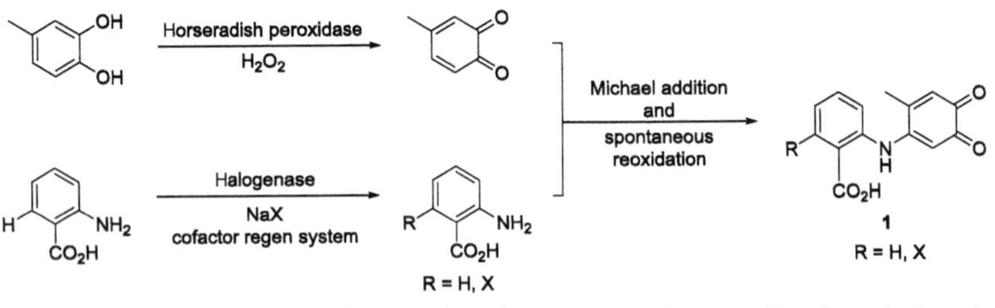

Scheme 2 Method for peroxidase-mediated quinone–amine coupling for colorimetric scheme for arylamine halogenation. *Adapted from Hosford, J., Shepherd, S. A., Micklefield, J., & Wong, L. S. (2014). A high-throughput assay for arylamine halogenation based on a peroxidase-mediated quinone-amine coupling with applications in the screening of enzymatic halogenations. Chemistry A European Journal, 20, 16759–16763.*

2-aminonaphthalene. This substrate undergoes a marked decrease in the absorbance at 516 nm upon chlorination. Thus, the response at a given wavelength can be inversely correlated to the percent conversion of the halogenation reaction, and if a response factor is calculated, this correlation can be quantified.

Such assays offer several distinct advantages when applied to enzyme engineering. Each measurement requires only seconds, allowing a 96-well plate of halogenase variants to be analyzed in a matter of minutes. This should allow access to much larger library sizes, thus allowing for the discovery of more beneficial mutations per round of evolution. Furthermore, it only requires a UV/vis plate reader for analysis, which is more frequently available than UPLC or SFC instruments. The key limitation of such methods, however, is the requirement that the target reaction result in a unique spectroscopic observable, which is not typical for most substrates or for instances in which isomeric products are formed.

Researchers have also explored different methods to replace enzyme screening altogether with selection processes in which higher performing variants have an increased rate of survival and are thus enriched (Lin & Cornish, 2002; Taylor, Kast, & Hilvert, 2001). Selections have been widely used to engineer enzyme properties that can be readily linked to the genetic circuitry of the host organism (Collins, Leadbetter, & Arnold, 2006). For example, Esvelt et al. reported the development of a phage-assisted continuous evolution system that could enable the continuous directed evolution of any enzyme that can be directly linked to the host genetic circuitry and demonstrated its use by rapidly evolving RNA polymerase (Esvelt, Carlson, & Liu, 2011). Unfortunately, the requirement to link a trait of interest to cell survival can be difficult to satisfy when the trait of interest is selective functionalization of a small molecule. Three-hybrid systems have been developed to enable selections for chemical reactions, but these systems also impose limitations on the types of reactions that can be examined (Lin, Tao, & Cornish, 2004).

Selections can also be enabled by differential biological activities of potential reaction products. For example, a selection was used to improve the enantioselectivity of a lipase for hydrolysis of a mixture of the acetate ester of (S)-isopropylidene glycerol and the fluoroacetate ester of (R)-isopropylidene glycerol, since the product of the hydrolysis of the acetate is an energy source while that of the fluoroacetate is toxic (Reetz, Höbenreich, Soni, & Fernández, 2008). While fortuitous, this activity, like the cases noted earlier, is not readily generalizable to the vast majority of substrates, enzymes, or reactions. Neither the evolution of RebH substrate scope (Section 4.2) nor the evolution of RebH regioselectivity (Section 3.1) involves products that can be readily linked to *E. coli* survival. No selection has, as of yet, been reported to have been successfully employed for the evolution of FDHs.

Despite the limitations of high-throughput screens and selections, the high-throughput nature of these processes makes them attractive options for halogenase engineering, particularly for applications involving a single substrate that does not change over the course of the engineering project. Efforts to improve the thermostability, organic solvent tolerance, activity at a particular pH, or to optimize turnover frequency/number on a particular substrate may potentially be compatible with these screens. The full procedure for the colorimetric assay described by Hosford et al. (2014) or any other high-throughput assay would be substituted for steps 20–22 of the procedure described in Section 2.1.1.

2.3 Improving Enzyme Stability Through Immobilization

Enzyme immobilization has been reported to increase enzyme stability without significantly compromising activity, for example, in improving the organic solvent tolerance (Truppo, Strotman, & Hughes, 2012) and pH range (Koszelewski, Müller, Schrittwieser, Faber, & Kroutil, 2010) of transaminases. Frese and Sewald recently reported on the enzymatic halogenation of L-tryptophan on a gram scale by forming cross-linked enzyme aggregates (CLEAs) from crude RebH lysate (Frese & Sewald, 2015). The authors reported that RebH combiCLEAs had significantly increased retention of activity relative to free, purified RebH after extended storage at 4°C, and that RebH combiCLEAs also displayed significantly increased catalyst lifetime in an active biohalogenation reaction (although relative total turnover numbers were not reported). Furthermore, RebH combiCLEAs could be reused up to 10 times while still displaying significant halogenation activity. CombiCLEAs, which have been demonstrated to be used successfully with RebH, and other enzyme immobilization methods, which as of yet lack specific demonstration with FDHs, are most likely to be useful in preparative-scale halogenation reactions, rather than high-throughput screening applications. The reported procedure for the particular immobilization strategy employed would thus be substituted for steps 31–34 of the procedure described in Section 2.1.1.

3. ALTERING THE REGIOSELECTIVITY OF FDHs

The ability to change the site on a substrate that an enzyme functionalizes is a powerful tool in protein engineering. Functionalizing specific C–H bonds is a particularly challenging problem (Lewis et al., 2011). Fishman et al. demonstrated that structure-guided mutagenesis of toluene *para*-monooxygenase TpMO could be used to generate variants for *ortho*-, *meta*-, and (improved) *para*-hydroxylation of toluene (Fishman, Tao, Rui, & Wood, 2005). Cytochromes P450 have also been engineered to catalyze hydroxylation of different C–H bonds on a given substrate (Fasan, 2012; Lewis et al., 2009). The unique selectivities of tryptophan FDHs have led researchers to investigate the possibility of altering the functionalization site of these halogenation catalysts as well. For example, the crystal structure of a tryptophan 5-halogenase, PyrH (Zehner et al., 2005) was used to identify a single residue that, when introduced into the tryptophan 7-halogenase PrnA (Keller et al., 2000), imparted some tryptophan 5-halogenation ability to PrnA (Lang et al., 2011). However, this variant still produced a 2:1 ratio of regioisomers with mostly 7-halotryptophan produced and displayed severely decreased activity (from 0.316 to 0.020 min/μM).

As mentioned in Section 1, Shepherd et al. demonstrated that the tryptophan 7-halogenase PrnA and the tryptophan 5-halogenase PyrH display activity on the tryptophan metabolite kynurenine and several anthranilamides and anthranilic acid (Shepherd et al., 2015). Halogenated anthranilic acids are present in many pharmaceuticals and agrochemicals so the authors sought to increase the halogenation activity of PrnA on anthranilic acid through protein engineering. By examining an available crystal structure of PrnA (Dong et al., 2005), the authors reasoned that selected residues could be mutated to the positively charged residues lysine or arginine to create a salt bridge contact with the carboxylate moiety of anthranilic acid. PrnA E450K not only increased anthranilic acid conversion eightfold (from less than 2% to 14% conversion after 1 h at 2 mol% enzyme loading) but also increased the ratio of 3- to 5-halogenation relative to wild-type PrnA (93% 3-halogenation compared with 86% for wild-type PrnA). Conversely, F454K increased conversion fourfold on the same substrate but had the opposite effect on regioselectivity, increasing the ratio of 5- to 3-halogenation (38% 5-halogenation compared with 14% for wild-type PrnA). A double mutant incorporating both of these mutations furnished a 16-fold increase in conversion (to 27% conversion) and an even greater preference for 5-halogenation (54% 5-halogenation).

The work by Payne et al. discussed in Section 4.2 (vide infra) led to the RebH variant 3-SS which not only had significantly improved activity on the substrate tryptoline but also changed the regioselectivity from a nearly 1:1 mixture of 6- and 7-chlorotryptoline with the wild-type enzyme to >95% selectivity for 6-chlorotrypotoline. While these examples demonstrate the ability to tune enzyme selectivity via rational design and directed evolution, respectively, no screening pressure was applied to enable discovery of variants with a specific regioselectivity. Libraries were screened for activity improvements and selectivity was determined through further characterization of activity hits. Directly screening for a specific change in regioselectivity can greatly expedite the development of halogenase variants that produce a specific halometabolite; a procedure for doing so is discussed later.

3.1 Altered Regioselectivity via Iterative Mutagenesis and Screening

Recently, Andorfer et al. demonstrated that selectivity of FDHs could be systematically tuned using a MALDI-MS assay (Scheme 3) (Andorfer et al., 2016). By using a combination of random and targeted mutagenesis, variants capable of halogenating the substrate tryptamine with 90% selectivity for C-6 and 95% selectivity for C-5 were developed from the 7-halogenase RebH. Starting from the point mutant RebH-N470S, 10 rounds of iterative mutagenesis and screening were performed to obtain these 5- and 6-halogenases (named 10S and 8F, respectively, Fig. 2). Libraries were screened using a selectively deuterated tryptamine probe as substrate and analyzed by MALDI-MS. In this way, variants were directly screened for

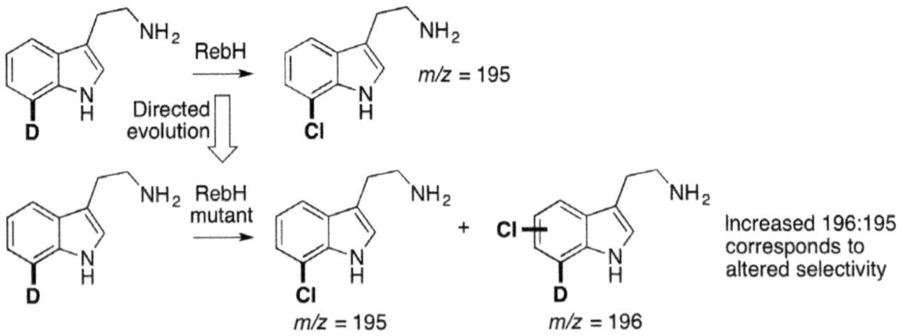

Scheme 3 Deuterium labeling of substrates to determine regioselectivity by mass spectrometry. *Adapted from Andorfer, M. C., Park, H. J., Vergara-Coll, J., & Lewis, J. C. (2016). Directed evolution of RebH for catalyst-controlled halogenation of indole C–H bonds,* Chemical Science, Advance article.

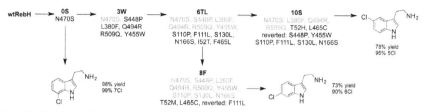

Fig. 2 Mutations introduced and relative conversions seen along lineage of altered regioselectivity RebH variants. *Adapted from Andorfer, M. C., Park, H. J., Vergara-Coll, J., & Lewis, J. C. (2016). Directed evolution of RebH for catalyst-controlled halogenation of indole C–H bonds, Chemical Science, Advance article.*

altered selectivity. The 5- and 6-halogenases developed in this work were found to change the selectivity on a variety of substrates, including tryptamine derivatives, tryptophol, simple arenes, and some of the indole and carbazole drugs discussed in Section 4.2.

The benefits of using random mutagenesis alongside targeted libraries are highlighted in this work. In the first rounds (1–5) of evolution, random mutagenesis was used to identify sites that led to large changes in selectivity. These sites found via random mutagenesis were targeted through libraries of variants containing the degenerate NNK codon. These targeted libraries led to the rapid identification of highly selective variants for C-5 and C-6 of tryptamine. Interestingly, when key mutations from targeted libraries were introduced into wild-type RebH, little to no change in selectivity was observed, suggested the advantageous mutagenesis of certain sites is template dependent.

3.1.1 Procedure for Evolving FDH Regioselectivity via Random Mutations

1. Follow steps 1–17 in the procedure described in Section 2.1.1 to generate the crude lysate of the FDH mutant library. A similar mutation rate and library size to those described for the evolution of increased thermostability were employed for each of the rounds of random mutagenesis.
2. Select a suitable substrate for screening. The primary considerations for selecting the proper substrate are:
   a. Sufficient activity on the substrate. For the MALDI-MS screening described later, at least 10% conversion on the substrate greatly decreased the quantity of false positives obtained (and thus the workload necessary for steps 25–35). In contrast to screening for increased activity on a particular substrate for which conversion

should be kept relatively low so as to highlight activity improvements, higher conversion is desirable when screening for altered selectivity. In many cases, variants with altered selectivity on a substrate also have significantly lowered activity (Andorfer et al., 2016; Lang et al., 2011).

   b. Solubility in a buffer:cosolvent system tolerated by the FDH. RebH shows very little loss of activity with 95% HEPES buffer: 5% organic solvent (MeOH/EtOH/iPrOH/acetone/DMSO), providing ample options for a suitable solvent system.

   c. Substrate can be selectively deuterated. Deuterium incorporation can be introduced either at the native site of halogenation or at the desired halogenation site. If halogenation at more than one site on a substrate is of interest, deuterium incorporation at the native halogenation site is an effective method to screen for altered selectivity at any position. However, if a particular site is of interest, deuteration at this site is most effective.

3. Add solutions of substrate (eg, 7-deutero-tryptamine, 30 mM in ethanol, 1.5 mM final concentration), halide (eg, NaCl, 1.5 M in HEPES buffer (25 mM, pH 7.4), 10 mM final concentration), FAD (10 mM in HEPES buffer (25 mM, pH 7.4), 100 µM final concentration), NAD (10 mM in HEPES buffer (25 mM, pH 7.4), 100 µM final concentration), FAD reductase (eg, MBP-RebF, 50–100 µM in HEPES storage buffer (25 mM, pH 7.4, 10% glycerol), 2.5 µM final concentration—other reductases have been employed for other FDHs), and glucose dehydrogenase (eg, GDH-105 from Codexis, 1800 U/mL in HEPES buffer (25 mM, pH 7.4), 9 U/mL final concentration) to the pretreated crude lysate. Add additional HEPES buffer (25 mM, pH 7.4) such that the final reaction volume is 75 µL and then add a solution of glucose (1 M in HEPES buffer (25 mM, pH 7.4), 20 mM final concentration) to initiate the reaction. Reseal the microtiter plates with AeraSeals from Research Products International and then shake overnight at 600 rpm at room temperature.

4. Quench the reactions with 75 µL of methanol to precipitate out the proteins. Add dilute acid to each reaction mixture to slightly acidify it (eg, for reactions with tryptamine probes, 10 µL of 75 mM HCl was sufficient). *Note*: Acidification is necessary when the substrate is basic. Without this step, the reaction mixture dissolves the prespotted matrix on the MALDI target. However, adding significantly more acid could cause loss of the deuterium label due to proton exchange.

5. Spin the quenched microtiter plates down at 3600 rpm for 15 min at 4°C.
6. Filter the supernatant directly into a 96-well microtiter plate using a 0.45-µm filter plate.
7. Spot matrix onto a 384-well MALDI target plate. Conditions can vary by substrate. For tryptamine, 2 µL of a solution of α-cyano-4-hydroxycinnamic acid was used per spot (7.5 mg/mL of α-cyano-4-hydroxycinnamic acid in 1:1 THF:H$_2$O). This can then be dried on the benchtop or in a vacuum oven.
8. Once the MALDI target is dry, spot 2 µL of filtered reaction mixture onto each spot using a liquid handling robot. Dry on the benchtop or in a vacuum oven.
9. Seal the 96-well plates containing the remaining reaction mixtures with aluminum foil using a plate sealer. These can be stored at 4°C and used as a secondary screen to eliminate false positives from the primary MALDI-MS screen.
10. Collect spectra for MALDI-MS. For tryptamine, spectra were collected on a Bruker Ultraflextreme MALDI-TOF-TOF using an automated method (AutoXecute tool of the Flex Control acquisition software). The spectra were generated in the positive ion reflector mode over a detector range of 160–200 Da (m/z of deutero-tryptamine = 162, m/z of chloro-deutero-tryptamine = 196). Final spectra constituted the average of 500 raster shots taken at 50 random positions.
11. Set all peaks besides those corresponding to starting material and products as background. Export data into Microsoft Excel. Use an Excel macro to insert "0" for all reactions that did not have product peak. Array these data into 96-well plate format using Excel and calculate the conversions and selectivities for each reaction.
12. The reactions that show the greatest degree of change in selectivity or the greatest degree of improvement in conversion for all products while maintaining parent selectivity can be rescreened on a UPLC using the procedure outlined in Section 2.1.1, step 22. This will often not show resolution of product isomers, but can remove false positives for improved conversions.
13. The reactions that show the greatest degree of improvement with MALDI-MS and UPLC screening should be grown on a larger scale and purified as described in steps 25–34 of Section 2.1.1 (using the saved primary culture in the 1 mL plates), plasmid isolated and sequenced to confirm the presence of nonsynonymous mutations,

and halogenase variants analyzed in vitro as described in step 35 (for FPLC purification of either non-His-tagged or His-tagged halogenases, see van Pée (2012)).

3.1.2 Procedure for Altering FDH Regioselectivity via Targeted Mutations

1. Generate a mutant library by introducing the degenerate NNK codon at the desired site via overlap extension PCR: for RebH, PCR conditions were as follows: 98°C 30 s (98°C 10 s, 72°C 50 s) for 35 cycles, 72°C 10 min.
2. Follow steps 3–17 in the procedure described in Section 2.1.1 to generate the crude lysate of the FDH mutant library.
   a. *Note*: For step 11, the number of colonies that should be picked for each library depends on the number of degenerate NNK codons that are introduced as well as the desired library coverage. For RebH, enough colonies were picked to ensure 95% library coverage. The CASTer tool was used to determine these numbers (http://www.kofo.mpg.de/en/research/biocatalysis).
3. Follow steps 2–13 in the procedure described in Section 3.1.1.

4. EXPANDING THE SUBSTRATE SCOPE OF FDHs

4.1 Expanding FDH Substrate Scope via Targeted Mutations

The narrow structural range of substrates known to be halogenated by wild-type FDHs, as described in Section 1, was a significant barrier to the application of FDHs in chemical synthesis. Glenn et al. determined that the L-tryptophan 7-halogenase RebH could also accept the nonnative substrate tryptamine, producing solely 7-chlorotryptamine as its product (Glenn et al., 2011). The authors had previously integrated RebH and another halogenase into biosynthetic pathways to produce halogenated metabolites (Runguphan et al., 2010), but found that the subsequent enzyme in the biosynthetic pathways, tryptophan decarboxylase, accepted 7-chlorotryptophan at only 3% of the efficiency with which it accepted L-tryptophan. They accordingly sought to instead halogenate tryptamine, the product of tryptophan decarboxylase, thus bypassing this bottleneck. Wild-type RebH was found to accept L-tryptophan and tryptamine roughly equally (although a later report indicated wild-type RebH prefers for L-tryptophan significantly over tryptamine), so the authors sought use the available crystal structure of RebH

(Yeh et al., 2007) to guide a protein engineering effort to increase selectivity for tryptamine over L-tryptophan.

Six residues were selected proximal to the carboxylate moiety of L-tryptophan with the intention to mutate these residues to larger amino acids. It was hoped that this would disrupt binding of L-tryptophan in the RebH active site without impacting the binding of tryptamine, which lacks this carboxylate moiety. A total of 17 mutants were explored, only two of which retained any activity for either L-tryptophan or tryptamine. One of these mutants, RebH Y455W, displayed significantly increased preference for tryptamine over tryptophan. In a competition assay, RebH Y455W displayed a 10-fold decrease in production of 7-chlorotryptophan, with a simultaneous threefold increase in the production of 7-chlorotryptamine, resulting in a net 30-fold increase in selectivity for tryptamine. More recently, Shepherd et al. performed point mutagenesis of select residues of PrnA based on its crystal structure with the intent of generating a salt bridge contact with the carboxylate moiety of anthranilic acid (Shepherd et al., 2015). Several of these point mutants displayed increased conversion on nonnative substrates; these variants were discussed in detail in Section 3.

4.2 Expanding FDH Substrate Scope via Random Mutations

The availability of crystal structures of several FDHs affords great potential for structure-guided protein engineering efforts. However, the effects of structure-guided mutations can often be surprising. For example, in the work of Glenn et al. (2011), a total of 17 variants were designed, for all of which it was intended that the enzyme would prefer tryptamine over L-tryptophan, but only one of these variants had this preference. A general engineering approach that can be applied to any FDH to improve activity on any particular substrate would also be very useful. In particular, all the FDHs, wild type or engineered, discussed thus far only accept substrates comparable in size to or smaller than the native substrate and do so with significantly diminished activities. An engineered FDH capable of site-selective halogenation of large, bioactive substrates would be of great utility in both synthetic chemistry and metabolic engineering applications like those described earlier. The engineered FDH would then act as the last step in the biosynthetic sequence, thus avoiding any disruption of the intermediate steps by creating bottlenecks from poor processing of halogenated intermediates (Glenn et al., 2011).

The FDHs discussed so far were not known to accept any particularly large, nonnative substrates, making an engineering effort to directly increase activity on a particular large substrate difficult. However, the activity of an FDH can be improved on a known substrate with structural homology to a target substrate, thus identifying FDH variants with improved activity on the target substrate, and then repeating this process iteratively in order to gradually evolve the substrate scope of the enzyme. This technique is known as substrate walking, and examples have been reported of its use to expand the substrate scopes of cytochrome P450 monooxygenases (Fasan, Chen, Crook, & Arnold, 2007), transaminases (Savile et al., 2010), tRNA synthetases (Xie, Liu, & Schultz, 2007), and monoamine oxidases (Ghislieri et al., 2013).

Payne et al. applied this technique to expand the substrate scope of RebH to a range of large, bioactive indoles (Scheme 4) (Payne, Poor, & Lewis, 2015). As described in Section 1, RebH showed some initial activity on the tricyclic indole tryptoline, though its conversion of this substrate is 50-fold lower than that seen with L-tryptophan (Payne et al., 2013). Starting with a thermostabilized RebH variant generated in the work described in Section 2.1, error-prone PCR was used to introduce mutations at random throughout the entirety of the RebH sequence, following the procedure described in Section 2.1.1. This first library was screened for increased activity on L-tryptophan, and one of the resulting variants also demonstrated nearly twofold increased activity on tryptoline. This increase provided sufficient activity to allow direct UPLC screening on tryptoline, and the subsequent round of evolution afforded a nearly 10-fold increase in conversion on tryptoline. This variant, 3-SS, possesses a total of six nonsynonymous mutations relative to wild-type RebH and displays over 60-fold increased activity on tryptoline. In addition, site-selective halogenation was afforded on a range of bioactive substituted tryptoline derivatives (Table 1).

The authors had chosen an analog of an inhibitor of bacterial biofilm formation, deformylflustrabromine, as a target for FDH halogenation. Deformylflustrabromine is an inhibitor of bacterial biofilm formation when the bromine substituent is present at the indole 6-position, but when this bromine is removed, no such activity is seen (Bunders et al., 2011). However, analogs with the bromine present at the indole 4-, 5-, or 7-positions have not been assayed, nor have any of the chlorinated analogs. Performing site-selective halogenations at these other positions would be extremely challenging using conventional chemical techniques, but FDHs seem well

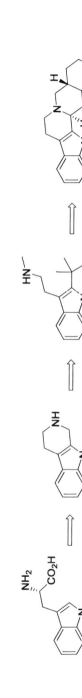

Scheme 4 Substrate walking used to expand the substrate scope of RebH. Adapted from Payne, J. T., Poor, C. B., & Lewis, J. C. (2015). Directed evolution of ReBH for site-selective halogenation of large biologically active molecules. Angewandte Chemie International Edition, 54, 4226–4230.

Table 1 Substrates Halogenated (with Selectivities Shown) by Engineered RebH Variants with Expanded Substrate Scope

equipped for this task. However, wild-type RebH displayed no quantifiable activity on unhalogenated deformylflustrabromine. Gratifyingly, the RebH variant engineered to have increased activity on tryptoline, 3-SS, displayed low activity on this substrate. The authors found that reverting one of the mutations present in 3-SS provided a fivefold increase in activity, enough to screen directly for further improvements on unhalogenated deformylflustrabromine.

An additional round of error-prone PCR furnished variant 4-V, which possesses a single addition A442V mutation and displays nearly twofold increased activity on unhalogenated deformylflustrabromine. Furthermore, it was found that 4-V provides significant increases in activity on a range of large, bioactive substrate, including the pentacyclic compounds yohimbine and evodiamine as well as a range of indole and carbazole drugs used clinically as beta blockers, which possess significant substituents at the indole 4-position. One of these substrates, carvedilol, has nearly twice the molecular weight of the native substrate of RebH, L-tryptophan. Especially remarkable is the fact that the A442V mutation that enabled these large substrates to be functionalized is distal from the enzyme active site, and thus would not have been a likely target of structure-guided engineering efforts. Such mutations have been reported to occur in other enzyme classes as well (Fasan et al., 2007; Romero & Arnold, 2009; Shimotohno, Oue, Yano, Kuramitsu, & Kagamiyama, 2001; Spiller, Gershenson, Arnold, & Stevens, 1999) and these examples highlight a unique advantage of random mutagenesis efforts.

4.2.1 Procedure for Evolving FDH Substrate Scope via Random Mutations

1. Follow steps 1–17 in the procedure described in Section 2.1.1 to generate the crude lysates of the FDH mutant library. A similar mutation rate and library size to those described for the evolution of increased thermostability was employed for each of the three rounds of evolution used to expand the substrate scope of RebH.
2. Select a suitable substrate for screening. The primary considerations for selecting the proper substrate are:
   a. Structural homology to the ultimate target substrate. For example, if the target substrate possesses an additional ring system or bulky substituent relative to the native substrate, an intermediate substrate might possess methyl groups in these locations. One reported example that sought to accommodate an additional bipyridyl group first

engineered activity on an intermediate biphenyl analog (Xie et al., 2007). They thus developed an enzyme variant that could accommodate the added steric bulk prior to engineering a final enzyme capable of recognizing the increased hydrophilicity of the bipyridyl group relative to the biphenyl group.
  b. Sufficient activity on the substrate. For the UPLC screening described later, at least 10% conversion on the substrate greatly decreased the quantity of false positives obtained (and thus the workload necessary for steps 25–35) and 20% conversion was optimal.
  c. Solubility in a buffer:cosolvent system tolerated by the FDH. RebH shows very little loss of activity with 95% HEPES buffer: 5% organic solvent (MeOH/EtOH/iPrOH/acetone/DMSO), providing ample options for a suitable solvent system.
3. Add solutions of halide (eg, NaCl, 1.5 M in HEPES buffer (25 mM, pH 7.4), 10 mM final concentration), FAD (10 mM in HEPES buffer (25 mM, pH 7.4), 100 μM final concentration), NAD (10 mM in HEPES buffer (25 mM, pH 7.4), 100 μM final concentration), FAD reductase (eg, MBP-RebF, 50–100 μM in HEPES storage buffer (25 mM, pH 7.4, 10% glycerol), 2.5 μM final concentration—other reductases have been employed for other FDHs), and glucose dehydrogenase (eg, GDH-105 from Codexis, 1800 U/mL in HEPES buffer (25 mM, pH 7.4), 9 U/mL final concentration) to the pretreated crude lysate. Add additional HEPES buffer (25 mM, pH 7.4) such that the final reaction volume is 75 μL, a solution of substrate in the organic cosolvent (eg, tryptoline for RebH 2-T, 10 mM in methanol), 0.5 mM final concentration), and then add a solution of glucose (1 M in HEPES buffer (25 mM, pH 7.4), 20 mM final concentration) to initiate the reaction. Reseal the microtiter plates with AeraSeals from Research Products International and then shake overnight at 600 rpm at room temperature.
  a. Note: the substrate in the organic cosolvent is added near the end of the procedure such that the enzymes are not subjected to a transiently higher than desired concentration of organic cosolvent.
4. Quench the reactions with 75 μL of methanol to precipitate out the proteins, add an internal standard to control for differences in handling (phenol at a final concentration of 0.5 mM is commonly employed, but a different standard may be required depending on the substrate and analysis method used), and spin the microtiter plates down at 3600 rpm for 15 min at 4°C.

5. Filter the supernatants into a 96-well plate using a 96-well 0.45-μm filter plate prior to UPLC analysis.
6. Seal the plates with aluminum foil using a plate sealer to minimize evaporation and prevent particulate contamination, then analyze the reactions (eg, by UPLC) for halogenated product production. The following methods were used for the different substrates employed in the expansion of the substrate scope of RebH:
    a. L-tryptophan: Agilent 1200 UHPLC with an Agilent Eclipse Plus C18 2.1 × 50 mm column, 1.8 μM particle size, 0.550 mL/min flow rate; solvent $A = H_2O/0.1\%$ TFA, solvent $B = CH_3CN$; 0–0.5 min, $B = 16\%$; 0.5–1.5 min, $B = 16–80\%$.
    b. Tryptoline: Agilent 1200 UHPLC with an Agilent Eclipse Plus C18 2.1 × 50 mm column, 1.8 μM particle size, 0.550 mL/min flow rate; solvent $A = H_2O/0.1\%$ TFA, solvent $B = CH_3CN$; 0–0.5 min, $B = 15\%$; 0.5–1.5 min, $B = 15–40\%$; 1.5–2.1 min, $B = 40\%$.
    c. Unhalogenated deformylflustrabromine: Agilent 1200 UHPLC with an Agilent Eclipse Plus C18 2.1 × 50 mm column, 1.8 μM particle size, 0.550 mL/min flow rate; solvent $A = H_2O/0.1\%$ TFA, solvent $B = CH_3CN$; 0–0.5 min, $B = 15\%$; 0.5–1.5 min, $B = 15–40\%$; 1.5–2.5 min, $B = 40\%$.
7. Analyze the reaction data to determine the relative improvement in halogenated product production relative to the parent wells added to each plate in step 11 in Section 2.1.1. The wells that show the greatest degree of improvement should be grown on a larger scale and purified as described in step 9 (using the saved primary culture in the 1-mL plates in step 13 in Section 2.1.1), plasmid isolated and sequenced to confirm the presence of nonsynonymous mutations, and halogenase variants analyzed in vitro as described in step 9 (for FPLC purification of either non-His-tagged or His-tagged halogenases, see van Pée (2012)).
8. If multiple improved variants are found displaying different nonsynonymous mutations, it is recommended that the beneficial mutations be analyzed both individually and in combination. Mutations can be recombined via SOE PCR; for RebH, PCR conditions were as follows: 98°C 30 s (98°C 10 s, 72°C 50 s) for 35 cycles, 72°C 10 min. Once plasmids are obtained containing the combined mutations, the halogenase variants should be immediately assayed in vitro as described in step 9.
9. To analyze the halogenase variants in vitro, follow steps 25–35 from Section 2.1.1.

5. CONCLUSIONS

While protein engineering of FDHs is a relatively new practice, significant progress has already been achieved. Using directed evolution techniques that should be generalizable to any FDH, the thermostability (Poor et al., 2014), substrate scope (Payne et al., 2015), and regioselectivity (Andorfer et al., 2016) of RebH have all been significantly altered. In addition to random mutagenesis, crystal structures are available for several FDHs that should assist in structure-guided mutagenesis efforts. Continued evolution of enhanced FDH variants will facilitate further biotechnology efforts, including expansions on the work that has already been done on metabolic engineering (Roy et al., 2010; Runguphan et al., 2010; Wang et al., 2015), combinatorial biosynthesis (Sánchez et al., 2005), and preparative-scale in vitro biocatalysis (Frese & Sewald, 2015) using wild-type FDHs.

REFERENCES

Andorfer, M. C., Park, H. J., Vergara-Coll, J., & Lewis, J. C. (2016). Directed evolution of RebH for catalyst-controlled halogenation of indole C–H bonds. *Chemical Science,* Advance article.

Bernhardt, P., Okino, T., Winter, J. M., Miyanaga, A., & Moore, B. S. (2011). A stereoselective vanadium-dependent chloroperoxidase in bacterial antibiotic biosynthesis. *Journal of the American Chemical Society, 133,* 4268–4270.

Bloom, J. D., Labthavikul, S. T., Otey, C. R., & Arnold, F. H. (2006). Protein stability promotes evolvability. *Proceedings of the National Academy of Sciences of the United States of America, 103,* 5869–5874.

Bunders, C. A., Minvielle, M. J., Worthington, R. J., Ortiz, M., Cavanagh, J., & Melander, C. (2011). Intercepting bacterial indole signaling with flustramine derivatives. *Journal of the American Chemical Society, 133,* 20160.

Carter-Franklin, J. N., & Butler, A. (2004). Vanadium bromoperoxidase-catalyzed biosynthesis of halogenated marine natural products. *Journal of the American Chemical Society, 126,* 15060–15066.

Collins, C. H., Leadbetter, J. R., & Arnold, F. H. (2006). Dual selection enhances the signaling specificity of a variant of the quorum-sensing transcriptional activator LuxR. *Nature Biotechnology, 24,* 708–712.

Dairi, T., Nakano, T., Aisaka, K., Katsumata, R., & Hasegawa, M. (1995). Cloning and nucleotide sequence of the gene responsible for chlorination of tetracycline. *Bioscience, Biotechnology, and Biochemistry, 59,* 1099–1106.

Derewenda, Z. S. (2010). Application of protein engineering to enhance crystallizability and improve crystal properties. *Acta Crystallographica. Section D, Biological Crystallography, 66,* 604–615.

Diethelm, S., Teufel, R., Kaysser, L., & Moore, B. S. (2014). A Multitasking vanadium-dependent chloroperoxidase as an inspiration for the chemical synthesis of the merochlorins. *Angewandte Chemie International Edition, 53,* 11023–11026.

Dong, C. J., Flecks, S., Unversucht, S., Haupt, C., van Pée, K.-H., & Naismith, J. H. (2005). Tryptophan 7-halogenase (PrnA) structure suggests a mechanism for regioselective chlorination. *Science, 309,* 2216–2219.

Durak, L. J., Payne, J. T., & Lewis, J. C. (2016). Late-stage diversification of biologically active molecules via chemoenzymatic C–H functionalization, *ACS Catalysis*, *6*, 1451–1454.
Esvelt, K. M., Carlson, J. C., & Liu, D. R. (2011). A system for the continuous directed evolution of biomolecules. *Nature*, *472*, 499–503.
Fasan, R. (2012). Tuning P450 enzymes as oxidation catalysts. *ACS Catalysis*, *2*, 647–666.
Fasan, R., Chen, M. M., Crook, N. C., & Arnold, F. H. (2007). Engineered alkane-hydroxylating cytochrome P450$_{BM3}$ exhibiting native like catalytic properties. *Angewandte Chemie International Edition*, *46*, 8414–8418.
Fishman, A., Tao, Y., Rui, L., & Wood, T. K. (2005). Controlling the regiospecific oxidation of aromatics via active site engineering of toluene para-monooxygenase of *Ralstonia pickettii* PKO1. *The Journal of Biological Chemistry*, *280*, 506–514.
Flecks, S., Patallo, E. P., Zhu, X., Ernyei, A. J., Seifert, G., Schneider, A., et al. (2008). New insights into the mechanism of enzymatic chlorination of tryptophan. *Angewandte Chemie International Edition*, *47*, 9533–9536.
Fox, R. J., & Huisman, G. W. (2008). Enzyme optimization: Moving from blind evolution to statistical exploration of sequence-function space. *Trends in Biotechnology*, *26*, 132–138.
Frese, M., Guzowska, P. H., Voß, H., & Sewald, N. (2014). Regioselective enzymatic halogenation of substituted tryptophan derivatives using the FAD-dependent halogenase RebH. *ChemCatChem*, *6*, 1270–1276.
Frese, M., & Sewald, N. (2015). Enzymatic halogenation of tryptophan on a gram scale. *Angewandte Chemie International Edition*, *54*, 298–301.
Ghislieri, D., Green, A. P., Pontini, M., Willies, S. C., Rowles, I., Frank, A., et al. (2013). Engineering an enantioselective amine oxidase for the synthesis of pharmaceutical building blocks and alkaloid natural products. *Journal of the American Chemical Society*, *135*, 10863–10869.
Glenn, W. S., Nims, E., & O'Connor, S. E. (2011). Reengineering a tryptophan halogenase to preferentially chlorinate a direct alkaloid precursor. *Journal of the American Chemical Society*, *133*, 19346–19349.
Gribble, G. W. (2010). *Naturally occurring organohalogen compounds—A comprehensive update*. Wien: Springer.
Harris, C. M., Kannan, R., Kopecka, H., & Harris, T. M. (1985). The role of the chlorine substituents in the antibiotic vancomycin: Preparation and characterization of mono- and didechlorovancomycin. *Journal of the American Chemical Society*, *107*, 6652–6658.
Hasan, Z., Renirie, R., Kerkman, R., Ruijssenaars, H. J., Hartog, A. F., & Wever, R. (2006). Laboratory-evolved vanadium chloroperoxidase exhibits 100-fold higher halogenating activity at alkaline pH: Catalytic effects from first and second coordination sphere mutations. *The Journal of Biological Chemistry*, *281*, 9738–9744.
Hernandes, M. Z., Cavalcanti, S. M. T., Moreira, D. R. M., de Azevedo Junior, W. F., & Leite, A. C. L. (2010). Halogen atoms in the modern medicinal chemistry: Hints for the drug design. *Current Drug Targets*, *11*, 303–314.
Herrera-Rodriguez, L. N., Khan, F., Robins, K. T., & Meyer, H.-P. (2011). Perspectives on biotechnological halogenation Part I: Halogenated products and enzymatic halogenation. *Chemistry Today*, *29*, 31.
Hölzer, M., Burd, W., Reißig, H.-U., & van Pée, K.-H. (2001). Substrate specificity and regioselectivity of tryptophan 7-halogenase from *Pseudomonas fluorescens* BL915. *Advanced Synthesis & Catalysis*, *343*, 591–595.
Hosford, J., Shepherd, S. A., Micklefield, J., & Wong, L. S. (2014). A high-throughput assay for arylamine halogenation based on a peroxidase-mediated quinone-amine coupling with applications in the screening of enzymatic halogenations. *Chemistry A European Journal*, *20*, 16759–16763.
For PrnA, see: Keller, S., Wage, T., Hohaus, K., Hölzer, M., Eichhorn, E., & van Pée, K.-H. (2000). Purification and partial characterization of tryptophan 7-halogenase (PrnA) from Pseudomonas fluorescens. *Angewandte Chemie International Edition*, *39*, 2300–2302.

Kirner, S., Hammer, P. E., Hill, D. S., Altmann, A., Fischer, I., Weislo, L. J., et al. (1998). Functions encoded by pyrrolnitrin biosynthetic genes from *Pseudomonas fluorescens*. *Journal of Bacteriology*, *180*, 1939–1943.

Koszelewski, D., Müller, N., Schrittwieser, J. H., Faber, K., & Kroutil, W. (2010). Immobilization of ω-transaminases by encapsulation in a sol-gel/celite matrix. *Journal of Molecular Catalysis B: Enzymatic*, *63*, 39–44.

Lang, A., Polnick, S., Nicke, T., William, P., Patallo, E. P., Naismith, J. H., et al. (2011). Changing the regioselectivity of the tryptophan 7-halogenase PrnA by site-directed mutagenesis. *Angewandte Chemie International Edition*, *50*, 2951–2953.

Lewis, J. C., Bastian, S., Bennett, C. S., Fu, Y., Mitsuda, Y., Chen, M. M., et al. (2009). Chemoenzymatic elaboration of monosaccharides using engineered cytochrome P450$_{BM3}$ demethylases. *Proceedings of the National Academy of Sciences of the United States of America*, *106*, 16550–16555.

Lewis, J. C., Coelho, P. S., & Arnold, F. H. (2011). Enzymatic functionalization of carbon-hydrogen bonds. *Chemical Society Reviews*, *40*, 2003–2021.

Lewis, J. C., Poor, C. B., Payne, J. T., & Andorfer, M. C. (2014). *Directed evolution of a regioselective halogenase for increased thermostability*. Washington DC: US Patent and Trademark Office. PCT/US14/67661.

Liao, H. H. (1993). Thermostable mutants of kanamycin nucleotidyltransferase are also more stable to proteinase K, urea, detergents, and water-miscible organic solvents. *Enzyme and Microbial Technology*, *15*, 286–292.

Lin, H., & Cornish, V. W. (2002). Screening and selection methods for large-scale analysis of protein function. *Angewandte Chemie International Edition*, *41*, 4402–4425.

Lin, H., Tao, H., & Cornish, V. W. (2004). Directed evolution of a glycosynthase via chemical complementation. *Journal of the American Chemical Society*, *126*, 15051–15059.

Manoj, K. M. (2006). Chlorinations catalyzed by chloroperoxidase occur via diffusible intermediate(s) and the reaction components play multiple roles in the overall process. *Biochimica et Biophysica Acta*, *1764*, 1325–1339.

Morris, D. R., & Hager, L. P. (1966). Chloroperoxidase I. Isolation and properties of the crystalline glycoprotein. *The Journal of Biological Chemistry*, *241*, 1763–1768.

Payne, J. T., Andorfer, M. C., & Lewis, J. C. (2013). Regioselective arene halogenation using the FAD-dependent halogenase RebH. *Angewandte Chemie International Edition*, *52*, 5271–5274.

Payne, J. T., Poor, C. B., & Lewis, J. C. (2015). Directed evolution of RebH for site-selective halogenation of large biologically active molecules. *Angewandte Chemie International Edition*, *54*, 4226–4230.

Petty, M. A. (1961). An introduction to the origin and biochemistry of microbial halometabolites. *Bacteriological Reviews*, *25*, 111–130.

Polizzi, K. M., Bommarius, A. S., Broering, J. S., & Chaparro-Riggers, J. F. (2007). Stability of biocatalysts. *Current Opinion in Chemical Biology*, *11*, 220–225.

Poor, C. B., Andorfer, M. C., & Lewis, J. C. (2014). Improving the stability and catalyst lifetime of the halogenase RebH by directed evolution. *ChemBioChem*, *15*, 1286–1289.

Reetz, M. T., Höbenreich, H., Soni, P., & Fernández, L. (2008). A genetic selection system for evolving enantioselectivity of enzymes. *Chemical Communications*, *43*, 5502–5504.

Romero, P. A., & Arnold, F. H. (2009). Exploring protein fitness landscapes by directed evolution. *Nature Reviews. Molecular Cell Biology*, *10*, 866–876.

Roy, A. D., Grüschow, S., Cairns, N., & Goss, R. J. M. (2010). Gene expression enabling synthetic diversification of natural products: Chemogenetic generation of pacidamycin analogs. *Journal of the American Chemical Society*, *132*, 12243–12245.

Runguphan, W., & O'Connor, S. E. (2013). Diversification of monoterpene indole alkaloid analogs through cross-coupling. *Organic Letters*, *15*, 2850–2853.
Runguphan, W., Qu, X., & O'Connor, S. E. (2010). Integrating carbon-halogen bond formation into medicinal plant metabolism. *Nature*, *468*, 461–464.
Sánchez, C., Zhu, L., Braña, A. F., Salas, A. P., Rohr, J., Méndez, C., et al. (2005). Combinatorial biosynthesis of antitumor indolocarbazole compounds. *Proceedings of the National Academy of Sciences of the United States of America*, *102*, 461–466.
Savile, C. K., Janey, J. M., Mundorff, E. C., Moore, J. C., Tam, S., Jarvis, W. R., et al. (2010). Biocatalytic asymmetric synthesis of chiral amines from ketones applied to sitagliptin manufacture. *Science*, *329*, 305–309.
For thal, see: Seibold, C., Schnerr, H., Rumpf, J., Kunzendorf, A., Hatscher, C., Wage, T., et al. (2006). A flavin-dependent tryptophan 6-halogenase and its use in modification of pyrrolnitrin biosynthesis. *Biocatalysis and Biotransformation*, *24*, 401–408.
Shepherd, S. A., Karthikeyan, C., Latham, J., Struck, A.-W., Thompson, M. L., Menon, B. R. K., et al. (2015). Extending the biocatalytic scope of regiocomplementary flavin-dependent halogenase enzymes. *Chemical Science*, *6*, 3454–3460.
Shimotohno, A., Oue, S., Yano, T., Kuramitsu, S., & Kagamiyama, H. (2001). Demonstration of the importance and usefulness of manipulating non-active-site residues in protein design. *Journal of Biochemistry*, *129*, 943–948.
Smith, K., & El-Hiti, G. A. (2004). Regioselective control of electrophilic aromatic substitution reactions. *Current Organic Synthesis*, *1*, 253–274.
Smith, B. M., Smith, J. M., Tsai, J. H., Schultz, J. A., Gilson, C. A., Estrada, S. A., et al. (2008). Discovery and structure-activity relationship of (1R)-8-chloro-2,3,4,5-tetrahydro-1-methyl-1H-3-benzazepine (lorcaserin), a selective serotonin 5-HT2C receptor agonist for the treatment of obesity. *Journal of Medicinal Chemistry*, *51*, 305–313.
Spiller, B., Gershenson, A., Arnold, F. H., & Stevens, R. C. (1999). A structural view of evolutionary divergence. *Proceedings of the National Academy of Sciences of the United States of America*, *96*, 12305–12310.
Sun, H., Keefer, C. E., & Scott, D. O. (2011). Systematic and pairwise analysis of the effects of aromatic halogenation and trifluoromethyl substitution on human liver microsomal clearance. *Drug Metabolism Letters*, *5*, 232–242.
Taylor, S. V., Kast, P., & Hilvert, D. (2001). Investigating and engineering enzymes by genetic selection. *Angewandte Chemie International Edition*, *40*, 3310–3335.
Truppo, M. D., Strotman, H., & Hughes, G. (2012). Development of an immobilized transaminase capable of operating in organic solvent. *ChemCatChem*, *4*, 1071–1074.
Vaillancourt, F. H., Yeh, E., Vosburg, D. A., Garneau-Tsodikova, S., & Walsh, C. T. (2006). nature's inventory of halogenation catalysts: Oxidative strategies predominate. *Chemical Reviews*, *106*, 3364–3378.
van Pée, K.-H. (1990). Bacterial haloperoxidases and their role in secondary metabolism. *Biotechnology Advances*, *8*, 185–205.
van Pée, K.-H. (2012). Enzymatic chlorination and bromination. *Methods in Enzymology*, *516*, 237–257.
van Pée, K.-H. (2013). Halogenating enzymes for selective halogenation reactions. *Current Organic Chemistry*, *16*, 2583–2597.
van Pée, K.-H., & Patallo, E. P. (2006). Flavin-dependent halogenases involved in secondary metabolism in bacteria. *Applied Microbiology and Biotechnology*, *70*, 631–641.
Wagner, C., & König, G. M. (2012). Handbook of marine natural products: Mechanisms of halogenation of marine secondary metabolites. In E. Fattorusso, W. H. Gerwick, & O. Taglialatela-Scafati (Eds.), Netherlands: Wiley-Springer.

Wang, S., Zhang, S., Xiao, A., Rasmussen, M., Skidmore, C., & Zhan, J. (2015). Metabolic engineering of *Escherichia coli* for the biosynthesis of various phenylpropanoid derivatives. *Metabolic Engineering, 29*, 153–159.

Williams, P. G., Buchanan, G. O., Feling, R. H., Kauffman, C. A., Jensen, P. R., & Fenical, W. (2005). New cytotoxic salinosporamides from the marine actinomycete *Salinispora tropica*. *Journal of Organic Chemistry, 70*, 6196–6203.

Wu, I., & Arnold, F. H. (2013). Engineered thermostable fungal Cel6A and Cel7A cellobiohydrolases hydrolyze cellulose efficiently at elevated temperatures. *Biotechnology and Bioengineering, 110*, 1874–1883.

Xie, J., Liu, W., & Schultz, P. G. (2007). A genetically encoded bidentate, metal-binding amino acids. *Angewandte Chemie International Edition, 46*, 9239–9242.

Yamada, R., Higo, T., Yoshikawa, C., China, H., & Ogino, H. (2014). Improvement of the stability and activity of the BPO-A1 haloperoxidase from Streptomyces aureofaciens by directed evolution. *Journal of Biotechnology, 192*, 248–254.

Yeh, E., Blasiak, L. C., Koglin, A., Drennan, C. L., & Walsh, C. T. (2007). Chlorination by a long-lived intermediate in the mechanism of flavin-dependent halogenases. *Biochemistry, 46*, 1284–1292.

For RebH, see: Yeh, E., Garneau, S., & Walsh, C. T. (2005). Robust in vitro activity of RebF and RebH, a two component reductase/halogenase, generating 7-chlorotryptophan during rebeccamycin biosynthesis. *Proceedings of the National Academy of Sciences of the United States of America, 102*, 3960–3965.

You, L., & Arnold, F. H. (1994). Directed evolution of subtilisin E in Bacillus subtilis to enhance total activity in aqueous dimethylformamide. *Protein Engineering, 9*, 77–83.

For PyrH, see: Zehner, S., Kotzsch, A., Bister, B., Süssmuth, R. D., Méndez, C., Salas, J. A., et al. (2005). A regioselective tryptophan 5-halogenase is involved in pyrroindomycin biosynthesis in streptomyces rugosporus LL-42D005. *Chemistry & Biology, 12*, 445–452.

Zeng, J., Lutle, A. K., Gage, D., Johnson, S. J., & Zhan, J. (2013). Specific chlorination of isoquinolines by a fungal flavin-dependent halogenase. *Bioorganic & Medicinal Chemistry Letters, 23*, 1001–1003.

Zhao, H., & Arnold, F. H. (1999). Directed evolution converts subtilisin E into a functional equivalent of thermitase. *Protein Engineering, 12*, 47–53.

CHAPTER SIX

Heterologous Expression of Fungal Secondary Metabolite Pathways in the *Aspergillus nidulans* Host System

J.W.A. van Dijk*, C.C.C. Wang*,†,1

*School of Pharmacy, University of Southern California, Los Angeles, CA, United States
†Dornsife College of Letters, Arts, and Sciences, University of Southern California, Los Angeles, CA, United States
[1]Corresponding author: e-mail address: clayw@usc.edu

Contents

1. Introduction	128
2. Identification of Secondary Metabolite Genes in Fungal Genomes	130
3. Design Primers for Fusion PCR	130
4. Obtain Genomic DNA for PCR Template	131
4.1 Materials	132
4.2 DNA Extraction from Hyphae	132
4.3 Materials	133
4.4 DNA Extraction from Spores	133
5. Fusion PCR Construction	134
5.1 Materials	134
5.2 Genomic PCR	134
5.3 One Pot Fusion Reaction	135
5.4 Two Pot Fusion Reaction	135
6. Transformation	136
6.1 Materials	136
6.2 Protoplasting	136
7. Diagnostic PCR	138
8. Liquid Culturing of Mutant Strains	138
8.1 Materials	138
8.2 Culturing Strains in Glucose Minimal Media	139
9. Recipes	139
10. Conclusions	140
Acknowledgments	141
References	141

Abstract

Heterologous expression of fungal secondary metabolite genes allows for the product formation of otherwise silent secondary metabolite biosynthesis pathways. It also allows facile expression of mutants or combinations of genes not found in nature. This capability makes model fungi an ideal platform for synthetic biology. In this chapter a detailed description is provided of how to heterologously express any fungal secondary metabolite gene(s) in a well-developed host strain of *Aspergillus nidulans*. It covers all the necessary steps from identifying a gene(s) of interest to culturing mutant strains to produce secondary metabolites.

1. INTRODUCTION

Fungal species have been widely used to study fundamental biology of eukaryotes since they grow faster and are easier to grow compared to higher order mammalian cells. Several species such as *Aspergillus fumigatus* and *Aspergillus flavus* are of particular interest because of their pathogenicity due to both invasiveness in immunocompromised patients and the toxic compounds they produce, for example, gliotoxin and aflatoxin. These compounds are often the product of the secondary metabolite (SM) pathways in these fungi. SM pathways offer a myriad of organic molecules and can also sometimes be medicinally useful such as most famously penicillin G, cyclosporine A, and lovastatin. Since many of these biosynthetic pathways present in the genomes of fungal species have not been linked to their product, this provides an opportunity for discovery as more (fungal) genomes are sequenced.

Fungal secondary metabolite biosynthesis pathways involve one core set of enzymes that is responsible for the backbone of the compound. These core enzymes could either be polyketide synthases (PKS), nonribosomal synthetases (NRPS), a hybrid PKS–NRPS, or a terpene cyclase (TC). Discovery of new secondary metabolite genes is often accomplished by homology searches using known genes of these core enzymes (Inglis et al., 2013; Khaldi et al., 2010; Takeda, Umemura, Koike, Asai, & Machida, 2014). Conveniently, in many cases, the other genes (coding for tailoring enzymes, resistance proteins, transcription factors, or transporters) in the biosynthetic pathway are clustered in the genome so they can be identified easily once the core enzymes have been identified. There are several ways to link the identified secondary metabolite pathways with their products. One way is to create knockouts in the producing organism (Chiang et al., 2008). In this approach the disappearance of one or more compounds in the secondary metabolite profile in the mutant strain compared to a control strain would indicate that the deleted gene in the mutant is necessary for product formation. This method is quite straightforward but would require that a culture condition can be identified under which the gene product is produced. In many cases secondary metabolite clusters identified from a sequenced genome are either silent or produce very small amounts, so that a condition to turn on the pathway cannot be identified easily. Alternatively, our lab and others have used a heterologous expression approach to link secondary metabolism genes with their respective compounds (Chiang et al., 2013, 2015). The heterologous expression approach is useful particularly when the secondary pathways are silent in the producing organism or when no genetic system is available.

The three commonly used host system in the lab are *Saccharomyces cerevisiae* (yeast) (Tsunematsu, Ishiuchi, Hotta, & Watanabe, 2013), *Aspergillus oryzae* (Sakai, Kinoshita, & Nihira, 2012), and *Aspergillus nidulans*. Gene targeting is very efficient in yeast, but wild-type yeast is not rich in SMs and thus relevant building blocks may be absent. Introns also differ between yeast and filamentous fungi, so cDNA is necessary for heterologous expression. Both *A. nidulans* and *A. oryzae* are more suited in these respects since intron splicing is not an issue. In our lab we have focused on using *A. nidulans* as a host. *A. nidulans* has the advantage that its genetic system is well developed. Host strains with multiple nutritional markers are available. In collaboration with Berl Oakley's group at University of Kansas an *A. nidulans* strain, LO8030, with four mutations (*pyroA4*, *riboB2*, *pyrG89*, and *nkuA::argB*) and eight gene clusters knocked out (sterigmatocystin AN7804-7825, emericellamide AN2545-2549, asperfuranone AN1039-1029, monodictyphenone AN10023-10021, terrequinone AN8512-8520, austinol AN8379-8384/AN9246-9259, F9775 AN7906-7915, and asperthecin AN6000-6002) has been created (Chiang et al., 2015). The strain LO8030 wastes fewer resources on known secondary metabolites and has a clean background ideal for detecting new compounds as the result of heterologous expression. This chapter will explain step by step how to rapidly heterologously express SM genes in this highly optimized host without the need for cosmids, plasmids, or vectors. Parts of this protocol have also been described elsewhere in detail (Lim, Sanchez, Wang, & Keller, 2012), but will be included to enable readers to perform the experiments using this chapter alone.

2. IDENTIFICATION OF SECONDARY METABOLITE GENES IN FUNGAL GENOMES

The first step is to identify the gene(s) you want to express. The Joint Genome Institute (JGI) is a great resource for (newly) sequenced genomes, but NCBI and AspGD also have genomes available. Genes of interest are usually found by blast homology analysis with known biosynthetic genes. The predicted function of hits and their surrounding genes can be used as an indication of validity. Once a gene is identified, the genomic sequence can be downloaded with introns included, since the *A. nidulans* system can recognize them.

Note: Since JGI provides genome sequences prior to publication, consult the JGI website for their most recent policy regarding data usage and acknowledgments.

3. DESIGN PRIMERS FOR FUSION PCR

For genes smaller than 4000 base pairs, one forward and one reverse primer can be designed. The forward primer typically starts at the start codon and has a 5′ sticky end of ~20 base pairs for fusion PCR to an upstream promoter. The powerful alcohol inducible promoter *alcA* (Waring, May, & Morris, 1989) is most often used in our lab, but other (noninducible) promoters like Tet-on system (Meyer et al., 2011) or *gpdA* (Punt, Kramer, Kuyvenhoven, Pouwels, & Vandenhondel, 1992) are also possibilities. The reverse primer is designed at roughly 300 base pairs after the stop codon to allow proper transcription termination. It has a 3′ sticky end of ~20 base pairs for a selection marker. The gene *pyrG* from *A. fumigatus* is most often used to complement the *A. nidulans pyrG89* background strain. *AfriboB* and *AfpyroA* markers can be used for subsequent transformations. More genes can be inserted by marker recycling where a marker of the first construct is replaced by a second construct with a different marker. *AfpyrG* deletion can also be selected for with 5-fluoroorotic acid. Genes larger than 4000 base pairs require two or more DNA fragments for transformation due to PCR limits for large (8 + kb) constructs. The 5′ and the 3′ primers are designed in the same way and additional primers are designed to generate fragments with a 1000 base pair overlap. Additionally a second marker is fused to the 5′ fragment before the promoter (see Fig. 1B).

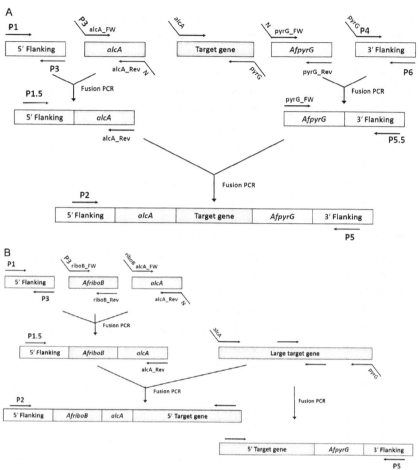

Fig. 1 Fusion PCR overview for genes smaller (A) and larger than 4000 base pairs (B). (A) Individual fragments are amplified from genomic DNA. Flanking fragments and the *alcA* promoter from the *A. nidulans* host, *AfpyrG* from *A. fumigatus* genomic DNA, and the target gene from its respective source. The "N" sticky ends of the alcA_Rev and pyrG_FW primers on the first line are an arbitrary number of nucleotides added to make the fusion PCRs on the middle line have nested primers on both sides. (B) The target gene is amplified in two separate pieces with 1000 base pairs overlap. In addition to the promoter, an additional marker is added to the 5′ fragment. The two fragments are mixed and added to the protoplasts for transformation. (See the color plate.)

4. OBTAIN GENOMIC DNA FOR PCR TEMPLATE

The ability of *A. nidulans* to recognize introns, allows the use of unmodified genomic DNA for gene amplification. Genomic DNA is preferably extracted from hyphae, since they presumably contain less metabolites that may inhibit polymerase. Usually genomic DNA from spores also works though. For hazardous species, it is possible to purchase only the genomic DNA from commercial sources such as ATCC. Since many fungal species are either plant or human pathogens sometimes obtaining the necessary permits will take several months and purchasing only genomic DNA can then greatly speed up the project. Lastly, with the significant improvements in commercial gene synthesis, it is also possible to obtain secondary metabolite genes via commercial DNA synthesis.

4.1 Materials

Note: In this section we provide the supplier and the item number we use in the Wang lab. Equivalent items can be used instead unless specifically noted below.
1. Miracloth (EMD Millipore, #475855)
2. Mortar and pestle
3. Sterile toothpick (Forster)
4. Lysis buffer hyphae (see Section 9)
5. Phenol/chloroform/isoamyl alcohol (25:24:1) adjusted to pH 8 (EMD Millipore, #516726)
6. 3 M NaOAc, pH 5 (AMRESCO, #0602)
7. Isopropanol (J.T. Baker, #9084-01)
8. Ethanol, biotechnology grade (IBI scientific, # IB15720)
9. TE buffer (see Section 9)

4.2 DNA Extraction from Hyphae

1. Harvest the hyphae by filtration through Miracloth.
2. Transfer to Eppendorf tube, lyophilize, and grind dry mycelia with a sterile toothpick or transfer to mortar and freeze with liquid nitrogen and grind to powder.
3. Add 700 µL lysis buffer and vortex for 30 s.
4. Incubate at 65°C for 1 h (tube with cap-lock).
5. Add 700 µL (an equal amount of lysis buffer in step 3) phenol/chloroform/isoamyl alcohol (25:24:1) adjusted to pH 8 and vortex for 1 min.
6. Microcentrifuge at max speed (or at least $10,000 \times g$) for 15 min at RT.
7. Transfer aqueous phase (~400 µL in 100 µL steps to avoid disturbing separation) to a new Eppendorf tube.

8. Add 13.3 μL of 3 M NaOAc, pH 5, followed by 223 μL (0.54 × V) of isopropanol and invert gently. DNA clots may be visible.
9. Microcentrifuge at max speed (or at least 10,000 × g) for 3 min at RT.
10. Remove supernatant with a pipette and rinse the pellet twice with 70% ethanol.
11. Remove supernatant with a pipette and place the tube in a dry bath at 50°C for ~10 min.
12. Resuspend in 100 μL TE buffer. If the DNA does not dissolve, place in dry bath at 65°C for 15 min.
13. Take 0.5 μL for a 50 μL PCR reaction.
14. Genomic DNA can be stored at −20°C for months or at −80°C for longer.

4.3 Materials

1. Miracloth (EMD Millipore, #475855)
2. Lysis buffer spores (see Section 9)
3. Glass beads, acid washed (Sigma, #G8772)
4. Phenol/chloroform/isoamyl alcohol (25:24:1) adjusted to pH 8 (EMD Millipore, #516726)
5. 3 M NaOAc, pH 5 (AMRESCO, #0602)
6. Isopropanol (J.T. Baker, #9084-01)
7. Ethanol, biotechnology grade (IBI scientific, # IB15720)
8. TE buffer (see Section 9)

4.4 DNA Extraction from Spores

1. Take 500 μL of spore suspension in an Eppendorf tube and centrifuge at 8000 × g for 5 min.
2. Remove supernatant and resuspend in 100 μL lysis buffer, then add 150 mg/100 μL of 0.45–0.5-mm glass beads.
3. Vortex for 30 s.
4. Incubate for 30 min at 65°C vortexing 30 s every 10 min (tube with cap-lock).
5. Add 100 μL of phenol/chloroform/isoamyl alcohol (25:24:1).
6. Vortex for 5 min.
7. Microcentrifuge at max speed (or at least 10,000 × g) for 5 min at RT.
8. Collect 50–60 μL from the upper phase and transfer to a new tube.
9. Add 5 μL of 3 M NaOAc, pH 5, mix, add 100 μL ethanol, mix. Put in −20°C for 20 min.

10. Microcentrifuge at max speed (or at least 10,000 × g) for 30 min at RT.
11. Discard supernatant and wash with 100 μL of 70% ethanol.
12. Microcentrifuge at max speed (or at least 10,000 × g) for 15 min at RT.
13. Remove supernatant and place the tube in a dry bath at 50°C for ~5 min.
14. Dissolve in 30 μL TE buffer.
15. Take 0.5 μL for a 50 μL PCR reaction.

5. FUSION PCR CONSTRUCTION

Individual fragments are amplified using the designed primers in 50 μL PCR reactions (Fig. 2). Note: Accuprime™ high-fidelity *Taq* is the commercial polymerase we use due to its accuracy, high yield, and ability to generate large DNA fragments. Even though it is more expensive than other polymerases commercially available, in our experience those have performed significantly worse in terms of both success and yield. Therefore, we highly recommend using the same polymerase we use in the Wang lab for the fusion PCR construction.

5.1 Materials

1. Techne Prime thermocycler (Techne, #5PRIMEG/02)
2. Accuprime™ *Taq* high fidelity with buffer I and II (Thermo Fisher Scientific, #12346094)
3. Agarose (IBI scientific, #IB70071)
4. Ethidium bromide (IBI scientific, #40075)
5. QIAquick gel extraction kit (Qiagen, #28704)
6. Zymo-Spin I columns (Zymo Research, #C1003)

5.2 Genomic PCR

PCR reactions are analyzed on a 1% agarose gel with 0.01% ethidium bromide and correct bands are cut out and purified using a gel extraction kit (Qiagen) and eluted in 30 μL elution buffer. The concentration of eluted DNA is determined by nanodrop. For fusion PCR, each fragment is mixed at a fixed molar ratio (1:1, 1:2:1, or 1:2:2:1). Fusion PCR with more than four fragments should be split up into two separate fusions of which the products can be fused again. The amount of each fragment needed depends on each size, but a minimum of 30 ng of each fragment is recommended. Fusion PCR can be done in one or two pots. In one pot the primers are

Reaction components

Genomic DNA	0.5 μL
Forward primer (10 μM)	2 μL
Reverse primer (10 μM)	2 μL
10× buffer II (accuprime™)	5 μL
Accuprime™ Taq high fidelity (5u/μL)	0.2 μL
ddH₂O	40.3 μL

Thermocycler program

T	Time	Ramp rate	
94°C	5 min	3.4 °C/s	
94°C	20 s	3.4 °C/s	
70°C	1 s	3.4 °C/s	30×
55°C	30 s	0.1 °C/s	
68°C	1 min/kb + 1 min	0.2 °C/s	
68°C	20 min	–	
4°C	∞	1 °C/s	

Fig. 2 Genomic PCR conditions.

added with the separate fragments, which is the quickest way of obtaining the fusion construct (Fig. 3). In two pots the fragments are first fused without primers and then an aliquot is amplified in a second round with primers, which takes more time than the one pot reaction, but it is easier to generate more fusion construct, since the amplification step can be scaled up easily (Fig. 4).

5.3 One Pot Fusion Reaction

Reaction components

5′ Flanking fragment	30+ ng
DNA fragment (gene)	30+ ng
3′ Flanking fragment	30+ ng
Forward primer (10 μM)	2 μL
Reverse primer (10 μM)	2 μL
10× buffer I (accuprime™)	5 μL
Accuprime™ Taq high fidelity (5u/μL)	0.2 μL
ddH₂O	up to 50 μL

Thermocycler program

T	Time	Ramp rate	
94°C	5 min	3.4 °C/s	
94°C	20 s	3.4 °C/s	
70°C	1 s	3.4 °C/s	10×
55°C	30 s	0.1 °C/s	
68°C	1 min/kb + 1 min	0.2 °C/s	
94°C	20 s	3.4 °C/s	
70°C	1 s	3.4 °C/s	5×
55°C	30 s	0.1 °C/s	
68°C	1 min/kb + 1 min + 5 s/cycle	0.2 °C/s	
94°C	20 s	3.4 °C/s	
70°C	1 s	3.4 °C/s	15×
55°C	30 s	0.1 °C/s	
68°C	1 min/kb + 1 min + 20 s+ 15 s/cycle	0.2 °C/s	
68°C	20 min	–	
4°C	∞	1 °C/s	

Fig. 3 One pot fusion PCR conditions.

5.4 Two Pot Fusion Reaction

The fusion PCR is again analyzed on 1% agarose gel and correct product is cut out and purified using a gel extraction kit. Fusion PCR construct is eluted in 10–15 μL elution buffer or ddH₂O from Zymo-Spin I columns that allow elution with volumes as low as 6 μL and can be directly used for transformation or stored at −20°C. The fusion PCR product is not quantified, but the amount of DNA is correlated with transformation success: more DNA gives more transformed colonies. Also, for larger constructs

Reaction components

DNA fragment 1	30+ ng
DNA fragment 2	30+ ng
10× buffer I (accuprime™)	2.5 μL
Accuprime™ Taq high fidelity (5 u/μL)	0.1 μL
ddH₂O	up to 25 μL

Fusion template	2 μL
Forward primer (10 μM)	2 μL
Reverse primer (10 μM)	2 μL
10× buffer II (accuprime™)	5 μL
Accuprime™ Taq high fidelity (5 u/μL)	0.2 μL
ddH₂O	38.8 μL

Thermocycler program

T	Time	Ramp rate	
94°C	5 min	3.4 °C/s	
94°C	30 s	3.4 °C/s	15×
55°C	10 min	0.1 °C/s	
68°C	1 min/kb + 1 min	0.2 °C/s	
68°C	20 min	–	
4°C	∞	1 °C/s	
94°C	5 min	3.4 °C/s	
94°C	20 s	3.4 °C/s	
70°C	1 s	3.4 °C/s	30×
55°C	30 s	0.1 °C/s	
68°C	1 min/kb + 1 min	0.2 °C/s	
68°C	20 min	–	
4°C		1 °C/s	

Fig. 4 Two pot fusion PCR conditions.

and transformation with more than one fragment, DNA amounts need to be higher to be successful.

6. TRANSFORMATION

Protoplasts of the *A. nidulans* background strain are always freshly prepared for transformation (Szewczyk et al., 2006).

6.1 Materials

1. Supplemented yeast glucose (YG) media (see Section 9)
2. Gyratory shaker (Barnstead 4000 MaxQ)
3. 2 × Protoplasting solution (see Section 9)
4. Miracloth (EMD Millipore, #475855)
5. 1.2 M sucrose (sterile) (Sucrose, Omnipur, VWR, #EM-8550)
6. 0.6 M KCl (sterile) (VWR, #EM-PX1404-5)
7. 0.6 M KCl, 50 mM CaCl$_2$ (sterile) (CaCl$_2$·H$_2$O, VWR, #CX0130-2)
8. PEG solution (see Section 9)
9. Glucose minimal media agar plates (see Section 9)
10. Sterile cotton swab (Puritan, #25-806 1WC)
11. S/T buffer (see Section 9)

6.2 Protoplasting

1. Fresh spores (1e8, less than 1 week old) are inoculated in 20 mL YG media with supplements pyridoxine, riboflavin, uridine, and uracil.
2. Spores are shaken at 30°C, 135 rpm for 14 h.
3. 2 × Protoplasting solution made 20 min before.

4. The hyphae are filtered with sterile Miracloth.
5. Hyphae are washed with 2 mL of YG media and transferred with a sterile spatula to a sterile 50-mL conical flask with 8 mL YG media.
6. 8 mL of 2× protoplasting solution is filtered into the flask.
7. Protoplasting at 30°C, 135 rpm for 2–2.5 h, beware of overdigestion.
8. 4 mL of protoplasting solution is carefully layered on 8 mL of 1.2 M sucrose in a 15-mL tube and centrifuged with the brake off at 1800 × g for 10 min.
9. Protoplasts are collected from slightly above the resulting interface with a sterile dropper. Avoid disturbing the interface or collecting from below the interface. Collect 1–1.5 mL from each tube in a new 15-mL tube.
10. Add 2× volume of 0.6 M KCl and centrifuge with brake at 1800 × g for 10 min.
11. Remove supernatant with a serological pipette and resuspend in 2 × 1 mL of 0.6 M KCl. Resuspend by pipetting the first 300 µL slowly up and down a few times, then add the rest and transfer to a sterile Eppendorf tube (1 mL in each tube).
12. Centrifuge at 2400 × g for 3 min. Then wash each tube with 1 mL of 0.6 M KCl two more times. Material that does not resuspend after a few times of pipetting up and down should be removed by keeping the suspension sideways in the tip for 20 s. Unsuspended clogs will stick to the tip wall.
13. All protoplasts are resuspended in 1 mL of 0.6 M KCl, 50 mM CaCl$_2$ and centrifuged at 2400 × g for 3 min.
14. Supernatant is discarded and protoplasts are resuspended in 100 µL of 0.6 M KCl, 50 mM CaCl$_2$ per transformation. Up to 10 transformations can be done with one batch.
15. Protoplast solution is added to the DNA solution (10–15 µL) and vortexed once for 1 s.
16. 50 µL of freshly filtered PEG solution is added and vortexed once for 1 s.
17. Put on ice for 25 min.
18. Add 100 µL PEG solution and mix by pipetting up and down a minimum of 10 times.
19. Keep at RT for 25 min.
20. Plate each transformation mixture gently with a sterile spreader on two 10-cm glucose minimal media (GMM) plates with 0.6 M KCl, riboflavin, and pyridoxin (roughly 130 µL per plate).

21. Incubate face up at 30°C overnight.
22. Flip plates and transfer to 37°C.
23. Colonies should be visible 2 days after transfer.

By choosing the flanking DNA of the fusion construct (~1000 base pairs each side) to target the *yA* or the *wA* locus, the visible colonies after transformation have changed color from green to either yellow or white. When picking colonies for diagnostic PCR, the green ones, if any, can be disregarded. Three colonies per transformation are picked and scratched on selective plates. After 2 days a single colony is picked with a sterile cotton swab and replated on selective plates to ensure genetic homogeneity. Spores are harvested by covering the plate in 7 mL S/T buffer by rubbing a sterile cotton swab over the surface and collecting the suspension.

7. DIAGNOSTIC PCR

Genomic DNA of the transformants is extracted as described above. Diagnostic PCR (same as genomic PCR except 10 μL reactions) is performed by using primers outside of the insert on both background and new strains. An increase or decrease in size according to the fragment inserted shows correct transformation. For large, multiple fragment insertions the whole insertion is too large for a single PCR. Therefore, shorter fragments both within and on the border are chosen and compared with absence in the background. Diagnostic PCR usually shows success rates near 90%. Southern blots can be performed if necessary to double confirm. In addition (part of) the genes can be sequenced to confirm insertion and identify potential mutations not detected by PCR.

8. LIQUID CULTURING OF MUTANT STRAINS

8.1 Materials

1. Glucose minimal media (see Section 9)
2. Methylethyl ketone (or 2-butanone) (Sigma-Aldrich, #34861)
3. Ethyl acetate (EMD Millipore, #EX0240-5)
4. 6 M HCl (J.T. Baker, #5619-03)
5. Dimethyl sulfoxide (DMSO) (J.T. Baker, #9224-06)
6. Methanol (EMD Millipore, #MX0488-1)
7. 0.2-μm PTFE filter (VWR, #28145-491)

8.2 Culturing Strains in Glucose Minimal Media

Genetically correct strains are cultured alongside the background strain in 30 or 50 mL of liquid GMM in 125-mL conical flasks at 37°C, 180 rpm for 42 h. This ensures most if not all glucose is used up and will not inhibit subsequent *alcA* induction. The temperature is lowered to 30°C and the *alcA* promoter is induced by adding 134.4 or 224 μL methylethylketone (MEK). Cyclopentanone can also be used, but is more toxic to the fungus. After 3 days of induction, the cultures are filtered with filter paper and extracted with an equal volume of ethyl acetate. Aqueous phase is acidified to pH 2 with 6 *M* HCl and again extracted with an equal volume of ethyl acetate. Ethyl acetate extractions are dried, weighed, and redissolved in 20% DMSO in methanol at a concentration of roughly 1 mg/mL or as desired. After filtration with a 0.2-μm PTFE filter, 10 μL is injected for LC MS analysis. New peaks compared to the background are identified. Hit strains will be scaled up to liter scale. Extracts are first purified by flash chromatography followed by semipreparative HPLC. Purified compounds are characterized by HighRes MS and NMR. Detailed protocols for secondary metabolite analysis can be found in our earlier *Methods in Enzymology* review (Lim et al., 2012).

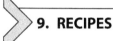

9. RECIPES

Lysis buffer hyphae (200 mL)
Add 1.211 g Tris, 2.922 g EDTA, and 6 g SDS to a final volume of 200 mL ddH$_2$O. Adjust pH to 7.2 with NaOH. Add 1% 2-mercaptoethanol right before use.

Lysis buffer spores (200 mL)
Add 4 mL Triton X-100, 2 g SDS, 1.1688 g NaCl, 0.0584 g EDTA, and 0.3152 g Tris–HCl to a final volume of 200 mL ddH$_2$O. Adjust to pH 8.0 with NaOH.

TE buffer (200 mL)
Add 0.3152 g Tris–HCl and 5.84 mg EDTA to a final volume of 200 mL ddH$_2$O. Adjust to pH 8.0 with NaOH. Autoclave.

Yeast glucose (YG) rich media (300 mL)
Add 1.5 g yeast extract, 6 g dextrose, 300 μL of Hutner's trace element solution to a final volume of 300 mL. Pyridoxine-dependent strains can grow on this medium. Riboflavin-dependent strain requires the addition

of 750 µg riboflavin (from stock solution). Uridine[a] and uracil-dependent strains require 0.7326 and 0.3 g, respectively. Autoclave.

Glucose minimal media (500 mL)
Add 5 g dextrose, 500 µL KOH (5.5 M), 500 µL Hutner's trace element solution and 25 mL of 20 × salt solution (see below) to a final volume of 500 mL. Optional supplements: 0.25 mg pyridoxine, 1.25 mg riboflavin, 500 mg uracil, and 1.221 g uridine.[a] For agar plates, add 7.5 g agar. For 0.6 M KCl plates, add 22.36 g KCl. Autoclave. Agar media is kept at 55°C in a water bath before pouring 25 mL per 10-cm plate.

20× salt solution (2000 mL)
Add 240 g $NaNO_3$, 20.8 g KCl, 20.8 g $MgSO_4·7H_2O$, and 60.8 g KH_2PO_4 to a final volume of 2000 mL.

2 × Protoplasting solution (10 mL)
Dissolve 8.2 g KCl and 2.1 g citric acid monohydrate in 50 mL ddH_2O. Adjust pH to 5.8 with 1.1 M KOH solution. Make the volume up to 100 mL. Autoclave. Take 10 mL of KCl/citric acid and add 2 g VinoTaste® Pro. Vortex and let it dissolve for 20 min. Filter sterilize into flask for immediate use (see step 6 of protoplasting).

PEG solution (100 mL)
Add 4.47 g KCl and 0.74 g $CaCl·2H_2O$ to about 20 mL ddH_2O and 0.802 mL 1 M Tris–HCl and 0.196 mL 1 M Tris. Add 25 g PEG (average molecular weight 3350). Add ddH_2O approximately up to 90 mL. Use a magnetic stirrer to dissolve PEG. Remove stirrer and make up to 100 mL. Autoclave. Before use, filter necessary amount with 0.2-µm filter to remove detrimental PEG precipitates.

S/T buffer (1000 mL)
Add 8.5 g NaCl and 1 mL Tween 80 to 1000 mL ddH_2O. Autoclave.

10. CONCLUSIONS

Gene transfer using this method is very facile and mutant strains with correct gene inserts are easily obtained. The functionality of the heterologously expressed genes is hard to predict and often requires additional experiments, like the addition of genes that only work in tandem. However, the ability to generate 10 or more different mutants per transformation and the use of several orthogonal markers, make it easy to screen many gene mutants and combinations. The use of the powerful *alcA* promoter in this system

---

[a] Add filter-sterilized uridine stock solution (1 M) after autoclaving.

allows for the detection of secondary metabolites from less than optimal enzymes and even protein purification for in vitro studies. The artificial recreation and subsequent manipulation of biosynthetic pathways that this system enables, will greatly contribute to synthetic biology and drug discovery in the next decade.

ACKNOWLEDGMENTS
Research in the Wang group is funded by the following grants: NIH Grant POIGM084077 and NSF Emerging Frontiers in Research and Innovation-MIKS (Grant # 1136903).

REFERENCES
Chiang, Y.-M., Ahuja, M., Oakley, C. E., Entwistle, R., Asokan, A., Zutz, C., ... Oakley, B. R. (2015). Development of genetic dereplication strains in *Aspergillus nidulans* results in the discovery of aspercryptin. *Angewandte Chemie, 127*, 1–5.

Chiang, Y. M., Oakley, C. E., Ahuja, M., Entwistle, R., Schultz, A., Chang, S. L., ... Oakley, B. R. (2013). An efficient system for heterologous expression of secondary metabolite genes in Aspergillus nidulans. *Journal of the American Chemical Society, 135*(20), 7720–7731. http://dx.doi.org/10.1021/ja401945a.

Chiang, Y. M., Szewczyk, E., Nayak, T., Davidson, A. D., Sanchez, J. F., Lo, H. C., ... Wang, C. C. C. (2008). Molecular genetic mining of the Aspergillus secondary metabolome: Discovery of the emericellamide biosynthetic pathway. *Chemistry & Biology, 15*(6), 527–532. http://dx.doi.org/10.1016/j.chembiol.2008.05.010.

Inglis, D. O., Binkley, J., Skrzypek, M. S., Arnaud, M. B., Cerqueira, G. C., Shah, P., ... Sherlock, G. (2013). Comprehensive annotation of secondary metabolite biosynthetic genes and gene clusters of Aspergillus nidulans, A. fumigatus, A. niger and A. oryzae. *Bmc Microbiology, 13*, 23. http://dx.doi.org/10.1186/1471-2180-13-91.

Khaldi, N., Seifuddin, F. T., Turner, G., Haft, D., Nierman, W. C., Wolfe, K. H., & Fedorova, N. D. (2010). SMURF: Genomic mapping of fungal secondary metabolite clusters. *Fungal Genetics and Biology, 47*(9), 736–741. http://dx.doi.org/10.1016/j.fgb.2010.06.003.

Lim, F. Y., Sanchez, J. F., Wang, C. C. C., & Keller, N. P. (2012). Toward awakening cryptic secondary metabolite gene clusters in filamentous fungi. *Methods in Enzymology, 517*, 303–324.

Meyer, V., Wanka, F., van Gent, J., Arentshorst, M., van den Hondel, C. A. M. J. J., & Ram, A. F. J. (2011). Fungal gene expression on demand: An inducible, tunable, and metabolism-independent expression system for Aspergillus niger. *Applied and Environmental Microbiology, 77*(9), 2975–2983. http://dx.doi.org/10.1128/aem.02740-10.

Punt, P. J., Kramer, C., Kuyvenhoven, A., Pouwels, P. H., & Vandenhondel, C. (1992). An upstream activating sequence from the aspergillus-nidulans gpda gene. *Gene, 120*(1), 67–73. http://dx.doi.org/10.1016/0378-1119(92)90010-m.

Sakai, K., Kinoshita, H., & Nihira, T. (2012). Heterologous expression system in Aspergillus oryzae for fungal biosynthetic gene clusters of secondary metabolites. *Applied Microbiology and Biotechnology, 93*(5), 2011–2022. http://dx.doi.org/10.1007/s00253-011-3657-9.

Szewczyk, E., Nayak, T., Oakley, C. E., Edgerton, H., Xiong, Y., Taheri-Talesh, N., ... Oakley, B. R. (2006). Fusion PCR and gene targeting in Aspergillus nidulans. *Nature Protocols, 1*(6), 3111–3120.

Takeda, I., Umemura, M., Koike, H., Asai, K., & Machida, M. (2014). Motif-independent prediction of a secondary metabolism gene cluster using comparative genomics:

Application to sequenced genomes of aspergillus and ten other filamentous fungal species. *DNA Research*, *21*(4), 447–457. http://dx.doi.org/10.1093/dnares/dsu010.

Tsunematsu, Y., Ishiuchi, K., Hotta, K., & Watanabe, K. (2013). Yeast-based genome mining, production and mechanistic studies of the biosynthesis of fungal polyketide and peptide natural products. *Natural Product Reports*, *30*(8), 1139–1149. http://dx.doi.org/10.1039/c3np70037b.

Waring, R. B., May, G. S., & Morris, N. R. (1989). Characterization of an inducible expression system in aspergillus-nidulans using alca and tubulin-coding genes. *Gene*, *79*(1), 119–130. http://dx.doi.org/10.1016/0378-1119(89)90097-8.

CHAPTER SEVEN

Plug-and-Play Benzylisoquinoline Alkaloid Biosynthetic Gene Discovery in Engineered Yeast

J.S. Morris, M. Dastmalchi, J. Li, L. Chang, X. Chen, J.M. Hagel[1], P.J. Facchini[1]

University of Calgary, Calgary, AB, Canada
[1]Corresponding authors: e-mail address: hagelj@ucalgary.ca; pfacchin@ucalgary.ca

Contents

1. Introduction	144
1.1 Benzylisoquinoline Alkaloid Metabolic Biochemistry	144
1.2 Functional Genomics in BIA-Producing Plants	150
1.3 Engineering BIA Pathways in Yeast	152
2. Transcriptome Resources and Mining Candidate Genes	155
2.1 Selection of Plants and Tissues	155
2.2 RNA Extraction and DNA Sequencing	155
2.3 Selection of Candidate Genes	159
3. Building Yeast Platform Strains	161
3.1 Overview of USER Cloning	161
3.2 Preparing USER-Compatible Vectors	163
3.3 Simultaneous Cloning of Multiple PCR Fragments	163
3.4 Yeast Transformation	164
3.5 Recombinant Protein Analysis	165
3.6 Ura Marker excision	165
4. Functional Testing of Candidate Genes	166
4.1 Transient Expression Constructs	166
4.2 Yeast Transformation, Culture, and Substrate Feeding	166
5. Liquid Chromatography—Tandem Mass Spectrometry	167
5.1 Analytical Strategy Overview	167
5.2 List of Instrumentation and Software	168
5.3 List of Consumables, Solvents, and Reagents	168
5.4 Analytical Methods	169
5.5 Technical Notes	172
6. Summary and Future Prospects	172
Acknowledgments	173
References	173

Abstract

Benzylisoquinoline alkaloid (BIA) metabolism has been the focus of a considerable research effort over the past half-century, primarily because of the pharmaceutical importance of several compounds produced by opium poppy (*Papaver somniferum*). Advancements in genomics technologies have substantially accelerated the rate of gene discovery over the past decade, such that most biosynthetic enzymes involved in the formation of the major alkaloids of opium poppy have now been isolated and partially characterized. Not unexpectedly, the availability of all perceived biosynthetic genes has facilitated the reconstitution of several BIA pathways in microbial hosts, including yeast (*Saccharomyces cerevisiae*). Product yields are currently insufficient to consider the commercial production of high-value BIAs, such as morphine. However, the rudimentary success demonstrated by the uncomplicated and routine assembly of a multitude of characterized BIA biosynthetic genes provides a valuable gene discovery tool for the rapid functional identification of the plethora of gene candidates available through increasingly accessible genomic, transcriptomic, and proteomic databases. BIA biosynthetic gene discovery represents a substantial research opportunity largely owing to the wealth of existing enzyme data mostly obtained from a single plant species. Functionally novel enzymes and variants with potential metabolic engineering applications can be considered the primary targets. Selection of candidates from sequence repositories is facilitated by the monophyletic relationship among biosynthetic genes belonging to a wide range of enzyme families, such as the numerous cytochromes P450 and AdoMet-dependent O- and N-methyltransferases that operate in BIA metabolism. We describe methods for the rapid functional screening of uncharacterized gene candidates encoding potential BIA biosynthetic enzymes using yeast strains engineered to perform selected metabolic conversions. As an initial screening tool, the approach is superior to the in vitro characterization of recombinant enzyme candidates, and provides a standardized functional genomics opportunity for otherwise recalcitrant exotic plant species.

1. INTRODUCTION
1.1 Benzylisoquinoline Alkaloid Metabolic Biochemistry

Benzylisoquinoline alkaloids (BIAs) are a large and structurally diverse class of nitrogen-containing plant specialized metabolites found predominantly in members of the order Ranunculales, such as opium poppy (*Papaver somniferum*), goldenseal (*Hydrastis canadensis*), and the tropical vine source for a curare arrow poison (*Chondrodendron tomentosum*) (Hagel & Facchini, 2013). BIAs display a remarkable range of biological and pharmacological properties, including the narcotic analgesic morphine, the antitussive and mitotic inhibitor noscapine, the vasodilator papaverine, and the antimicrobial sanguinarine, all of which accumulate in opium poppy. Despite the impressive biosynthetic capacity of opium poppy, these compounds represent only a fraction of the estimated 2500 BIAs occurring naturally in plants. The importance of opium poppy as our only commercial source for morphine, codeine, and thebaine, which is used for the semisynthesis of oxycodone, buprenorphine, naltrexone, and naloxone (Berényi, Csutorás, & Sipos, 2009), accounts for the longstanding research focus on BIA metabolism in this plant. Moreover, the establishment of transcriptomic resources (Desgagné-Penix & Facchini, 2012), coupled with the development of effective functional genomics methodologies in opium poppy (Dang, Onoyovwe, Farrow, & Facchini, 2012), have facilitated remarkable progress with respect to the isolation of most BIA biosynthetic genes involved in the formation of morphine (Farrow, Hagel, Beaudoin, Burns, & Facchini, 2015; Hagel & Facchini, 2010), noscapine (Chen, Dang, & Facchini, 2015), papaverine (Desgagné-Penix & Facchini, 2012), and sanguinarine (Beaudoin & Facchini, 2013). Additional progress has been made on the isolation and characterization of BIA biosynthetic genes from other plant species, such as Japanese goldthread (*Coptis japonica*), meadow rue (*Thalictrum flavum*), California poppy (*Eschscholzia californica*), and Wilson's barberry (*Berberis wilsoniae*) (Hagel & Facchini, 2013). Expanding the current collection of several dozen characterized BIA biosynthetic genes to a more complete catalogue encompassing the multitude of specific enzymes responsible for the remarkable diversity of natural BIAs requires a relatively generic functional genomics platform to test candidate genes from various undomesticated and generally recalcitrant plant species. Biosynthetic gene catalogues will be valuable as a source of parts for the burgeoning field of synthetic biology, in particular with respect to the metabolic engineering of plant specialized metabolism in microbial systems. We describe methods for the efficient and effective functional characterization of putative BIA biosynthetic genes from any species, citing examples of gene discovery in model organisms such as opium poppy.

1.1.1 BIA Biosynthesis in Plants

Understanding BIA metabolism, in general, and the capacity of individual plant species to produce different compounds are central considerations when devising a generic method for biosynthetic gene discovery. Despite impressive structural diversity, all BIAs originate from two L-tyrosine derivatives, dopamine and 4-hydroxyphenylacetaldehyde (4-HPAA), which provide the isoquinoline and benzyl moieties of the basic 1-benzylisoquinoline core structure (Facchini & De Luca, 1994; Hagel & Facchini, 2013; Lee &

Facchini, 2010) (Fig. 1). Elaboration of the 1-benzylisoquinoline scaffold, via reconfiguration of the backbone structure and the addition of functional groups, is mediated by a variety of enzymes belonging to different families, including cytochromes P450, FAD-linked oxidoreductases, 2-oxoglutarate/Fe(II)-dependent dioxygenases, AdoMet-dependent O-methyltransferases (OMTs), and N-methyltransferases (NMTs). The condensation of dopamine and 4-HPAA is catalyzed by norcoclaurine synthase (NCS) and yields the central intermediate (S)-norcoclaurine (Hagel & Facchini, 2013; Lee & Facchini, 2010). A metabolic lattice composed of norcoclaurine-6-O-methyltransferase (6OMT) (Morishige, Tsujita, Yamada, & Sato, 2000; Ounaroon, Decker, Schmidt, Lottspeich, & Kutchan, 2003), coclaurine N-methyltransferase (CNMT) (Choi, Morishige, & Sato, 2001), N-methylcoclaurine 3′-hydroxylase (NCMH) (Pauli & Kutchan, 1998), and 3′-hydroxy-N-methylcoclaurine 4′-O-methyltransferase (4′OMT) (Morishige et al., 2000) converts (S)-norcoclaurine to (S)-reticuline (Fig. 1). Considered a key branch-point intermediate, (S)-reticuline is known to undergo two major structural conversions, methylene bridge formation and stereochemical inversion, and various O-methylations (Chang, Hagel, & Facchini, 2015).

Methylene bridge formation in (S)-reticuline catalyzed by berberine bridge enzyme (BBE) yields a second important branch-point intermediate, the protoberberine (S)-scoulerine (Dittrich & Kutchan, 1991; Facchini, Penzes, Johnson, & Bull, 1996; Kutchan & Dittrich, 1995; Winkler, Hartner, Kutchan, Glieder, & Macheroux, 2006). (S)-Scoulerine can be converted to other protoberberines (eg, berberine), phthalideisoquinolines (eg, noscapine), to protopines (eg, protopine), and benzo[c]phenanthridines (eg, sanguinarine) (Fig. 1). A variety of additional alkaloid classes likely arise from further remodeling of these structural types, although the biochemical steps have generally not yet been elucidated. The stereochemical inversion of (S)-reticuline to (R)-reticuline is the gateway to the morphinan pathway and is catalyzed by a unique enzyme fusion, reticuline epimerase (REPI), consisting of the cytochromes P450 1,2-dehydroreticuline synthase (DRS), which oxidizes the tertiary amine in (S)-reticuline to the corresponding iminium, and the aldo–keto reductase 1,2-dehydroreticuline reductase (DRR), which performs NADPH-dependent reduction of the iminium to (R)-reticuline (Farrow et al., 2015; Winzer et al., 2015) (Fig. 1). (R)-Reticuline is converted by the cytochrome P450 salutaridine

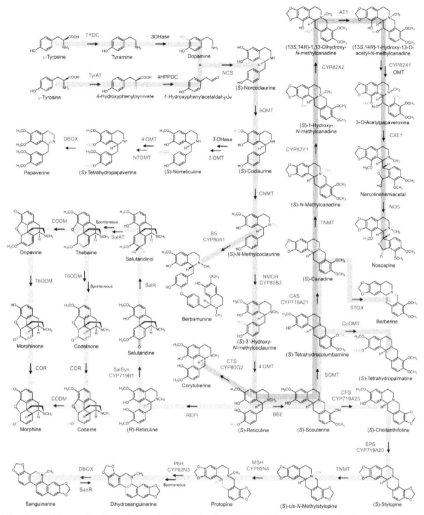

Fig. 1 BIA biosynthetic pathways for (S)-reticuline (*light pink*), papaverine (*yellow*), morphine (*green*), sanguinarine (*orange*), berberine (*blue*), noscapine (*purple*), berbamunine (*olive*), and corytuberine (*dark pink*). *Red* within each alkaloid highlights enzyme catalyzed. Structural changes in each compound are indicated in *red*. Cognate genes have been isolated for enzymes abbreviated in *blue*. *Abbreviations*: 3′-*OHase*, 3′-hydroxylase; *3′OMT*, 3′-O-methyltransferase; *3OHase*, 3-hydroxylase; *4HPPDC*, 4-hydroxyphenylpyruvate decarboxylase; *4′OMT*, 3′-hydroxy-N-methylcoclaurine 4′-O-methyltransferase; *6OMT*, norcoclaurine 6-O-methyltransferase; *AT1*, 1,13-dihydroxy-N-methylcanadine 13-O-acetyltransferase; *BBE*, berberine bridge enzyme; *BS*, berbamunine synthase; *CAS*, canadine synthase;

(Continued)

synthase (SalSyn) to the tetracyclic promorphinan, salutaridine (Gesell et al., 2009), which is then converted to salutaridinol by the short-chain dehydrogenase/reductase salutaridine reductase (SalR) (Ziegler et al., 2006). Salutaridinol is subsequently O-acetylated by salutaridinol 7-O-acetyltransferase (SalAT) (Grothe, Lenz, & Kutchan, 2001) yielding the first pentacyclic morphinan, thebaine (Theuns, Janssen, Biessels, Menichini, & Salemink, 1984; Theuns, Theuns, & Lousberg, 1986). Two 2-oxoglutarate/Fe(II)-dependent dioxygenases, thebaine 6-O-demethylase (T6ODM) and codeine O-demethylase (CODM), catalyze the regiospecific O-demethylation of various morphinan intermediates (Hagel & Facchini, 2010). T6ODM converts thebaine to neopinone, which spontaneously rearranges to form codeinone, whereas CODM and T6ODM sequentially convert thebaine through oripavine to morphinone. Codeinone and morphinone are substrates for the aldo–keto reductase codeinone reductase (COR), yielding codeine and morphine, respectively (Unterlinner, Lenz, & Kutchan, 1999). Finally, CODM also converts codeine to morphine to complete the bifurcated pathway from thebaine (Hagel & Facchini, 2010). The restricted number of protein families implicated in BIA metabolism facilitates the gene discovery process, as the type of enzyme catalyzing uncharacterized reactions can be predicted based catalytically similar conversions. For example, genes encoding biosynthetic enzymes from opium poppy can be used to query transcriptomes of related species. The feasibility of this gene discovery method is due in part to the monophyletic origin of BIA metabolism (Liscombe, Macleod, Loukanina, Nandi, & Facchini, 2005).

Fig. 1—Cont'd *CFS*, cheilanthifoline synthase; *CNMT*, coclaurine *N*-methyltransferase; *CODM*, codeine *O*-demethylase; *CoOMT*, columbamine *O*-methyltransferase; *COR*, codeinone reductase; *CTS*, corytuberine synthase; *CYP82X1*, 1-hydroxy-13-*O*-acetyl-*N*-methylcanadine 8-hydroxylase; *CYP82X2*, 1-hydroxy-*N*-methylcanadine 13-hydroxylase; *CYP82Y1*, *N*-methylcanadine 1-hydroxylase; *CDBOX*, dihydrobenzophenanthridine oxidase; *CXE1*, 3-*O*-acetylpapaveroxine carboxylesterase; *MSH*, *N*-methylstylopine hydroxylase; *N7OMT*, norreticuline 7-*O*-methyltransferase; *NCS*, norcoclaurine synthase; *NMCanH*, *N*-methylcanadine 1-hydroxylase; *NMCH*, *N*-methylcoclaurine 3'-hydroxylase; *NOS*, noscapine synthase; *P6H*, protopine 6-hydroxylase; *REPI*, reticuline epimerase; *SalAT*, salutaridinol 7-*O*-acetyltransferase; *SalR*, salutaridine reductase; *SalSyn*, salutaridine synthase; *SanR*, sanguinarine reductase; *SOMT*, scoulerine 9-*O*-methyltransferase; *SPS*, stylopine synthase; *STOX*, (*S*)-tetrahydroprotoberberine oxidase; *T6ODM*, thebaine 6-*O*-demethylase; *TNMT*, tetrahydroprotoberberine *N*-methyltransferase; *TYDC*, tyrosine decarboxylase; *TyrAT*, tyrosine aminotransferase. (See the color plate.)

1.1.2 Localization of BIA Biosynthetic Pathways in Plants

An important consideration for gene discovery is the location of relevant transcripts, enzymes or metabolites within a plant. Information on localization guides tissue selection and the design of comparative studies, and provides a framework to understand the overall functionality of the pathway. BIA biosynthetic gene expression generally appears to correlate with the spatio-temporal accumulation of alkaloids; however, transport between tissues and cell types is known to occur (Onoyovwe et al., 2013; Samanani, Park, & Facchini, 2005). The localization of BIA biosynthesis and accumulation are best characterized in opium poppy (Bird, Franceschi, & Facchini, 2003; Lee & Facchini, 2010; Onoyovwe et al., 2013; Samanani, Alcantara, Bourgault, Zulak, & Facchini, 2006). Morphine, noscapine, papaverine, along with other end products and relevant pathway intermediates accumulate in latex, which is the cytoplasm of specialized laticifers found in all opium poppy organs. In contrast, sanguinarine and certain protoberberine, protopine and benzo[c]phenanthridine pathway intermediates accumulate constitutively in roots, possibly in a cell type other than laticifers (Alcantara, Bird, Franceschi, & Facchini, 2005; Frick, Kramell, Schmidt, Fist, & Kutchan, 2005). Gene transcript levels are generally coordinated with BIA accumulation profiles in opium poppy organs. For example, noscapine and gene transcripts encoding noscapine biosynthetic enzymes occur in all plant organs, but are most abundant in stems and unripe seed capsules (Dang & Facchini, 2012). Noscapine synthase (NOS) catalyzes the final step in the pathway, converting narcotine hemiacetal to noscapine, and is the only pathway enzyme found in latex (Chen et al., 2015; Chen & Facchini, 2014). Similarly, thebaine formation is catalyzed by enzymes primarily localized in sieve elements of the phloem, whereas T6ODM, COR, and CODM occur predominantly in laticifers (Onoyovwe et al., 2013). In meadow rue (*T. flavum*), which lacks laticifers, BIA biosynthetic gene transcripts are associated with specific cell types in roots and rhizomes (Samanani et al., 2005). The accumulation of protoberberines is restricted to endodermis in roots, but occurs more broadly in the cortex of rhizomes. Strategies for gene discovery can be based on the specific localization of the biosynthetic machinery within a plant. For example, the establishment of transcriptomic and metabolomic resources for 20 BIA-producing plants relied on strategic selection of tissue type initially guided by alkaloid content (Hagel, Morris, et al., 2015; Hagel, Mandal, et al., 2015).

1.2 Functional Genomics in BIA-Producing Plants

The recently expanded availability of transcriptomic and proteomic resources has facilitated the application of high-throughput functional genomics in diverse plant species. Several plant genes have been identified through the integration of predicted transcript or protein annotations with relevant spatio-temporal coordinates and metabolomic information. However, in general, the functional characterization of plant biosynthetic genes remains a slow and challenging process. For nonmodel species, including BIA-producing plants, a major hurdle in the evaluation of candidate genes is the general lack of reproducible and robust protocols for stable or transient heterologous gene expression *in planta*. Virus-induced gene silencing (VIGS; Ruiz, Voinnet, & Baulcombe, 1998) in opium poppy has allowed the relatively rapid, transient functional analysis of candidates, including biosynthetic genes involved in a variety of BIA pathways (Desgagné-Penix & Facchini, 2012; Hagel & Facchini, 2010; Wijekoon & Facchini, 2012; Winzer et al., 2012). Coupled with comparative metabolite or transcript profiling, the role of a biosynthetic gene in a metabolic pathway can be deduced from the phenotype resulting from the targeted knockdown of a specific gene. VIGS has also been used successfully in California poppy (*E. californica*), but many other BIA-producing species exhibit recalcitrant features that limit widespread application of the technology. These limitations include a relatively low efficiency of transformation, the frequent visualization of only partial phenotypes, considerable phenotypic variability between individual plants, the lack of heritability, and an inability to investigate events that occur during early plant development. With respect to the latter, for example, opium poppy seedlings are most effectively infiltrated with *Agrobacterium tumefaciens* to initiate VIGS after emergence of the first true leaves, approximately 1 week after seed germination, but effects on BIA metabolism become apparent only after several additional weeks of growth. When available, the use of plant chemotypes displaying distinct BIA profiles is complementary to VIGS. For example, an opium poppy-specific DNA microarray was used to compare a high thebaine/oripavine, low-codeine/morphine opium poppy chemotype with high-codeine/morphine cultivars to discover the *T6ODM* gene (Hagel & Facchini, 2010). Similarly, RNA-seq analysis of high- and low-papaverine cultivars of opium poppy revealed differentially expressed genes involved in papaverine biosynthesis (Desgagné-Penix & Facchini, 2012) and numerous candidate genes with uncharacterized functions (Pathak et al., 2013). Although effective, functional genomic strategies such as DNA microarrays and RNA-seq require substantial investment to establish necessary tools and resources, and are dependent on the availability of appropriate chemotypes. Beyond opium poppy, inbred chemotypes are rarely available for BIA-producing species, most of which are not domesticated. Nevertheless, a wealth of transcriptomic and metabolite data has been made available recently for several nonmodel species. The 1000 Plants (1KP; Matasci et al., 2014) and PhytoMetaSyn (Xiao et al., 2013) projects are notable examples. Data for the 1KP and PhytoMetaSyn projects are accessible at https://sites.google.com/a/ualberta.ca/onekp/ and http://www.phytometasyn.ca, respectively.

BIA biosynthetic gene discovery via functional genomics has become highly dependent on VIGS. As a demonstration of the effectiveness of VIGS in opium poppy, systematic knockdown of six biosynthetic genes involved in the conversion of (R)-reticuline to morphine resulted in profound and generally predictable changes in the profile of pathway intermediates and end products (Wijekoon & Facchini, 2012). Notably, (i) the silencing of *SalSyn* enhanced reticuline accumulation; (ii) suppression of *SalR* increased salutaridine and decreased morphine levels; (iii) silencing *SalAT* resulted in elevated salutaridine and reduced codeine accumulation; (iv) suppression of *T6ODM* increased thebaine levels at the expense of morphine; and (v) silencing *CODM* increased the ratio of codeine to morphine accumulation. Interestingly, suppression of *COR* increased the accumulation of reticuline. RNAi-mediated silencing of *COR* produced the same phenotype owing to the co-silencing of *REPI* (Farrow et al., 2015), which had not yet been identified (Allen et al., 2004). *COR* and the *DRR* component of the *REPI* fusion show considerable nucleotide sequence identity (Farrow et al., 2015). Another functional genomics strategy, the screening of chemically mutagenized populations of a high-morphine variety of opium poppy, was used to independently isolate *REPI* (Winzer et al., 2015). In this case, genetic analysis of F2 populations, segregating for mutations conferring a deficiency in the conversion of (S)-reticuline to (R)-reticuline, facilitated detection of the COR paralog. However, functional genomic strategies such as RNAi-mediated gene silencing and induced mutagenesis are not applicable to most BIA-producing species owing to their recalcitrance to stable genetic transformation or large-scale genetic crossing and generational propagation.

The most widely used method for the functional characterization of unknown genes is the in vitro analysis of cognate recombinant enzymes produced in bacterial or yeast hosts, with a focus on the determination of substrate preferences. Native or codon-optimized gene candidates are typically cloned into an expression vector and fused to a peptide tag to facilitate detection and/or affinity purification of the recombinant enzyme. Crude or partially purified protein preparations are then assayed for catalytic activity using available substrates and predicted optimal enzymatic conditions (eg, cofactors, pH, and temperature).

1.3 Engineering BIA Pathways in Yeast

Recently, the use of yeast or bacteria for the purposes of BIA biosynthetic enzyme characterization has been extended to the realm of whole pathway reconstruction. Although strains of *Escherichia coli* capable of producing several BIA intermediates and end products have been reported (Minami et al., 2008; Nakagawa et al., 2016, 2014, 2011), the utility of prokaryotes for metabolic engineering of eukaryotic pathways is limited by the lack of cellular features required by cytochromes P450 and other membrane-associated enzymes. A eukaryotic microbial chassis, in particular baker's yeast (*S. cerevisiae*), is also preferred owing to the availability of metabolic modeling tools and various strain repositories.

Efforts to reassemble BIA metabolism in yeast can be divided into two major achievements: (i) production of key "scaffold" structures, such as the branch-point intermediate (*S*)-reticuline, and (ii) subsequent "tailoring" of scaffold molecules to produce reconfigured end products. The first report of an engineered BIA pathway in yeast involved the introduction of three biosynthetic genes, which conferred an ability to convert exogenous (*R,S*)-norlaudanosoline to (*S*)-reticuline (Hawkins & Smolke, 2008). Recently, de novo production of (*S*)-reticuline from endogenous L-tyrosine was achieved using a heterologous tyrosine hydroxylase from beet (*Beta vulgaris*) to generate L-DOPA, which was converted to dopamine by a prokaryotic DOPA decarboxylase (DeLoache et al., 2015). With the inclusion of NCS, dopamine condensed with endogenous 4-HPAA yielding (*S*)-norcoclaurine. In a related effort, a mammalian tyrosine hydroxylase was used in yeast engineered to produce the required tetrahydrobiopterin cofactor (Galanie, Thodey, Trenchard, Filsinger Interrante, & Smolke, 2015). In both cases, the reported yield of (*S*)-reticuline was only 0.2 μM, representing less than 1% of the endogenous dopamine pool. In contrast, de novo production of (*S*)-reticuline in *E. coli* has been reported at 140 μM, corresponding to a more substantial 4% conversion from dopamine (Nakagawa et al., 2012).

The extension of BIA metabolism beyond (S)-reticuline in yeast initially targeted the biosynthesis of the protoberberines scoulerine and canadine, and the aporphine magnoflorine (Hawkins & Smolke, 2008; Minami et al., 2008). A more extensive 10-step conversion of norlaudanosoline to the benzophenanthridine sanguinarine was also reported (Fossati et al., 2014). Recently, the production of morphinans, including the semisynthetic derivatives oxycodone and hydrocodone (Fossati, Narcross, Ekins, Falgueyret, & Martin, 2015; Galanie et al., 2015; Thodey, Galanie, & Smolke, 2014), and protoberberines has been achieved (Galanie & Smolke, 2015). Although de novo BIA production in yeast is associated with remarkably low yields, the relatively straightforward ability to assemble multistep pathways in a tractable microbial system provides opportunities for both research and future industrial applications.

Metabolic engineering strategies to improve product yields and specificity have generally focused on variations in yeast culture conditions. Factors such as carbon source, supplementation with antioxidants, and control of pH and aeration rate have been shown to impact yield. Optimization of codon usage in heterologous genes derived from plants and other organisms, to match tRNA abundance in yeast, can also improve pathway performance. In this regard, improved strategies for "tuning" codon usage have also been described (Boël et al., 2016; Quax, Claassens, Söll, & van der Oost, 2015). The regulation of heterologous gene expression is also an important consideration, and can be affected through alterations in gene dosage (eg, plasmid copy number and the number of genome integration events) and transcriptional or translational rates (eg, promoters with a range of expression capacity and the use of untranslated regions conferring differential mRNA stability). Although most metabolic engineering efforts targeting BIA production in yeast have been based on the simplistic notion that elevated biosynthetic enzyme levels should increase product yield, more sophisticated approaches will undoubtedly be required, including the sequential induction of incompatible enzyme types and the "balancing" of enzyme copy number to optimize flux (Thodey et al., 2014). Approaches to re-establish the spatial regulation of engineered metabolic pathways in yeast to mimic the compartmentalization of BIA biosynthetic enzymes in the native plant cell have also been proposed to enhance flux and product specificity (Conrado, Varner, & DeLisa, 2008). The spatial separation of enzymes catalyzing consecutive steps in a metabolic pathway can be improved by: (i) the creation of translational fusions of two or more biosynthetic genes, (ii) the inclusion of peptide signals targeting specific organelles or membranes, or (iii) the use of synthetic scaffolding systems (Lee, DeLoache, & Dueber, 2012). Such engineering strategies generally assume that biological systems are predictable. Alternatively, the screening of large expression libraries or the use of yeast strains developed in a semirandom manner can provide unexpected improvements (Kim, Du, Eriksen, & Zhao, 2013; Naesby et al., 2009). Undoubtedly, more information concerning local enzyme environments, modes of transport, and key enzyme–enzyme interactions must be derived from BIA-accumulating plants to inform successful engineering strategies in yeast. As such, gene discovery in BIA-producing plants could extend beyond biosynthetic enzymes to include ancillary pathway components. Enzyme interactions within alkaloid pathways should also be considered (Dastmalchi, Bernards, & Dhaubhadel, 2016).

Product yield can also be improved via engineering the endogenous metabolism of yeast, with the primary goal of providing sufficient precursor and co-factor molecules for the introduced pathway. For example, simultaneously increasing the endogenous supply of L-tyrosine and limiting the diversion of metabolic flux toward undesirable side products. Preliminary work has targeted the knockout or overexpression of host genes selected through the logical consideration of relevant biosynthetic networks (Trenchard & Smolke, 2015). More sophisticated methods have utilized integrated computational and experimental approaches to identify metabolic gene targets (Gold et al., 2015), whereas others involve "nonrational" methods such as directed evolution (Abatemarco, Hill, & Alper, 2013).

Although BIA-producing yeast strains are not yet amenable to industrial applications owing to the low yield of high-value compounds such as morphinans, the ability to routinely assemble BIA biosynthetic pathways in yeast provides a powerful experimental system for gene discovery. The reconstitution of an operational BIA biosynthetic pathway consisting of two or more enzymes in a "baseline strain" provides access to a suite of metabolic intermediates that are often not available as authentic standards to perform in vitro enzyme assays. Moreover, the identification of the specific reaction conditions required for the successful analysis of enzymes in vitro is not required since yeast have been shown to provide a competent cellular environment for the activity of many different enzyme types. Uncharacterized biosynthetic gene candidates can be introduced into the baseline yeast strains, which are subsequently analyzed for qualitative or quantitative alterations in BIA profiles. The "plug-and-play" strategy described in Section 3 is particularly suitable for the functional characterization of both novel biosynthetic genes and variants of known plant genes with potentially superior performance in a yeast cellular environment. The approach is scalable to allow the high-throughput screening of large numbers of candidate genes and can substantially reduce time to discovery.

2. TRANSCRIPTOME RESOURCES AND MINING CANDIDATE GENES

2.1 Selection of Plants and Tissues

A key initial consideration involves the selection of appropriate plants and tissues that will be used to establish transcriptomic resources for the selection of candidate biosynthetic genes. An impressive number of transcriptome databases are already available for a variety of BIA-producing species (Table 1). These include transcriptomes established using Illumina and/or 454 pyrosequencing for various organs and tissues from dozens of species (Hagel, Morris, et al., 2015; Liao et al., 2016; Matasci et al., 2014), and several chemotype-specific databases for opium poppy (Desgagné-Penix et al., 2012). For approximately two-thirds of samples used to generate transcriptomic databases, corresponding metabolomics resources or targeted metabolite profiles are available (Desgagné-Penix et al., 2012, 2010; Hagel, Mandal, et al., 2015). The availability of corresponding metabolomics databases adds an additional dimension to the strategy for selection of candidate biosynthetic genes. However, the phylogenetic relationship among BIA biosynthetic enzymes that belong to a distinct number of protein families facilitates the rational selection of high-priority candidate genes even in the absence of complementary metabolomics information (Hagel & Facchini, 2013). The establishment of new transcriptomic resources is also a cost-effective option, as described later.

2.2 RNA Extraction and DNA Sequencing

2.2.1. Collect fresh plant material and freeze immediately in liquid nitrogen. Samples can be stored in −80°C until total RNA is extracted.

2.2.2. Grind the plant material to a fine powder under liquid nitrogen using a motor and pestle. Alternatively a Tissue Lyzer (Qiagen, Venlo, The Netherlands) with prechilled blocks can be used.

2.2.3. Extract total RNA using a CTAB method (Meisel et al., 2005). Briefly, add 2.5 mL of CTAB buffer [2% (w/v) CTAB, 2.5% (w/v) PVP-40000, 2 M NaCl, 100 mM Tris–HCl, pH 8.0, 25 mM EDTA, pH 8.0, and 2% (v/v) β-mercaptoethanol added

Table 1 Public Transcriptome Resources for Benzylisoquinoline Alkaloid-Producing Plant Species

Species	Common Name	Family (Tribe)	Organ/Tissue	Reference
Argemone mexicana	Mexican Prickly Poppy	Papaveraceae	Stem	Hagel, Morris, et al. (2015)
Argemone mexicana	Mexican Prickly Poppy	Papaveraceae	Various	Matasci et al. (2014)
Berberis thunbergii	Japanese Barberry	Berberidaceae	Root	Hagel, Morris, et al. (2015)
Chelidonium majus	Greater Celandine	Papaveraceae	Stem	Hagel, Morris, et al. (2015)
Chelidonium majus	Greater Celandine	Papaveraceae	Leaf	Matasci et al. (2014)
Cissampelos mucronata	Abuta	Menispermaceae	Callus	Hagel, Morris, et al. (2015)
Corydalis chelanthifolia	Ferny Fumewort	Papaveraceae	Root	Hagel, Morris, et al. (2015)
Corydalis linstowiana	False Bleeding Heart	Papaveraceae	Leaf	Matasci et al. (2014)
Corydalis yanhusuo	Yan Hu So	Papaveraceae	Root	Liao et al. (2016)
Cocculus laurifolius	Snailseed	Menispermaceae	Leaf	Matasci et al. (2014)
Cocculus trilobus	Korean Moonseed	Menispermaceae	Callus	Hagel, Morris, et al. (2015)
Colchicum autumnale	Autumn Crocus	Colchicaceae	Various	Matasci et al. (2014)
Eschscholzia californica	California Poppy	Papaveraceae	Root	Hagel, Morris, et al. (2015)
Eschscholzia californica	California Poppy	Papaveraceae	Various	Matasci et al. (2014)
Glaucium flavum	Yellow Horn Poppy	Papaveraceae	Root	Hagel, Morris, et al. (2015)

Table 1 Public Transcriptome Resources for Benzylisoquinoline Alkaloid-Producing Plant Species—cont'd

Species	Common Name	Family (Tribe)	Organ/ Tissue	Reference
Hydrastis canadensis	Goldenseal	Ranunculaceae	Rhizome	Hagel, Morris, et al. (2015)
Hydrastis canadensis	Goldenseal	Ranunculaceae	Young shoot, flower	Hagel, Morris, et al. (2015)
Jeffersonia diphylla	Rheumatism Root	Berberidaceae	Root	Hagel, Morris, et al. (2015)
Mahonia aquifolium	Oregon Grape	Berberidaceae	Bark	Hagel, Morris, et al. (2015)
Menispermum canadense	Canadian Moonseed	Menispermaceae	Rhizome	Hagel, Morris, et al. (2015)
Nandina domestica	Sacred Bamboo	Berberidaceae	Root	Hagel, Morris, et al. (2015)
Nandina domestica	Sacred Bamboo	Berberidaceae	Young leaf, flower bud	Matasci et al. (2014)
Nigella sativa	Black Cumin	Ranunculaceae	Root	Hagel, Morris, et al. (2015)
Papaver bracteatum	Persian Poppy	Papaveraceae	Stem	Hagel, Morris, et al. (2015)
Papaver bracteatum	Persian Poppy	Papaveraceae	Various	Matasci et al. (2014)
Papaver rhoeas	Field Poppy	Papaveraceae	Various	Matasci et al. (2014)
Papaver setigerum	Poppy of Troy	Papaveraceae	Various	Matasci et al. (2014)
Papaver somniferum	Opium Poppy	Papaveraceae	Stem	Desgagné-Penix, Farrow, Cram, Nowak, and Facchini (2012)
Papaver somniferum	Opium Poppy	Papaveraceae	Various	Matasci et al. (2014)

Continued

Table 1 Public Transcriptome Resources for Benzylisoquinoline Alkaloid-Producing Plant Species—cont'd

Species	Common Name	Family (Tribe)	Organ/Tissue	Reference
Papaver somniferum	Opium Poppy	Papaveraceae	Cultured cells	Desgagné-Penix et al. (2010)
Podophyllum peltatum	Mayapple	Berberidaceae	Various	Matasci et al. (2014)
Stylophorum diphyllum	Celandine Poppy	Papaveraceae	Stem	Hagel, Morris, et al. (2015)
Sanguinaria canadensis	Bloodroot	Papaveraceae	Rhizome	Hagel, Morris, et al. (2015)
Sanguinaria canadensis	Bloodroot	Papavcraceae	Young shoot, flower bud	Matasci et al. (2014)
Thalictrum flavum	Meadow Rue	Ranunculaceae	Root	Hagel, Morris, et al. (2015)
Thalictrum thalictroides	Rue Anemone	Ranunculaceae	Floral bud	Matasci et al. (2014)
Tinospora cordifolia	Heartleaf Moonseed	Menispermaceae	Callus	Hagel, Morris, et al. (2015)
Xanthorhiza simplicissima	Yellowroot	Ranunculaceae	Root	Hagel, Morris, et al. (2015)

Access information is available in the indicated references.

immediately prior to use], to approximately 500 mg of frozen plant material. Vortex the slurry several times and incubate the samples at 65°C for 10 min.

2.2.4. Extract the aqueous phase twice with an equal volume of chloroform:isoamyl alcohol (24:1, v/v) at $14,000 \times g$ for 10 min at 4°C.

2.2.5. Add 10 M LiCl to the aqueous extract to a final concentration of 3 M. Precipitate total RNA for at least 1 h at 4°C.

2.2.6. Centrifuge at $14,000 \times g$ for 20 min at 4°C and resuspend the total RNA pellet in 0.5 mL of SSTE buffer [10 mM Tris–HCl, pH 8.0, 1 mM EDTA, pH 8.0, 1% (w/v) SDS, 1 M NaCl].

2.2.7. Extract the aqueous phase with an equal volume of chloroform: isoamyl alcohol (24:1, v/v) at $14,000 \times g$ for 10 min at 4°C.

2.2.8. Transfer the aqueous phase to a new tube and add double the volume of ethanol. Incubate the tube on ice for 1 h.
2.2.9. Centrifuge the tube at $14,000 \times g$ for 20 min at 4°C. Wash the RNA pellet twice with 70% (v/v) ethanol and air dry. Dissolve the total RNA in RNase-free water.
2.2.10. A high-quality total RNA sample should show a 260/280 ratio between 2.0 and 2.5, and a 230/280 ratio greater than 2.0. Assess total RNA quality and quantity using NanoDrop ND-1000 (Thermo Fisher Scientific, Waltham, MA) or BioAnalyzer 2100 (Agilent Technologies, Santa Clara, CA) instruments.
2.2.11. Poly(A)+RNA purification, cDNA library synthesis, Illumina HiSeq, and sequence assembly and annotation are routinely performed by a service provider, such as the McGill University and Génome Québec Innovation Center (www.gqinnovationcenter.com).

2.3 Selection of Candidate Genes

A variety of strategies can be used to select candidate genes depending on the availability of transcriptomic and metabolomic resources. When metabolomic or targeted metabolite (ie, BIA) profile data is available for two or more plant samples displaying different alkaloid chemotypes, comparative transcriptomics can be used to mine candidate genes potentially involved in the biosynthesis of differentially represented compounds (Chen & Facchini, 2014; Chen et al., 2015; Dang, Chen, & Facchini, 2015; Dang & Facchini, 2012, 2014; Desgagné-Penix et al., 2012; Hagel & Facchini, 2010). A comparative transcriptomics approach is most effective using databases representing different chemotypes, organs and/or tissues of the same species. However, most of the different plant species listed in Table 1 can also be compared owing to the monophyletic origin of BIA biosynthesis in members of the Ranunculales (Liscombe et al., 2005). In this case, transcriptomes from different species showing a differential representation of specific alkaloids can be mined for biosynthetic gene candidates with sequence similarity to known protein families implicated in BIA metabolism (Farrow, Hagel, & Facchini, 2012; Hagel & Facchini, 2013; Hagel, Morris, et al., 2015). Candidate genes encoding enzymes potentially capable of catalyzing specific metabolic conversions can then be aligned, and variants with relatively unique amino acid sequences compared with characterized protein family members can be prioritized. Ultimately, the strength of a plug-and-play functional genomics strategy using engineered yeast with the ability to perform discrete segments of BIA metabolism, either de novo or using exogenous substrates, is the relatively high-throughput capacity of the system; thus, the prioritization of candidate genes is not essential. In fact, any selection introduces the risk of overlooking functionally interesting candidates owing to the lack of a reliable relationship between primary structure (ie, amino acid sequence) and enzyme function (eg, substrate specificity) of BIA biosynthetic enzymes. In other words, the basic catalytic mechanism (eg, oxidation, reduction, methylation) can be accurately assumed, but the specific enzymatic function cannot be reliably predicted. Mining orthologs and paralogs of characterized BIA biosynthetic enzymes has proven effective as a strategy for the nontargeted discovery of functionally novel variants, including OMTs (Chang et al., 2015) and NMTs (Hagel, Morris, et al., 2015). In these examples, six OMT and six NMT candidates with greater than 40% amino acid sequence identity compared with characterized orthologs were detected in a transcriptome database for *Glaucium flavum* roots. When assayed in vitro, four of six OMT candidates (Chang et al., 2015) and all six NMT candidates (Hagel, Morris, et al., 2015) showed activity with one or more BIA, including previously unreported substrate- or regio-specificity. Similar mining of transcriptome databases established for 20 BIA-producing plant species yielded over 850 full-length candidate genes, including at least 230 OMT and 110 NMT orthologs (Hagel, Morris, et al., 2015). Owing to the rich collection of functionally characterized BIA biosynthetic enzymes belonging to a large, yet defined, number of protein families, the selection of candidate genes is not a bottleneck. However, the basic functional analysis of candidate genes requires a generic and high-throughput option beyond the in vitro testing of recombinant enzymes. To illustrate, a simple strategy could involve the stable introduction of 6OMT, CNMT, and 4′OMT into yeast fed exogenously with (R,S)-norlaudanosoline. Such a strain would allow plug-and-play analysis of various uncharacterized *OMT* and *NMT* orthologs with the potential ability to convert one or more of the intermediates or end products of this metabolically latticed pathway into a different BIA compound. The procedures below describe the construction of platform strains to establish known BIA biosynthetic capacity in a genetically stable yeast system (Section 3), the plug-and-play testing of uncharacterized biosynthetic gene candidates by transient expression in yeast platform strains (Section 4), and an effective analytical method to monitor changes in the BIA profile of a yeast platform strain (Section 5). A summary of the workflow is presented in Fig. 2.

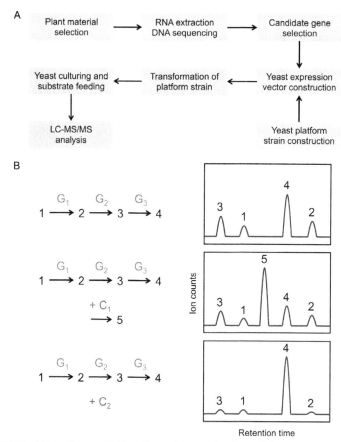

Fig. 2 (A) Workflow diagram linking the various methods used for acquisition and plug-and-play functional analysis of biosynthetic gene candidates in yeast platform strains with an operational BIA metabolic pathway, or parts thereof. (B) Example of a yeast platform strain able to convert a hypothetical BIA (compound 1), via two pathway intermediates (compounds 2 and 3), to an end product (compound 4). Transient expression in the yeast platform strain with a candidate gene (C_1) encoding an uncharacterized enzyme results in the accumulation of a new BIA (compound 5). Alternatively, a candidate gene (C_2) encoding an enzyme functionally similar to one already part of the platform strain increases accumulation of the end product. (See the color plate.)

3. BUILDING YEAST PLATFORM STRAINS

3.1 Overview of USER Cloning

Reconstruction of a BIA metabolic pathway in yeast requires the simultaneous introduction of numerous biosynthetic genes. Episomal expression systems depend on the availability of a large number of selection markers, and are associated with increased risks of gene loss and gene re-arrangement. Integration of foreign biosynthetic gene cassettes into the yeast genome is a more reliable strategy for the stable assembly of BIA pathways for which biosynthetic genes have been previously isolated and characterized. USER (*u*racil-*s*pecific *e*xcision *r*eaction)-based cloning has been used for the integration of multiple genes into the yeast genome owing to its relatively straightforward application and independence from the enzyme-based ligation of DNA fragments (Nour-Eldin, Hansen, Norholm, Jensen, & Halkier, 2006). A comprehensive review of USER-based cloning methods, including a one-step procedure for multipart DNA construct assembly termed "USER fusion," is available (Nour-Eldin, Geu-Flores, & Halkier, 2010). The development of several USER-based yeast integrative vectors, most notably the EasyClone system, has also been reported (Jensen et al., 2014). The EasyClone system includes Ura, Leu, and His selection markers flanked by LoxP sequences, and allows stable integration of up to 24 autonomous genes into 12 independent sites within the *S. cerevisiae* genome. With three different auxotrophic markers, up to six genes can be integrated simultaneously. Marker recycling based on the uracil excision reaction and Cre-LoxP-mediated recombination can be used to introduce additional genes. A previous version of this vector system allowed the engineering of a multistep, plant-derived, indolylglucosinolate production pathway into yeast (Mikkelsen et al., 2012). We have extended the application of the EasyClone vector set to the assembly of BIA biosynthetic pathways in yeast, with a focus on plug-and-play applications for the function characterization of novel genes.

EasyClone vectors are initially nicked with *Asi*SI and Nb.*Bsm*I to generate 8-nucleotide long 3′-overhangs (ie, fusion tails). Complementary fusion tails are generated via recognition and excision achieved using DNA glycosylase and DNA glycosylase-lyase Endo VIII, both of which are included in commercial USER™ enzyme mix available from New England Biolabs. The USER™ enzyme mix removes single deoxyuridine residues and dissociates single-stranded fragments lying upstream of the cleavage sites (Nour-Eldin et al., 2006). Ultimately, hybridization between the gene cassette inserts, and between inserts and the EasyClone vector, occurs spontaneously, and the corresponding circularized plasmid is transformed into *E. coli* without the need for enzymatic ligation. If more than two DNA fragments require fusion, primer design must include the creation of unique fusion tails for the fusion to occur in the correct orientation. Conveniently, PHUSER (primer help for USER) software is available to assist with primer design (Olsen et al., 2011). PHUSER considers the order of fusion and designs primers with 8–20 nucleotide fusion tails containing deoxyuridine residues. Multiple fragments can be theoretically processed using PHUSER until the availability of unique fusion tails is exhausted. For the plug-and-play synthetic biology applications described herein, a bidirectional Gal1/Gal10-promoter cassette is introduced along with candidate biosynthetic genes into the EasyClone vector.

3.2 Preparing USER-Compatible Vectors

3.2.1. Inoculate an overnight 3 mL culture with *E. coli* harboring the selected USER-compatible vector (Jensen et al., 2014). Purify plasmid DNA using a GeneJET Plasmid Miniprep Kit (Thermo Fisher Scientific).

3.2.2. Digest 10 µg of the purified USER-compatible vector with 10 U of the restriction enzyme mix *As*iSI/Nb.*Bsm*I (Thermo Fisher Scientific) at 37°C and in a total volume of 100 µL for 1 h.

3.2.3. Analyze 2 µL of digested plasmid by agarose gel electrophoresis to verify complete linearization.

3.2.4 Purify the digested vector using the GeneJET Gel Extraction Kit and elute in 100 µL TE buffer.

3.3 Simultaneous Cloning of Multiple PCR Fragments

3.3.1. Perform PCR reactions using specific uracil-containing primers and PfuTurbo Cx Hotstart DNA Polymerase (Agilent Technologies) following the manufacturer's instructions. Verify amplification of desired PCR amplicons by analyzing 2 µL of each reaction by agarose gel electrophoresis.

3.3.2. Purify amplicons using the GeneJET Gel Extraction Kit and elute in 100 µL TE buffer.

3.3.3. Mix multiple amplicons, linearized vectors, CutSmart buffer (New England Biolabs, Ipswitch, MA) with 0.75 U of the USER™ enzyme mix (New England Biolabs), and add water to a final volume of 12 µL. Prepare a negative control reaction with 12 µL of water instead of the PCR product. Incubate at 37°C for 15 min, followed by an additional 15 min at 25°C.

3.3.4. Transform 50 μL of competent *E. coli* cells by heat shock at 42°C for 30 s using the entire ligation reaction.

3.3.5. Add 500 μL of lysogeny broth (LB) to the transformation reaction, incubate the cells for 1 h at 37°C and plate on ampicillin-selective LB medium. Incubate the plates overnight at 37°C.

3.3.6. Screen for insertion of the correct amplicon by colony PCR. Grow selected colonies overnight as liquid cultures and isolate plasmid DNA.

3.3.7. Sequence to confirm correct assembly of the construct.

3.4 Yeast Transformation

To facilitate chromosomal integration, two yeast chromosomal-targeting sequences (named "UP" and "DOWN") flanked by two *Not*I restriction sites allow linearization of USER-compatible vector constructs prior to transformation of a yeast host strain (Orrweaver, Szostak, & Rothstein, 1981). We subsequently transform the CENPK102-5B yeast host strain (Jensen et al., 2014) using the LiAc/PEG/single-stranded carrier DNA (ssDNA) transformation method (Gietz & Schiestl, 2007).

3.4.1. Inoculate the CENPK102-5B host strain (MATa ura3-52 his3Δ1 leu2-3/112 MAL2-8c SUC2) in 2–5 mL of yeast extract-peptone-dextrose (YPD) medium (Bioshop, Burlington, ON) and incubate overnight at 200 rpm on a gyratory shaker at 30°C.

3.4.2. Transfer the overnight culture to a 250 mL flask containing 50 mL of YPD medium and grow the yeast for an additional 3–5 h.

3.4.3. Prepare carrier DNA by boiling 2 mg mL^{-1} of single-stranded salmon testes DNA (Sigma-Aldrich, St. Louis, MO) for 5 min and rapidly chill on ice.

3.4.4. After 3–5 h, transfer the culture to a 50 mL Falcon tube, centrifuge at 3000 × g for 5 min, and discard the supernatant.

3.4.5. Wash the cell pellet by adding 20 mL of sterile water, vortex, centrifuge at 3000 × g for 5 min, and discard the supernatant. Repeat this step twice.

3.4.6. Add 1 mL of sterile water to the cell pellet and mix by vortexing.

3.4.7. Transfer 100 μL of the resuspended cells to a 1.5 mL microcentrifuge tube, add 240 μL of 50% (v/v) polyethylene glycol (PEG3350, Sigma-Aldrich), 36 μL of 1 M lithium acetate (Sigma-Aldrich), and 50 μL of 2 mg mL^{-1} carrier DNA and mix by vortexing.

3.4.8. Add 300–500 ng of *Not*I-linearized USER-compatible vector constructs to the competent yeast cells and mix by vortexing.
3.4.9. Incubate the microcentrifuge tube at 42°C for 30–60 min.
3.4.10. Centrifuge at 20,000 × g for 30 s and discard the supernatant.
3.4.11. Add 200 μL of sterile water and mix well by vortexing.
3.4.12. Spread the cells on a plate containing synthetic defined (SD)-drop out medium.
3.4.13. Incubate the plates at 30°C for 2–4 days.
3.4.14. Confirm the production of target proteins in yeast by immunoblot analysis.

3.5 Recombinant Protein Analysis

3.5.1. Inoculate a single yeast colony from step 3.4.13 into 2 mL of synthetic defined (SD) medium and incubate overnight at 200 rpm on a gyratory shaker at 30°C.
3.5.2. Inoculate 5 mL of YPG medium [1% (w/v) yeast extract, 2% (w/v) bacto-peptone, 3% (v/v) glycerol] with 100 μL of the overnight yeast culture, and incubate overnight at 200 rpm on a gyratory shaker at 30°C.
3.5.3. Lyse 0.6 mL of the overnight yeast culture as described previously (Zhang et al., 2011) and confirm the production of target proteins in yeast by immunoblot analysis.

3.6 Ura Marker excision

3.6.1. Inoculate 2 mL of SD–Leu–His medium (synthetic defined medium lacking leucine and histidine) containing 5-fluoroorotic acid (5-FOA, 740 mg L^{-1}) with a single yeast colony produced on a SD–Leu–His–Ura (synthetic defined medium lacking leucine, histidine, and uracil) plate. Incubate the culture for 2–3 days at 200 rpm on a gyratory shaker at 30°C.
3.6.2. Streak out yeast from the 2 mL culture on a new SD–Leu–His plate containing 5-FOA and incubate for 2–3 days at 30°C.
3.6.3. To confirm deletion of Ura maker, inoculate 2 mL of SD–Leu–His medium and 2 mL of SD–Leu–His–Ura medium from a single colony and incubate for 24 h at 30°C. If no yeast colonies grow on the SD–Leu–His–Ura medium after 24 h, the Ura marker has been deleted.

4. FUNCTIONAL TESTING OF CANDIDATE GENES
4.1 Transient Expression Constructs

The high-copy number pESC-Ura (or, alternatively, pESC-Leu or pESC-His) vector is used to express gene candidates using the *Gal10* promoter.

4.1.1. Amplify candidate genes from cDNA using primers flanked with *Spe*I and *Not*I restriction sites.

4.1.2. Digest the PCR amplicons and the pESC-Ura vector with *Spe*I and *Not*I.

4.1.3. Gel purify the DNA fragments and ligate the PCR amplicons to the pESC-Ura vector to generate transient expressing constructs.

4.1.4. Transform 50 µL of competent *E. coli* cells by heat shock at 42 °C for 30 s using the entire ligation reaction.

4.1.5. Add 500 µL of LB to the transformation, incubate the cells for 1 h at 37 °C, and plate on ampicillin-selective LB medium. Incubate the plates overnight at 37 °C.

4.1.6. Screen for insertion of the correct PCR amplicon by colony PCR. Grow selected colonies overnight as liquid cultures and isolate plasmid DNA.

4.1.7. Sequence to confirm correct assembly of the construct.

4.2 Yeast Transformation, Culture, and Substrate Feeding

4.2.1. Transform the yeast baseline strain established in Section 3 with transient expression constructs generated in Section 4.1 using the LiAc/PEG/single-stranded carrier DNA (ssDNA) transformation method (Gietz & Schiestl, 2007). Use yeast baseline strains transformed with the empty pESC-Ura (Agilent Technologies) vector as negative controls.

4.2.2. Inoculate each yeast strain transiently expressing a different candidate gene in individual wells of a 96-well plate, with each containing 600 µL of SD–His–Ura medium supplemented with 2% (w/v) glucose.

4.2.3. Grow yeast cultures at 900 rpm on a gyratory microplate shaker overnight at 30 °C.

4.2.4. Transfer 30 µL of the original culture to 600 µL of SD–His–Ura containing 2% (w/v) galactose and 200 µM of the BIA suitable for conversion by the baseline yeast strain and/or the transient expression construct.

4.2.5. Grow yeast cultures at 900 rpm on a gyratory microplate shaker at 30°C.

4.2.6. Sample 100 μL aliquots of each culture at 24 and 48 h. Centrifuge to remove yeast cells.

4.2.7. Subject 5 μL of each supernatant, containing BIAs secreted by the yeast cells into the culture medium, to high-resolution mass spectrometry (MS) analysis as described in Section 5.

5. LIQUID CHROMATOGRAPHY—TANDEM MASS SPECTROMETRY

5.1 Analytical Strategy Overview

For more than a decade, high-resolution mass spectrometry (MS) has been widely used in metabolomics applications owing to an enhanced capacity for metabolite detection and identification compared with low-resolution instruments (Allwood & Goodacre, 2010; Kueger, Steinhauser, Willmitzer, & Giavalisco, 2012). In particular, high-resolution MS has revolutionized the study of specialized metabolite biosynthesis in plants, fungi, and marine organisms. High-resolution mass analyzers can be used to predict empirical formulae of unknown compounds, allowing de-replication during screens for novel natural products (Nielsen & Larsen, 2015). The inclusion of chromatography-based separation prior to mass analysis permits a second dimension for resolving complex mixtures. Liquid chromatography (LC) is most commonly used for metabolite analysis, although coupled systems employing gas chromatography (GC) are also emerging (eg, GC-Quadrupole-Orbitrap; Peterson et al., 2014). Choice of ion formation conditions, desolvation strategies, and subsequent tuning for specific analytes are important considerations in the development of MS-based analytical strategies. High-resolution LC-MS using electrospray ionization (ESI) in positive mode is particularly suited to the analysis of BIAs, which all contain at least one basic nitrogen atom. The use of exact molecular mass is often insufficient for the identification of BIAs since many structurally diverse compounds possess the same empirical formula. Chromatographic separation and the association of retention times with those of authentic standards are essential in BIA analysis, although high-throughput applications typically demand the development of relatively abbreviated methods. Additional structural information can be obtained using collision-induced fragmentation, which can yield diagnostic spectra. Comparing the collision-induced dissociation (CID) spectra of unknown analytes with those of authentic standards is a generally acceptable means to confirm compound identity with the inclusion of retention time and empirical formulae determinations. Standard CID spectra for many compounds are also available online at MassBank (http://www.massbank.jp) and MassBank of North America (http://mona.fiehnlab.ucdavis.edu), and in published journal articles.

Targeted alkaloid profiling must often be performed rapidly using complex mixtures of metabolites. Recent efforts to introduce BIA metabolic pathways into heterologous hosts, such as yeast or *E. coli*, have demonstrated the requirement for high-throughput analytical platforms (Brown, Clastre, Courdavault, & O'Connor, 2015; DeLoache et al., 2015; Fossati et al., 2015; Galanie et al., 2015; Nakagawa et al., 2012; Qu et al., 2015). High-resolution, LC-coupled tandem MS (MS/MS) is well suited for the high-throughput analysis of alkaloids in engineered microorganisms. We use an LTQ-Orbitrap-XL mass analyzer formatted with an LC system injecting samples arrayed in 96-well plates (Thermo Fisher Scientific). To cope with high flow and to enhance desolvation, we include an adjustable and heated electrospray ionization (HESI) source, and perform all analyses in positive mode using a single tune file. The data-dependent feature of this instrument permits the simultaneous acquisition of CID spectra along with exact mass data. Our method draws from a custom BIA parent ion mass list, and CID are acquired only for analytes with exact masses corresponding to those of alkaloids. Dynamic exclusion features are used to ensure that no alkaloids are missed during relatively short runs, which are approximately 10 min each.

5.2 List of Instrumentation and Software

1. LTQ-Orbitrap-XL equipped with an Accela HPLC and autosampler, HESI-II probe, and syringe pump (Thermo Fisher Scientific). All instruments are operated using LTQ Tune Plus v. 2.5.5 SP1 and Xcalibur v. 2.1.0.1140. Basic analysis is performed using the Qual Browser feature of Xcalibur and Microsoft Excel.
2. ProteoWizard 3.0.7414.
3. R i386 v. 3.2.0.
4. Package XCMS v.1.44.0 (Tautenhahn, Boettcher, & Neumann, 2008), updated versions of which are available at https://masspec.scripps.edu.

5.3 List of Consumables, Solvents, and Reagents

1. Agilent Poroshell 120 SB-C18 HPLC column.
2. Solvent-resistant 96-well sampling plates and sealing mats.

3. HPLC solvents. For analysis of most BIAs, we use a mobile phase composed of Solvent A [10 mM ammonium acetate, pH 5.5, 5% (v/v) acetonitrile (ACN)] and Solvent B [100% ACN].
4. Wash solvents [methanol and 50:50 (v/v) methanol:water].
5. Pierce LTQ ESI Positive Ion Calibration Solution (Thermo Fisher Scientific).
6. Alkaloid standards as internal calibrants.

5.4 Analytical Methods

5.4.1. Perform external calibration using Pierce LTQ ESI Positive Ion Calibration Solution (Thermo Fisher Scientific) according to the manufacturer's instructions (*LTQ Orbitrap XL™ Getting Started 2.4*). The heated feature of the HESI source should be turned off to avoid heat-degrading MRFA peptide.

5.4.2. Maintain internal calibration by adding approximately 1 mg of calibrant to 1000 mL of HPLC solvent. We use one or more of the following as calibrants for BIA analysis: dextromethorphan, 1-methyl-6,7-dihydroxy-1,2,3,4-tetrahydroisoquinoline hydrobromide, 6,7-dimethoxy-1,2,3,4-tetrahydroisoquinoline hydrochloride, or caffeine (Sigma-Aldrich).

5.4.3. Develop at least one tune method using relatively abundant and inexpensive BIAs. For example, papaverine, noscapine, and morphine can be used for tune methods targeting 1-benzylisoquinoline, phthalideisoquinoline, and morphinan alkaloids, respectively (see Note 5.5.1 in Section 5.5). Most BIAs are heat-stable, so the heated feature of the source can be applied in conjunction with high LC flow (0.5 mL min^{-1} of 50:50 (v/v) Solvent A:Solvent B) and syringe-pump infusion (5 µL min^{-1} of a 1 µg mL^{-1} alkaloid solution) via a Tee union. Tune conditions will vary from instrument to instrument. Typical source and interface conditions are as follows:
   a. Capillary temperature and voltage: 380°C and 6.00 V, respectively
   b. Vaporizer temperature: 400°C
   c. Sheath and auxiliary gas flows: 60 and 20 arbitrary units (AU), respectively
   d. Source voltage and current: 3.00 kV and 100 µA, respectively
   e. Tube lens voltage: 45 V

5.4.4. Perform high-performance LC (HPLC) at a flow rate of 0.5 mL min^{-1}. The following gradient of Solvent A and Solvent B can be used: 100% (Solvent A) to 80% over 5 min; 80–50% over 3 min; 50–0% over 3 min; isocratic 0% for 2 min; 0–100% over 0.1 min; isocratic 100% for 1.9 min. Total run time is 15 min, but data should not be collected past 10 min to reduce the file size.

5.4.5. Inject 5–10 μL from approximately 200 μL of each sample in short-depth 96-well plates. Wash the needle and port with methanol between samples to avoid contamination. When including relatively water-soluble analytes (eg, reticuline and morphine), the wash protocol can include 50:50 (v/v) methanol:water.

5.4.6. BIA standards should be included in each analysis for qualitative and quantitative purposes. Typical strategies include:
  a. Addition of a standard mix consisting of up to five BIAs (1 μg mL^{-1} each) to one or more wells of each 96-well sample plate. Each BIA in the mix must have a different exact mass, or run at a sufficiently unique retention time compared with other standards in the mix with the same exact mass. LTQ-Orbitrap-XL analysis cannot distinguish between structural isomers if they elute at the same time.
  b. Addition of alkaloids of known molarity for the construction of standard curves. Typical BIA concentrations for standard curve development are from mid-range nM to approximately 100 μM (see Note 5.5.2 in Section 5.5).

5.4.7. Mass analyzer instrumentation is programmed using Xcalibur support as follows:
  a. One segment with three scan events:
    i. First scan: high-resolution FTMS (60,000 setting) from 200 to 700 m/z with ion injection time of 500 ms and scan time approximately 1.5 s.
    ii. Second and third scans: lower resolution ITMS for CID of alkaloids detected by the first scan event. Scan time is approximately 0.5 s each. Parent ions should represent the first- and second-most abundant alkaloid masses, respectively, as determined by FFT preview using an ion mass list (>200 members) corresponding to exact masses of known BIAs.
  b. For scan events 2 and 3:
    i. Enable the dynamic exclusion feature with a repeat count of 2, repeat duration of 10 s, and exclusion duration of 20 s.

ii. Enable preview mode for FTMS master scans.
iii. Select CID at NCE 35.
c. Unless the instrument is running continuously, wash and shutdown procedures should be included in the method sequence to avoid salt build-up in the needle of the HESI and to remove salt and contaminants from the LC column. The last runs of the method sequence should include the following, after which the instrument should be put in standby mode:
   i. Flush with 50:50 (v/v) methanol:water without salts using the previous tune method.
   ii. Flush with 50:50 (v/v) methanol:water without salts, using a new tune method with lowered source temperatures, reduced drying gas, and reduced solvent flow (5 µL min^{-1}).
d. For routine analysis, .raw file data is quantified in batches according to the manual *Xcalibur: Getting Productive* (Thermo Fisher Scientific). To generate complete mass lists along with peak areas and retention times, .raw files generated using identical LC-MS methods must be converted to .mzXML format using the MSConvert feature of ProteoWizard. Select the Peak Picking filter feature during conversion.
e. Open Package XCMS using R and process .mzXML data using the following:
   i. centWave, with ppm=2.0, peakwidth=c(10,60), prefilter=c(3,5000), sleep=0
   ii. retcor and obiwarp, with plottype=c('deviation')
   iii. group, with bw=5, minfrac=0.5, mzwid=0.015
   iv. fillPeaks
f. Create a table (peaklist <-peakTable(xset, value="into") and write to a .csv file (write.table(peaklist, "Name", sep="\t") that can be opened using Microsoft Excel.
g. Calculate molar quantities of individual BIAs using peak areas of unknowns in each sample and standard curves based on peak areas of BIA standards (step 5.4.6.b).
h. Confirm the identities of unknowns using CID spectra. ITMS data can be accessed using the Qual Browser feature of Xcalibur. CID scans corresponding to individual BIAs should be averaged (see Note 5.5.3 in Section 5.5) and exported to Microsoft Excel as nominal mass lists, which can be plotted. Alternatively, CID data can be written to .raw files.

5.5 Technical Notes

5.5.1. Our tuning method is based on papaverine and is sufficient to analyze most BIAs. For high-throughput analyses it is inconvenient to switch tune methods, so we advise using only one.

5.5.2. For FTMS data acquired as described above, the BIA concentration range is appropriate. Concentration ranges require adjustment depending on the method. For example, lower BIA concentrations should be employed when acquiring full-scan data using the LTQ (ITMS mode) or performing SIM experiments.

5.5.3. Only two CID scans are acquired per BIA analyzed when the dynamic exclusion feature of Xcalibur is used as described.

6. SUMMARY AND FUTURE PROSPECTS

The routine accessibility of massively parallel DNA sequencing technologies has yielded an substantial collection of public transcriptome databases (Hagel, Morris, et al., 2015; Matasci et al., 2014) for BIA-producing plant species, and provides an unprecedented opportunity to establish custom resources within and beyond currently available systems. Impressive achievements in the identification of most BIA biosynthetic genes in opium poppy and other "model" systems, such as *Coptis japonica*, have highlighted the diverse roles of a limited, yet broad, number of enzyme families. The monophyletic origin of BIA biosynthetic genes in the Ranunculales (Liscombe et al., 2005), coupled with the catalytically diverse functionality of individual members within each enzyme family, allows the rational selection of orthologs and paralogs from any taxonomically related plant (Hagel & Facchini, 2013; Hagel, Morris, et al., 2015). The major bottleneck for establishing an extensive catalogue of BIA biosynthetic genes is the availability of a relatively high-throughput and generic functional genomics technology to determine the basic catalytic capacity of candidate genes. The widely used in vitro analysis of recombinant enzymes is relatively low throughput and largely limited by the availability of authentic BIA substrates. Strategies such as VIGS have been highly effective for gene discovery in opium poppy (Desgagné-Penix & Facchini, 2012; Farrow et al., 2015; Hagel & Facchini, 2010; Winzer et al., 2012), but plant functional genomics technologies are generally difficult to implement in most undomesticated BIA-producing species. The ongoing routine introduction of BIA biosynthetic pathways, or parts thereof, into yeast (DeLoache et al., 2015; Fossati et al., 2015; Galanie & Smolke, 2015; Galanie et al., 2015; Hawkins & Smolke, 2008; Thodey et al., 2014) can be exploited as a common biosynthetic baseline to test possible functional roles for candidate genes. Although dramatically inefficient with respect to industrial applications, yeast strains engineered to produce various BIA pathway intermediate or end products nevertheless represent a key functional genomics tool for biosynthetic gene discovery. More extensive functional characterization of new genes still requires work with recombinant enzymes and, if possible, in the host plant. However, a plug-and-play strategy based on engineered yeast has the potential to expedite the basic identification of novel biosynthetic genes and variants of known genes.

ACKNOWLEDGMENTS
We are grateful to Dr. Irina Borodina (Technical University of Copenhagen) for the gift of the USER vector system. Research funding in the Facchini Laboratory is provided in part by a Natural Sciences and Engineering Research Council of Canada Discovery Grant.

REFERENCES
Abatemarco, J., Hill, A., & Alper, H. S. (2013). Expanding the metabolic engineering toolbox with directed evolution. *Biotechnology Journal, 8*(12), 1397–1410.

Alcantara, J., Bird, D. A., Franceschi, V. R., & Facchini, P. J. (2005). Sanguinarine biosynthesis is associated with the endoplasmic reticulum in cultured opium poppy cells after elicitor treatment. *Plant Physiology, 138*(1), 173–183.

Allen, R. S., Millgate, A. G., Chitty, J. A., Thisleton, J., Miller, J. A. C., Fist, A. J., et al. (2004). RNAi-mediated replacement of morphine with the nonnarcotic alkaloid reticuline in opium poppy. *Nature Biotechnology, 22*(12), 1559–1566.

Allwood, J. W., & Goodacre, R. (2010). An introduction to liquid chromatography-mass spectrometry instrumentation applied in plant metabolomic analyses. *Phytochemical Analysis, 21*(1), 33–47.

Beaudoin, G. A., & Facchini, P. J. (2013). Isolation and characterization of a cDNA encoding (S)-*cis*-N-methylstylopine 14-hydroxylase from opium poppy, a key enzyme in sanguinarine biosynthesis. *Biochemical and Biophysical Research Communications, 431*(3), 597–603.

Berényi, S., Csutorás, C., & Sipos, A. (2009). Recent developments in the chemistry of thebaine and its transformation products as pharmacological targets. *Current Medicinal Chemistry, 16*(25), 3215–3242.

Bird, D. A., Franceschi, V. R., & Facchini, P. J. (2003). A tale of three cell types: Alkaloid biosynthesis is localized to sieve elements in opium poppy. *Plant Cell, 15*(11), 2626–2635.

Boël, G., Letso, R., Neely, H., Price, W. N., Wong, K. H., Su, M., et al. (2016). Codon influence on protein expression in *E. coli* correlates with mRNA levels. *Nature, 529*, 358–363.

Brown, S., Clastre, M., Courdavault, V., & O'Connor, S. E. (2015). De novo production of the plant-derived alkaloid strictosidine in yeast. *Proceedings of the National Academy of Sciences of the United States of America, 112*(11), 3205–3210.

Chang, L., Hagel, J. M., & Facchini, P. J. (2015). Isolation and characterization of O-methyltransferases involved in the biosynthesis of glaucine in *Glaucium flavum*. *Plant Physiology*, *69*(2), 1127–1140.
Chen, X., Dang, T. T. T., & Facchini, P. J. (2015). Noscapine comes of age. *Phytochemistry*, *111*, 7–13.
Chen, X., & Facchini, P. J. (2014). Short-chain dehydrogenase/reductase catalyzing the final step of noscapine biosynthesis is localized to laticifers in opium poppy. *Plant Journal*, *77*(2), 173–184.
Choi, K. B., Morishige, T., & Sato, F. (2001). Purification and characterization of coclaurine N-methyltransferase from cultured *Coptis japonica* cells. *Phytochemistry*, *56*(7), 649–655.
Conrado, R. J., Varner, J. D., & DeLisa, M. P. (2008). Engineering the spatial organization of metabolic enzymes: Mimicking nature's synergy. *Current Opinion in Biotechnology*, *19*(5), 492–499.
Dang, T. T. T., Chen, X., & Facchini, P. J. (2015). Acetylation serves as a protective group in noscapine biosynthesis in opium poppy. *Nature Chemical Biology*, *11*(2), 104–106.
Dang, T. T. T., & Facchini, P. J. (2012). Characterization of three O-methyltransferases involved in noscapine biosynthesis in opium poppy. *Plant Physiology*, *159*(2), 618–631.
Dang, T. T. T., & Facchini, P. J. (2014). CYP82Y1 is N-methylcanadine 1-hydroxylase, a key noscapine biosynthetic enzyme in opium poppy. *Journal of Biological Chemistry*, *289*(4), 2013–2026.
Dang, T. T. T., Onoyovwe, A., Farrow, S. C., & Facchini, P. J. (2012). Biochemical genomics for gene discovery in benzylisoquinoline alkaloid biosynthesis in opium poppy and related species. *Methods in Enzymology*, *515*, 231–236.
Dastmalchi, M., Bernards, M., & Dhaubhadel, S. (2016). Twin anchors of the soybean isoflavonoid metabolon: Evidence for tethering of the complex to the endoplasmic reticulum by IFS and C4H. *Plant Journal*, *85*(6), 689–706.
DeLoache, W. C., Russ, Z. N., Narcross, L., Gonzales, A. M., Martin, V. J. J., & Dueber, J. E. (2015). An enzyme-coupled biosensor enables (S)-reticuline production in yeast from glucose. *Nature Chemical Biology*, *11*(7), 465–471.
Desgagné-Penix, I., & Facchini, P. J. (2012). Systematic silencing of benzylisoquinoline alkaloid biosynthetic genes reveals the major route to papaverine in opium poppy. *Plant Journal*, *72*(2), 331–344.
Desgagné-Penix, I., Farrow, S. C., Cram, D., Nowak, J., & Facchini, P. J. (2012). Integration of deep transcript and targeted metabolite profiles for eight cultivars of opium poppy. *Plant Molecular Biology*, *79*(3), 295–313.
Desgagné-Penix, I., Khan, M. F., Schriemer, D. C., Cram, D., Nowak, J., & Facchini, P. J. (2010). Integration of deep transcriptome and proteome analyses reveals the components of alkaloid metabolism in opium poppy cell cultures. *BMC Plant Biology*, *10*, 252.
Dittrich, H., & Kutchan, T. M. (1991). Molecular cloning, expression, and induction of berberine bridge enzyme, an enzyme essential to the formation of benzophenanthridine alkaloids in the response of plants to pathogenic attack. *Proceedings of the National Academy of Sciences of the United States of America*, *88*(22), 9969–9973.
Facchini, P. J., & De Luca, V. (1994). Differential and tissue-specific expression of a gene family for tyrosine/dopa decarboxylase in opium poppy. *Journal of Biological Chemistry*, *269*(43), 26684–26690.
Facchini, P., Penzes, C., Johnson, A. G., & Bull, D. (1996). Molecular characterization of berberine bridge enzyme genes from opium poppy. *Plant Physiology*, *112*(4), 1669–1677.
Farrow, S. C., Hagel, J. M., Beaudoin, G. A. W., Burns, D. C., & Facchini, P. J. (2015). Stereochemical inversion of (S)-reticuline by a cytochrome P450 fusion in opium poppy. *Nature Chemical Biology*, *11*(9), 728–732.

Farrow, S. C., Hagel, J. M., & Facchini, P. J. (2012). Transcript and metabolite profiling in cell cultures of 18 plant species that produce benzylisoquinoline alkaloids. *Phytochemistry*, 77, 79–88.
Fossati, E., Ekins, A., Narcross, L., Zhu, Y., Falgueyret, J.-P., Beaudoin, G. A. W., et al. (2014). Reconstitution of a 10-gene pathway for synthesis of the plant alkaloid dihydrosanguinarine in *Saccharomyces cerevisiae*. *Nature Communications*, 5, 3283.
Fossati, E., Narcross, L., Ekins, A., Falgueyret, J.-P., & Martin, V. J. J. (2015). Synthesis of morphinan alkaloids in *Saccharomyces cerevisiae*. *PloS One*, 10(4), e0124459.
Frick, S., Kramell, R., Schmidt, J., Fist, A. J., & Kutchan, T. M. (2005). Comparative qualitative and quantitative determination of alkaloids in narcotic and condiment *Papaver somniferum* cultivars. *Journal of Natural Products*, 68(5), 666–673.
Galanie, S., & Smolke, C. D. (2015). Optimization of yeast-based production of medicinal protoberberine alkaloids. *Microbial Cell Factories*, 14, 144.
Galanie, S., Thodey, K., Trenchard, I. J., Filsinger Interrante, M., & Smolke, C. D. (2015). Complete biosynthesis of opioids in yeast. *Science*, 349(6252), 1095–1100.
Gesell, A., Rolf, M., Ziegler, J., Díaz Chávez, M. L., Huang, F.-C., & Kutchan, T. M. (2009). CYP719B1 is salutaridine synthase, the C-C phenol-coupling enzyme of morphine biosynthesis in opium poppy. *Journal of Biological Chemistry*, 284(36), 24432–24442.
Gietz, R. D., & Schiestl, R. H. (2007). Quick and easy yeast transformation using the LiAc/SS carrier DNA/PEG method. *Nature Protocols*, 2(1), 35–37.
Gold, N. D., Gowen, C. M., Lussier, F.-X., Cautha, S. C., Mahadevan, R., & Martin, V. J. J. (2015). Metabolic engineering of a tyrosine-overproducing yeast platform using targeted metabolomics. *Microbial Cell Factories*, 14, 73.
Grothe, T., Lenz, R., & Kutchan, T. M. (2001). Molecular characterization of the salutaridinol 7-O-acetyltransferase involved in morphine biosynthesis in opium poppy *Papaver somniferum*. *Journal of Biological Chemistry*, 276(33), 30717–30723.
Hagel, J. M., & Facchini, P. J. (2010). Dioxygenases catalyze the O-demethylation steps of morphine biosynthesis in opium poppy. *Nature Chemical Biology*, 6, 273–275.
Hagel, J. M., & Facchini, P. J. (2013). Benzylisoquinoline alkaloid metabolism: A century of discovery and a brave new world. *Plant Cell Physiology*, 54(5), 647–672.
Hagel, J. M., Mandal, R., Han, B., Han, J., Dinsmore, D. R., Borchers, C. H., et al. (2015). Metabolome analysis of 20 taxonomically related benzylisoquinoline alkaloid-producing plants. *BMC Plant Biology*, 15, 220.
Hagel, J. M., Morris, J. S., Lee, E. J., Desgagné-Penix, I., Bross, C. D., Chang, L., et al. (2015). Transcriptome analysis of 20 taxonomically related benzylisoquinoline alkaloid-producing plants. *BMC Plant Biology*, 15, 227.
Hawkins, K. M., & Smolke, C. D. (2008). Production of benzylisoquinoline alkaloids in *Saccharomyces cerevisiae*. *Nature Chemical Biology*, 4, 564–573.
Jensen, N. B., Strucko, T., Kildegaard, K. R., David, F., Maury, J., Mortensen, U. H., et al. (2014). EasyClone: Method for iterative chromosomal integration of multiple genes in *Saccharomyces cerevisiae*. *FEMS Yeast Research*, 14(2), 238–248.
Kim, B., Du, J., Eriksen, D. T., & Zhao, H. (2013). Combinatorial design of a highly efficient xylose-utilizing pathway in *Saccharomyces cerevisiae* for the production of cellulosic biofuels. *Applied and Environmental Microbiology*, 79(3), 931–941.
Kueger, S., Steinhauser, D., Willmitzer, L., & Giavalisco, P. (2012). High-resolution plant metabolomics: From mass spectral features to metabolites and from whole-cell analysis to subcellular metabolite distributions. *Plant Journal*, 70(1), 39–50.
Kutchan, T. M., & Dittrich, H. (1995). Characterization and mechanism of the berberine bridge enzyme, a covalently flavinylated oxidase of benzophenanthridine alkaloid biosynthesis in plants. *Journal of Biological Chemistry*, 270(41), 24475–24481.
Lee, H., DeLoache, W. C., & Dueber, J. E. (2012). Spatial organization of enzymes for metabolic engineering. *Metabolic Engineering*, 14(3), 242–251.

Lee, E.-J., & Facchini, P. J. (2010). Norcoclaurine synthase is a member of the pathogenesis-related 10/Bet v1 protein family. *Plant Cell, 22*(10), 3489–3503.

Liao, D., Wang, P., Jia, C., Sun, P., Qi, J., Zhou, L., et al. (2016). Identification and developmental expression profiling of putative alkaloid biosynthetic genes in *Corydalis yanhusuo* bulbs. *Scientific Reports, 6*, 19460.

Liscombe, D. K., Macleod, B. P., Loukanina, N., Nandi, O. I., & Facchini, P. J. (2005). Evidence for the monophyletic evolution of benzylisoquinoline alkaloid biosynthesis in angiosperms. *Phytochemistry, 66*(11), 1374–1393.

Matasci, N., Hung, L. H., Yan, Z., Carpenter, E. J., Wickett, N. J., Mirarab, S., et al. (2014). Data access for the 1,000 Plants (1KP) project. *Gigascience, 3*, 17.

Meisel, L., Fonseca, B., González, S., Baeza-Yates, R., Cambiazo, V., Campos, R., et al. (2005). A rapid and efficient method for purifying high quality total RNA from peaches (*Prunus persica*) for functional genomics analyses. *Biological Research, 38*, 83–88.

Mikkelsen, M. D., Buron, L. D., Salomonsen, B., Olsen, C. E., Hansen, B. G., Mortensen, U. H., et al. (2012). Microbial production of indolylglucosinolate through engineering of a multi-gene pathway in a versatile yeast expression platform. *Metabolic Engineering, 14*(2), 104–111.

Minami, H., Kim, J.-S., Ikezawa, N., Takemura, T., Katayama, T., Kumagai, H., et al. (2008). Microbial production of plant benzylisoquinoline alkaloids. *Proceedings of the National Academy of Sciences of the United States of America, 105*(21), 7393–7398.

Morishige, T., Tsujita, T., Yamada, Y., & Sato, F. (2000). Molecular characterization of the S-adenosyl-L-methionine: 3'-hydroxy-N-methylcoclaurine 4'-O-methyltransferase involved in isoquinoline alkaloid biosynthesis in *Coptis japonica*. *Journal of Biological Chemistry, 275*(30), 23398–23405.

Naesby, M., Nielsen, S. V., Nielsen, C. A., Green, T., Tange, T. Ø., Simón, E., et al. (2009). Yeast artificial chromosomes employed for random assembly of biosynthetic pathways and production of diverse compounds in *Saccharomyces cerevisiae*. *Microbial Cell Factories, 8*, 45.

Nakagawa, A., Matsumura, E., Koyanagi, T., Katayama, T., Kawano, N., Yoshimatsu, K., et al. (2016). Total biosynthesis of opiates by stepwise fermentation using engineered *Escherichia coli*. *Nature Communications, 7*, 10390.

Nakagawa, A., Matsuzaki, C., Matsumura, E., Koyanagi, T., Katayama, T., Yamamoto, K., et al. (2014). (R,S)-Tetrahydropapaveroline production by stepwise fermentation using engineered Escherichia coli. *Scientific Reports, 4*, 6695.

Nakagawa, A., Minami, H., Kim, J. S., Koyanagi, T., Katayama, T., Sato, F., et al. (2011). A bacterial platform for fermentative production of plant alkaloids. *Nature Communications, 2*, 326.

Nakagawa, A., Minami, H., Kim, J. S., Koyanagi, T., Katayama, T., Sato, F., et al. (2012). Bench-top fermentative production of plant benzylisoquinoline alkaloids using a bacterial platform. *Bioengineered Bugs, 3*(1), 49–53.

Nielsen, K. F., & Larsen, T. (2015). The importance of mass spectrometric dereplication in fungal secondary metabolite analysis. *Frontiers in Microbiology, 6*, 71.

Nour-Eldin, H. H., Geu-Flores, F., & Halkier, B. A. (2010). USER cloning and USER fusion: The ideal cloning techniques for small and big laboratories. *Methods in Molecular Biology, 643*, 185–200.

Nour-Eldin, H. H., Hansen, B. G., Norholm, M. H. H., Jensen, J. K., & Halkier, B. A. (2006). Advancing uracil-excision based cloning towards an ideal technique for cloning PCR fragments. *Nucleic Acids Research, 34*(18), e122.

Olsen, L. R., Hansen, N. B., Bonde, M. T., Genee, H. J., Holm, D. K., Carlsen, S., et al. (2011). PHUSER (Primer Help for USER): A novel tool for USER fusion primer design. *Nucleic Acids Research, 39*(Suppl. 2), W61–W67.

Onoyovwe, A., Hagel, J. M., Chen, X., Khan, M. F., Schriemer, D. C., & Facchini, P. J. (2013). Morphine biosynthesis in opium poppy involves two cell types: Sieve elements and laticifers. *Plant Cell, 25*(10), 4110–4122.

Orrweaver, T. L., Szostak, J. W., & Rothstein, R. J. (1981). Yeast transformation—A model system for the study of recombination. *Proceedings of the National Academy of Sciences of the United States of America, 78*(10), 6354–6358.

Ounaroon, A., Decker, G., Schmidt, J., Lottspeich, F., & Kutchan, T. M. (2003). (R,S)-Reticuline 7-O-methyltransferase and (R,S)-norcoclaurine 6-O-methyltransferase of *Papaver somniferum*—cDNA cloning and characterization of methyl transfer enzymes of alkaloid biosynthesis in opium poppy. *Plant Journal, 36*(6), 808–819.

Pathak, S., Lakhwani, D., Gupta, P., Kishore Mishra, B., Shukla, S., Hasan Asif, M., et al. (2013). Comparative transcriptome analysis using high papaverine mutant of *Papaver somniferum* reveals pathway and uncharacterized steps of papaverine biosynthesis. *PLoS One, 8*(5), e65622.

Pauli, H. H., & Kutchan, T. M. (1998). Molecular cloning and functional heterologous expression of two alleles encoding (S)-N-methylcoclaurine 3'-hydroxylase (CYP80B1), a new methyl jasmonate-inducible cytochrome P-450-dependent mono-oxygenase of benzylisoquinoline alkaloid biosynthesis. *Plant Journal, 13*(6), 793–801.

Peterson, A. C., Hauschild, J.-P., Quarmby, S. T., Krumwiede, D., Lange, O., Lemke, R. A. S., et al. (2014). Development of a GC/Quadrupole-Orbitrap mass spectrometer, Part I: Design and characterization. *Analytical Chemistry, 86*(20), 10036–10043.

Qu, Y., Easson, M. L., Froese, J., Simionescu, R., Hudlicky, T., & De Luca, V. (2015). Completion of the seven-step pathway from tabersonine to the anticancer drug precursor vindoline and its assembly in yeast. *Proceedings of the National Academy of Sciences of the United States of America, 112*(19), 6224–6229.

Quax, T. E. F., Claassens, N. J., Söll, D., & van der Oost, J. (2015). Codon bias as a means to fine-tune gene expression. *Molecular Cell, 59*(2), 149–161.

Ruiz, M. T., Voinnet, O., & Baulcombe, D. C. (1998). Initiation and maintenance of virus-induced gene silencing. *Plant Cell, 10*(6), 937–946.

Samanani, N., Alcantara, J., Bourgault, R., Zulak, K. G., & Facchini, P. J. (2006). The role of phloem sieve elements and laticifers in the biosynthesis and accumulation of alkaloids in opium poppy. *Plant Journal, 47*(4), 547–563.

Samanani, N., Park, S. U., & Facchini, P. J. (2005). Cell type-specific localization of transcripts encoding nine consecutive enzymes involved in protoberberine alkaloid biosynthesis. *Plant Cell, 17*(3), 915–926.

Tautenhahn, R., Boettcher, C., & Neumann, S. (2008). Highly sensitive feature detection for high resolution LC/MS. *BMC Bioinformatics, 9*, 504.

Theuns, H. G., Janssen, R. H. A. M., Biessels, H. W. A., Menichini, F., & Salemink, C. A. (1984). A new rearrangement product of thebaine, isolated from *Papaver bracteatum* Lindl. Structural assignment of thebaine N-oxides. *Journal of the Chemical Society Perkin Transactions I Organic and Bioorganic Chemistry*, (8), 1701–1706.

Theuns, H. G., Theuns, H. L., & Lousberg, R. J. J. C. (1986). Search for new natural sources of morphinans. *Economic Botany, 40*(4), 485–497.

Thodey, K., Galanie, S., & Smolke, C. D. (2014). A microbial biomanufacturing platform for natural and semisynthetic opioids. *Nature Chemical Biology, 10*, 837–844.

Trenchard, I. J., & Smolke, C. D. (2015). Engineering strategies for the fermentative production of plant alkaloids in yeast. *Metabolic Engineering, 30*, 96–104.

Unterlinner, B., Lenz, R., & Kutchan, T. M. (1999). Molecular cloning and functional expression of codeinone reductase: The penultimate enzyme in morphine biosynthesis in the opium poppy *Papaver somniferum*. *Plant Journal, 18*(5), 465–475.

Wijekoon, C. P., & Facchini, P. J. (2012). Systematic knockdown of morphine pathway enzymes in opium poppy using virus-induced gene silencing. *Plant Journal*, *69*(6), 1052–1063.

Winkler, A., Hartner, F., Kutchan, T. M., Glieder, A., & Macheroux, P. (2006). Biochemical evidence that berberine bridge enzyme belongs to a novel family of flavoproteins containing a bi-covalently attached FAD cofactor. *Journal of Biological Chemistry*, *281*(30), 21276–21285.

Winzer, T., Gazda, V., He, Z., Kaminski, F., Kern, M., Larson, T. R., et al. (2012). A *Papaver somniferum* 10-gene cluster for synthesis of the anticancer alkaloid noscapine. *Science*, *336*(6089), 1704–1708.

Winzer, T., Kern, M., King, A. J., Larson, T. R., Teodor, T. R., Teodor, R. I., et al. (2015). Morphinan biosynthesis in opium poppy requires a P450-oxidoreductase fusion protein. *Science*, *349*(6245), 309–312.

Xiao, M., Zhang, Y., Chen, X., Lee, E. J., Barber, C. J., Chakrabarty, R., et al. (2013). Transcriptome analysis based on next-generation sequencing of non-model plants producing specialized metabolites of biotechnological interest. *Journal of Biotechnology*, *166*(3), 122–134.

Zhang, T. T., Lei, J., Yang, H. J., Xu, K., Wang, R., & Zhang, Z. Y. (2011). An improved method for whole protein extraction from yeast *Saccharomyces cerevisiae*. *Yeast*, *28*(11), 795–798.

Ziegler, J., Voigtländer, S., Schmidt, J., Kramell, R., Miersch, O., Ammer, C., et al. (2006). Comparative transcript and alkaloid profiling in *Papaver* species identifies a short chain dehydrogenase/reductase involved in morphine biosynthesis. *Plant Journal*, *48*(2), 177–192.

CHAPTER EIGHT

Optimizing Metabolic Pathways for the Improved Production of Natural Products

J.A. Jones*,†, M.A.G. Koffas*,†,1

*Rensselaer Polytechnic Institute, Troy, NY, United States
†Center for Biotechnology and Interdisciplinary Sciences, Rensselaer Polytechnic Institute, Troy, NY, United States
1Corresponding author: e-mail address: koffam@rpi.edu

Contents

1. Introduction 180
2. Genetic Optimization 180
   2.1 Gene Homolog Sourcing 181
   2.2 Selection of Expression Plasmid Backbone(s) 181
   2.3 DNA Copy Number Balancing 182
   2.4 Transcriptional Balancing 182
   2.5 Translational Balancing 183
   2.6 Posttranslational Balancing 183
   2.7 Dynamic Balancing 184
   2.8 Coculture Optimization 184
3. Fermentation Optimization 185
   3.1 Media Optimization 185
   3.2 Temperature Optimization 186
   3.3 Induction Optimization 186
4. Conclusion 189
Acknowledgments 189
References 189

Abstract

Metabolic engineering strives to develop microbial strains that are capable of high-titer production of a variety of industrially significant pharmaceuticals, nutraceuticals, commodity, and high-value compounds. Despite extensive success with many proof-of-concept systems there is still the need for optimization to achieve industrially relevant titers, yields, and productivities. The field of metabolic pathway optimization and balancing has formed to address this need using a scientific and systematic approach. In this chapter, we aim to outline various pathway optimization and system balancing strategies while giving insights and tips into the systems and procedures that have demonstrated recent success in the peer-reviewed literature.

1. INTRODUCTION

Metabolic pathway optimization is the synergistic combination of both genetic and fermentation optimization strategies. To achieve true global optimization of any microbially based natural product production process, optimization must be addressed from both regimes (Kim, Moore, & Yoon, 2015; Lee et al., 2012). An ever-increasing list of proof-of-concept pathways, producing novel compounds from a variety of carbon sources, supports the need for methods to expedite the transition from proof-of-concept to industrially viable production titers and yields (Bhan, Xu, & Koffas, 2013; Wang, Guleria, Koffas, & Yan, 2015). Advances in high-throughput selection and screening techniques have enabled great improvements to be achieved over the past several years.

In this chapter, we will begin by discussing methods for genetic optimization. In this section, we will summarize methods for pathway optimization at the DNA, RNA, and protein level, as well as posttranslational, dynamic, and coculture-based approaches. A section will follow on fermentation optimization with focus on media, temperature, and induction optimization. Although commonly ignored in the literature, fermentation optimization is a powerful tool that can rapidly improve titers, yields, and productivities. These methods are focused on applications in *Escherichia coli*, but most can be generalized for any genetically tractable production host. The purpose of the chapter is to highlight potential approaches, their success stories, and the possible pros and cons for many optimization methods. This chapter should enable you to decide which optimization methods are the best suited for your system or pathway of choice.

2. GENETIC OPTIMIZATION

Genetic optimization focuses on the development of the most efficient strain platform for the production of a given product of interest. This includes traditional metabolic engineering approaches to overexpress pathways leading toward the desired product and deleting or downregulating competing pathways. The scope of this section is to address the methods for fine-tuning the optimization of these overexpressions and downregulations to minimize metabolic burden and balance pathway fluxes to result in maximal production titers, yields, and productivities.

Genetic optimization strategies can be categorized as either static (Jones, Toparlak, & Koffas, 2015) or dynamic (Cress, Trantas, Ververidis, Linhardt, &

Koffas, 2015). Static approaches are characterized as a fixed enzyme expression level that is not designed to react or adjust depending on intercellular metabolite concentrations. These methods are viewed as traditional genetic optimization such as varying plasmid backbones, promoter strengths, and ribosome binding site (RBS) strengths. While these are often viewed as old-fashioned genetic engineering, they have seen significant success improving titer of a wide variety of products (Ajikumar et al., 2010; Huang, Lin, & Yan, 2013; Jones, Collins, Lachance, Vernacchio, & Koffas, 2016; Jones, Toparlak, et al., 2015; Jones, Vernacchio, et al., 2015; Xu et al., 2013). Dynamic balancing methods employ biosensors to sense and react to changing intercellular metabolite concentrations. The transient nature of these systems makes them a powerful tool for the metabolic engineering, but with increasing power comes increasing complexity, further complicating the task of experimental optimization.

2.1 Gene Homolog Sourcing

An important, yet oftentimes overlooked step in the microbial production of natural products is the selection of gene homologs. The natural product pathways are commonly mimicked from the original plant, mammalian, or microbial systems and therefore have not evolved for optimal activity in the genetically tractable production host of choice. The differences between original and desired production host can be quite significant and the effects of these differences on enzyme expression and activity are currently impossible to predict. Therefore, the combinatorial screening of several gene homologs from different sources has been shown to be an effective strategy to overcome this unpredictability (Hashimoto & Yamada, 2003; Leonard, Yan, Lim, & Koffas, 2005; Santos, Koffas, & Stephanopoulos, 2011; Thodey, Galanie, & Smolke, 2014; Yan, Huang, & Koffas, 2007; Yan, Li, & Koffas, 2008; Zhao et al., 2015). The ever-decreasing cost of DNA synthesis has significantly reduced the financial burden of this step. Codon optimization, or the act of adjusting the codon usage to match that of the desired production host, should be viewed as another method to affect enzyme expression and specific activity rather than a task that should be done for all synthesized constructs.

2.2 Selection of Expression Plasmid Backbone(s)

Traditionally the first step in pathway optimization is the selection of plasmid backbone from which your desired pathway genes will be expressed. Sets of compatible expression vectors having various copy numbers, expression levels, antibiotic resistances, and purification/solubilization tags have been developed to allow for many options in genetic optimization (Anderson et al., 2010; Lee et al., 2011; Shetty, Endy, & Knight, 2008; Xu, Vansiri, Bhan, & Koffas, 2012). Plasmids utilizing isocaudomer restriction enzyme sites such that entire pathways can be cloned in a single vector have recently become popular, but typically result in repetitive regions of DNA that can cause difficulty for homologous recombination-based chromosomal integration (CI) and should be avoided if CI strategies are planned. Recent advances in rapid one-step DNA assembly techniques such as Golden Gate (Engler, Kandzia, & Marillonnet, 2008), Ligase Cycling Reaction (LCR; De Kok et al., 2014), Gibson Isothermal Assembly (Gibson et al., 2009), and DNA Assembler (Shao, Zhao, & Zhao, 2009) have enabled plasmids containing entire pathways to be quickly assembled. It is important to make your plasmid backbone selection based on the type of optimization techniques you wish to employ (ie, transcriptional balancing requires easy incorporation of various strength promoters).

2.3 DNA Copy Number Balancing

Copy number balancing (CNB) was the first and is probably the most widely utilized form of genetic optimization. CNB is the expression of genes, modules, or pathways from plasmids with various expression levels. Generally, a plasmid's copy number is related to its expression level in vivo, although there are exceptions (Xu et al., 2012). Additionally, for metabolic engineering applications, it has been shown on many occasions that more is *not* always better when it comes to pathway enzyme overexpressions (Jones, Collins, et al., 2016; Jones, Kim, & Keasling, 2000; Jones, Toparlak, et al., 2015; Jones, Vernacchio, et al., 2015). The proper balance of pathway genes is necessary to minimize bottlenecks and promote proper resource allocation within the cell (Xu et al., 2013). Furthermore, in a deviation from traditional CNB, CI approaches as well as incorporation of multiple copies of the same gene on a single plasmid have been utilized effectively (Ajikumar et al., 2010; Tyo, Ajikumar, & Stephanopoulos, 2009; Zhao et al., 2015).

2.4 Transcriptional Balancing

Pathway balancing from the RNA level represents a powerful method for upfront optimization of enzyme expression level due to the significant experimental thrusts to develop and characterize promoter libraries (Brewster, Jones, & Phillips, 2012; De Mey, Maertens, Lequeux,

Soetaert, & Vandamme, 2007; Mutalik et al., 2013; Temme, Hill, Segall-Shapiro, Moser, & Voigt, 2012) and to understand transcriptional bottlenecks (Blazeck & Alper, 2013). The early balancing of mRNA levels enabled by transcriptional optimization lessens one of the major sources of metabolic burden giving an inherent advantage over other genetic optimization methods (Mairhofer, Scharl, Marisch, Cserjan-Puschmann, & Striedner, 2013). Methods such as ePathOptimize (Jones, Collins, et al., 2016; Jones, Toparlak, et al., 2015; Jones, Vernacchio, et al., 2015) and Golden Gate Assembly (Freestone & Zhao, 2015) have been used to quickly develop libraries of transcriptionally varied mutants, but advances in selection and screening tools are still needed to quickly identify lead mutants for further optimization. In addition to changing the promoter sequence, modification of the transcriptional operon configuration (operon, pseudo-operon, monocistronic, or any combination in between) and CRISPRi-based approaches are also considered transcriptional optimization, but are difficult to address from a combinatorial approach due to the unpredictable effects of such changes on the transcriptional profile (Cress, Toparlak, et al., 2015; He et al., 2015; Jakočiūnas, Jensen, & Keasling, 2015; Wu, Du, Chen, & Zhou, 2015).

2.5 Translational Balancing

Modification of the translation rate through mutation of the RBS sequence has recently become a very predictable tool due to research resulting in the RBS Calculator (Salis, Mirsky, & Voigt, 2009), RBSDesigner (Na & Lee, 2010), and others (Seo et al., 2013). This understanding allows for the prediction of dynamic ranges and sensitivity of RBS libraries that can be easily and efficiently created using degenerate primers and site-directed mutagenesis protocols (Ho, Hunt, Horton, Pullen, & Pease, 1989). Although this predictability is an advantage over other forms of genetic optimization, more still needs to be done to understand the protein–pathway flux relationship before true rational design can become a realistic reality (Farasat et al., 2014).

2.6 Posttranslational Balancing

Colocalization of pathway enzymes, through self-docking or the use of synthetic protein scaffolds, is used to facilitate substrate tunneling for diffusion-limited systems. To utilize this posttranslational balancing technique, a transcriptional fusion of desired enzyme(s) and binding ligand(s) must be created and coexpressed with a synthetic protein scaffold containing complementary binding domains. Several synthetic scaffolds with three binding domains in various ratios have been constructed and used to improve titers of mevalonate (Dueber et al., 2009), glucaric acid (Moon, Dueber, Shiue, & Prather, 2010), butyrate (Baek et al., 2013), gamma-aminobutyric acid (Pham, Lee, Park, & Hong, 2015), and (+)-catechin (Zhao et al., 2015) in *E. coli*. Although effective for several systems, the protein scaffold overexpression increases metabolic burden of the system and can oftentimes decrease growth rate and production if the optimal balance is not achieved (Jones, Toparlak, et al., 2015; Zhao et al., 2015).

2.7 Dynamic Balancing

Dynamic balancing systems have recently proven to be powerful in their ability to control metabolite fluxes according to intracellular metabolite concentrations in real time (Reizman et al., 2015; Solomon, Sanders, & Prather, 2012; Xu, Li, Zhang, Stephanopoulos, & Koffas, 2014; Zhang, Carothers, & Keasling, 2012). Biosensors have been developed to sense and react to a wide range of target molecules with generalizable strategies being developed for eukaryotic hosts (Feng et al., 2015; Siedler, Stahlhut, Malla, Maury, & Neves, 2014; Zhang, Jensen, & Keasling, 2015; Zhang, Li, Pereira, & Stephanopoulos, 2015; Zhang, Pereira, Li, & Stephanopoulos, 2015). In application, these systems can take many different forms and therefore analysis of these options is beyond the scope of this chapter. We refer readers to an excellent review on dynamic balancing in microbial systems (Cress, Trantas, et al., 2015).

2.8 Coculture Optimization

The utilization of co- and polyculture techniques has recently gained traction in the peer-reviewed literature for the production of compounds ranging from pharmaceutical precursors to biofuels (Jones, Vernacchio, et al., 2016; Saini, Hong Chen, Chiang, & Chao, 2015; Zhang, Li, et al., 2015; Zhang, Pereira, et al., 2015; Zhou, Qiao, Edgar, & Stephanopoulos, 2015). In addition to the aforementioned optimization methods, coculture techniques enable an additional level of optimization through balancing of the ratio of each strain of cells in the population. This parameter, strain ratio, allows for fluxes of each module to be adjusted through changes in the population distribution by varying the initial inoculation ratio. Additionally, coculture systems expedite process development by enabling previously optimized strains to be cultured together without the need for the genetic reoptimization that is commonly required when additional pathways are inserted in monoculture. Methods to analyze the strain ratio and decouple metabolic flux analysis of each strain without the need for physical cell separation have been recently developed (Gebreselassie & Antoniewicz, 2015).

3. FERMENTATION OPTIMIZATION

Fermentation optimization addresses parameters associated with process optimization rather than modification of the genetic elements within the microbe. This powerful, yet oftentimes overlooked tool has been shown to greatly improve pathway fluxes through careful optimization of media, temperature, and induction conditions. This field, previously championed by industrial partners, has the potential for rapid and significant improvements to product titer, yield, and productivity without the time delays associated with genetic optimization (ie, DNA synthesis, cloning, DNA sequencing, etc.). Here we will describe the main parameters that can be easily modified in shake flask experiments to result in significantly higher production.

To begin, it is useful to do some small-scale experiments to probe the production landscape such that you will have a starting point for the complete optimization. These experiments are typically not reported in the literature and therefore do not have to be reproduced or strategically planned; they exist to give a starting point for the optimization. Oftentimes, the preliminary fermentation optimization is done early in the genetic optimization process to determine initial screening conditions for genetically varied mutants. It is also important to revisit past optimizations, as this process is not linear but rather quite nonlinear with many factors affecting each other in smooth, yet unpredictable ways. The solution space or production landscape is incredibly large and it is impossible to screen all combinations of conditions. Therefore, when assumptions are made regarding less significant parameters such as overnight culture growth conditions, initial inoculation level, and media preparation protocols, care must be taken to use reproducible methods and properly document these details in the methods section to ensure reproducibility of the process.

3.1 Media Optimization

Media optimization varies significantly between batch shake flask and bioreactor-based studies, due to the scope of this chapter we will focus on shake flask systems and avoid the discussion of the challenges associated with bioreactor systems (Shiloach & Fass, 2005). As a good initial screen, we recommend selecting several different classes of base media to begin testing. Examples include: minimal media (M9, MOPS; Neidhardt, Bloch, & Smith, 1974), complex media (LB, TB, or SOC), and rich defined media (AMM; Jones, Collins, et al., 2016; Jones, Toparlak, et al., 2015; Jones, Vernacchio, et al., 2015, RDM—Teknova Inc.). To supplement these media we recommend the selection of a couple carbon sources that utilize different metabolic pathways for uptake and utilization (ie, glucose, xylose, glycerol, yeast extract; Bizzini et al., 2010; Da Silva, Mack, & Contiero, 2009; Shiloach & Fass, 2005). Combinatorial screening of these media and carbon sources will allow for selection of the best combination on which to base future optimization efforts.

3.2 Temperature Optimization

Temperature optimization attempts to hone in on the conditions under which the exogenous enzymes are the most active, while preserving the growth rate advantages of the host strain at elevated temperatures. It is recommended that several temperatures be selected at 5–10°C intervals to fully explore the solution space. Identification of the preferred growth temperature for the host strain (ie, *E. coli*, 37°C; *Saccharomyces cerevisiae*, 30°C) and in vitro temperature–activity relationship data for the exogenous enzymes from the peer-reviewed literature can give a base for the experimental design. In metabolic engineering studies, it is common to use a dual-temperature approach such that prior to induction the temperature is chosen to maximize growth, while postinduction it is chosen to optimize recombinant enzyme activity. Oftentimes, this temperature shift is applied slightly prior to induction to reduce simultaneous stresses to the cells. These insights allow for intelligent design of experiments, but currently it is still impossible to accurately predict the optimum conditions a priori.

3.3 Induction Optimization

Induction optimization is the most sensitive and therefore most important factor of fermentation optimization (Donovan, Robinson, & Glick, 1996). Improper selection of induction conditions can lead to less than desirable results, but until recently has been widely ignored in the metabolic engineering literature. The choice to use an inducible vs constitutive promoter system is case dependent, but proceeding with a constitutive system forfeits the ability to improve titers through induction optimization. Inducible systems allow for a rapid shift from the growth to the production phase of a fermentation enabling the majority of cellular resources to be focused on formation of desired product. Inducible systems have three main control points that are presented in detail later.

3.3.1 Induction Point

Without fail, the authors have observed high sensitivity to induction point for nearly every system we have worked with (Jones, Collins, et al., 2016; Jones, Toparlak, et al., 2015; Jones, Vernacchio, et al., 2015, 2016; Lim et al., 2015). These systems include the production of violacein, *n*-butanol, and various flavonoid and flavonoid-derived products, Fig. 1. Furthermore, these strains have observed differences in induction point optimum as well as slope of the traditional pyramid-shaped induction point optimization curve. These results lead us to believe that induction point choice cannot be

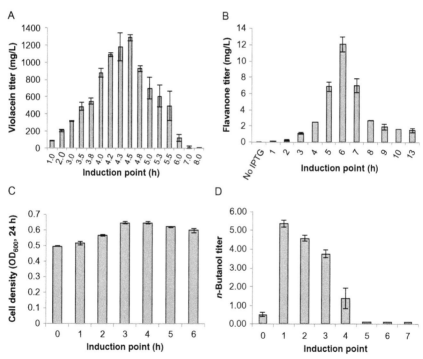

Fig. 1 Induction point sensitivity in various metabolic engineering applications. (A) De novo violacein production from *E. coli* in Andrew's Magic Media (AMM)—Optimum OD_{600}: 1.5. (B) Flavanone production from *E. coli* in minimal media—Optimum OD_{600}: 3.7. (C) Improved growth due to supplemental carbon source utilization in *E. coli*—Optimum OD_{600}: 0.45. (D) *n*-butanol production from *E. coli* in complex media—Optimum OD_{600}: 0.12.

assumed to be best at $OD_{600} = 0.6$ but rather should be carefully optimized to maximize production. Sensitivities on the order of 15 min have been observed (Jones, Collins, et al., 2016; Jones, Toparlak, et al., 2015; Jones, Vernacchio, et al., 2015), suggesting that induction point optimization involves more than just optimization of protein production, but is also intimately involved with the allocation of cellular resources associated with the transition from growth to production phase during the fermentation. To address induction point optimization, we suggest beginning with 1 h intervals to get the general shape of the induction point profile then fine-tuning the optimization near the peak. We recommend basing induction point on time-based inductions that can be converted to OD values based on a growth curve to allow for more reproducible experimental values.

3.3.2 Inducer Concentration

The choice of inducer concentration is another factor that is commonly ignored but oftentimes results in drastic changes in product titer. Modulation of inducer concentration is a way of globally adjusting transcriptional levels from all inducible promoters at the same time. This can be contrasted with transcriptional balancing (Section 2.4) where we discussed techniques to locally adjust transcription level through mutation of promoters controlling the expression of individual genes. While transcriptional optimization gives you fine-tuned control, it requires screening of many mutant strains to be effective. The coarse-tuned transcriptional control of changing inducer concentration allows for rapid analysis and is oftentimes a good indicator of transcriptional balance sensitivity. If high sensitivity is observed through varying inducer concentration, additional transcriptional optimization could result in further improvements in production.

3.3.3 Substrate Delay

Substrate delay refers to a time delay between the addition of inducer and precursor molecule supplementation. Therefore, this technique is only applicable in systems in which a precursor molecule is added to the fermentation broth to be further functionalized and converted to final product. This strategy is particularly effective when the precursor molecule is needed in concentrations that are inhibitory to cell growth and/or protein production. The time delay allows for the necessary enzymes to be expressed prior to precursor supplementation. This method has been applied in two cases for the production of caffeic acid from *p*-coumaric acid where a 2-h substrate delay resulted in up to 220% improvement over the no-delay control (Huang et al., 2013; Jones, Collins, et al., 2016).

4. CONCLUSION

When deciding on which form(s) of optimization to pursue for any given pathway it is important to realize that each case has its own system-specific sensitivities. Optimization protocols that have been shown to be successful for one pathway may not perform well for another. Some optimization strategies such as posttranslational balancing and substrate delay have narrow windows of applicability, while others such as induction point optimization are more generalizable. Currently, it is difficult to predict which strategies will yield the most fruitful results and methods to more accurately identify specific bottlenecks are greatly needed. Regardless, significant improvements have been achieved utilizing each of the optimization strategies mentioned above, and the synergistic effects of both genetic and fermentation optimization will lead the way in the pursuit of industrially viable microbial production processes.

ACKNOWLEDGMENTS

This work was supported by the ARPA-E REMOTE program grant number DE-AR0000432 awarded to M.K.

REFERENCES

Ajikumar, P. K., Xiao, W.-H., Tyo, K. E. J., Wang, Y., Simeon, F., Leonard, E., et al. (2010). Isoprenoid pathway optimization for Taxol precursor overproduction in Escherichia coli. Science, 330, 70–74. http://dx.doi.org/10.1126/science.1191652.

Anderson, J. C., Dueber, J. E., Leguia, M., Wu, G. C., Goler, J. A., Arkin, A. P., et al. (2010). BglBricks: A flexible standard for biological part assembly. Journal of Biological Engineering, 4, 1. http://dx.doi.org/10.1186/1754-1611-4-1.

Baek, J.-M., Mazumdar, S., Lee, S.-W., Jung, M.-Y., Lim, J.-H., Seo, S.-W., et al. (2013). Butyrate production in engineered Escherichia coli with synthetic scaffolds. Biotechnology and Bioengineering, 110, 2790–2794. http://dx.doi.org/10.1002/bit.24925.

Bhan, N., Xu, P., & Koffas, M. A. G. (2013). Pathway and protein engineering approaches to produce novel and commodity small molecules. Current Opinion in Biotechnology, 24, 1137–1143. http://dx.doi.org/10.1016/j.copbio.2013.02.019.

Bizzini, A., Zhao, C., Budin-Verneuil, A., Sauvageot, N., Giard, J.-C., Auffray, Y., et al. (2010). Glycerol is metabolized in a complex and strain-dependent manner in Enterococcus faecalis. Journal of Bacteriology, 192, 779–785. http://dx.doi.org/10.1128/JB.00959-09.

Blazeck, J., & Alper, H. S. (2013). Promoter engineering: Recent advances in controlling transcription at the most fundamental level. Biotechnology Journal, 8, 46–58. http://dx.doi.org/10.1002/biot.201200120.

Brewster, R. C., Jones, D. L., & Phillips, R. (2012). Tuning promoter strength through RNA polymerase binding site design in Escherichia coli. *PLoS Computational Biology*, *8*, 1–10. http://dx.doi.org/10.1371/journal.pcbi.1002811.

Cress, B. F., Toparlak, Ö. D., Guleria, S., Lebovich, M., Stieglitz, J. T., Englaender, J. A., et al. (2015). CRISPathBrick: Modular combinatorial assembly of type II-A CRISPR arrays for dCas9-mediated multiplex transcriptional repression in E. coli. *ACS Synthetic Biology*, *4*, 987–1000. http://dx.doi.org/10.1021/acssynbio.5b00012.

Cress, B. F., Trantas, E. A., Ververidis, F., Linhardt, R. J., & Koffas, M. A. (2015). Sensitive cells: Enabling tools for static and dynamic control of microbial metabolic pathways. *Current Opinion in Biotechnology*, *36*, 205–214. http://dx.doi.org/10.1016/j.copbio.2015.09.007.

Da Silva, G. P., Mack, M., & Contiero, J. (2009). Glycerol: A promising and abundant carbon source for industrial microbiology. *Biotechnology Advances*, *27*, 30–39. http://dx.doi.org/10.1016/j.biotechadv.2008.07.006.

De Kok, S., Stanton, L. H., Slaby, T., Durot, M., Holmes, V. F., Patel, K. G., et al. (2014). Rapid and reliable DNA assembly via ligase cycling reaction. *ACS Synthetic Biology*, *3*, 97–106. http://dx.doi.org/10.1021/sb4001992.

De Mey, M., Maertens, J., Lequeux, G. J., Soetaert, W. K., & Vandamme, E. J. (2007). Construction and model-based analysis of a promoter library for E. coli: An indispensable tool for metabolic engineering. *BMC Biotechnology*, *7*, 34. http://dx.doi.org/10.1186/1472-6750-7-34.

Donovan, R. S., Robinson, C. W., & Glick, B. R. (1996). Review: Optimizing inducer and culture conditions for expression of foreign proteins under the control of the lac promoter. *Journal of Industrial Microbiology*, *16*, 145–154. http://dx.doi.org/10.1007/BF01569997.

Dueber, J. E., Wu, G. C., Malmirchegini, G. R., Moon, T. S., Petzold, C. J., Ullal, A. V., et al. (2009). Synthetic protein scaffolds provide modular control over metabolic flux. *Nature Biotechnology*, *27*, 753–759. http://dx.doi.org/10.1038/nbt.1557.

Engler, C., Kandzia, R., & Marillonnet, S. (2008). A one pot, one step, precision cloning method with high throughput capability. *PloS One*, *3*, e3647. http://dx.doi.org/10.1371/journal.pone.0003647.

Farasat, I., Kushwaha, M., Collens, J., Easterbrook, M., Guido, M., & Salis, H. M. (2014). Efficient search, mapping, and optimization of multi-protein genetic systems in diverse bacteria. *Molecular Systems Biology*, *10*, 731. http://dx.doi.org/10.15252/msb.20134955.

Feng, J., Jester, B. W., Tinberg, C. E., Mandell, D. J., Antunes, M. S., Chari, R., et al. (2015). A general strategy to construct small molecule biosensors in eukaryotes. *Elife*, *4*, e10606. http://dx.doi.org/10.7554/eLife.10606.

Freestone, T. S., & Zhao, H. (2015). Combinatorial pathway engineering for optimized production of the anti-malarial FR900098. *Biotechnology and Bioengineering*, *113*, 384–392. http://dx.doi.org/10.1002/bit.25719.

Gebreselassie, N. A., & Antoniewicz, M. R. (2015). (13)C-metabolic flux analysis of co-cultures: A novel approach. *Metabolic Engineering*, *31*, 132–139. http://dx.doi.org/10.1016/j.ymben.2015.07.005.

Gibson, D. G., Young, L., Chuang, R.-Y., Venter, J. C., Hutchison, C. A., & Smith, H. O. (2009). Enzymatic assembly of DNA molecules up to several hundred kilobases. *Nature Methods*, *6*, 343–345. http://dx.doi.org/10.1038/nmeth.1318.

Hashimoto, T., & Yamada, Y. (2003). New genes in alkaloid metabolism and transport. *Current Opinion in Biotechnology*, *14*, 163–168. http://dx.doi.org/10.1016/S0958-1669(03)00027-2.

He, W., Fu, L., Li, G., Jones, J. A., Linhardt, R. J., & Koffas, M. (2015). Production of chondroitin in metabolically engineered E. coli. *Metabolic Engineering*, *27*, 92–100. http://dx.doi.org/10.1016/j.ymben.2014.11.003.

Ho, S. N., Hunt, H. D., Horton, R. M., Pullen, J. K., & Pease, L. R. (1989). Site-directed mutagenesis by overlap extension using the polymerase chain reaction. *Gene, 77*, 51–59. http://dx.doi.org/10.1016/0378-1119(89)90358-2.

Huang, Q., Lin, Y., & Yan, Y. (2013). Caffeic acid production enhancement by engineering a phenylalanine over-producing Escherichia coli strain. *Biotechnology and Bioengineering, 110*, 3188–3196. http://dx.doi.org/10.1002/bit.24988.

Jakočiūnas, T., Jensen, M. K., & Keasling, J. D. (2015). CRISPR/Cas9 advances engineering of microbial cell factories. *Metabolic Engineering, 34*, 44–59. http://dx.doi.org/10.1016/j.ymben.2015.12.003.

Jones, J. A., Collins, S. M., Lachance, D. M., Vernacchio, V. R., & Koffas, M. A. G. (2016). Optimization of naringenin and p-coumaric acid hydroxylation using the native E. coli hydroxylase complex, HpaBC. *Biotechnology Progress, 32*(1), 21–25. http://dx.doi.org/10.1002/btpr.2185.

Jones, K. L., Kim, S. W., & Keasling, J. D. (2000). Low-copy plasmids can perform as well as or better than high-copy plasmids for metabolic engineering of bacteria. *Metabolic Engineering, 2*, 328–338. http://dx.doi.org/10.1006/mben.2000.0161.

Jones, J. A., Toparlak, Ö. D., & Koffas, M. A. (2015). Metabolic pathway balancing and its role in the production of biofuels and chemicals. *Current Opinion in Biotechnology, 33*, 52–59. http://dx.doi.org/10.1016/j.copbio.2014.11.013.

Jones, J. A., Vernacchio, V. R., Lachance, D. M., Lebovich, M., Fu, L., Shirke, A. N., et al. (2015). ePathOptimize: A combinatorial approach for transcriptional balancing of metabolic pathways. *Scientific Reports, 5*, 11301. http://dx.doi.org/10.1038/srep11301.

Jones, J. A., Vernacchio, V. R., Sinkoe, A. L., Collins, S. M., Ibrahim, M. H. A., Lachance, D. M., et al. (2016). Experimental and computational optimization of an Escherichia coli co-culture for the efficient production of flavonoids. *Metabolic Engineering, 35*, 55–63.

Kim, E., Moore, B. S., & Yoon, Y. J. (2015). Reinvigorating natural product combinatorial biosynthesis with synthetic biology. *Nature Chemical Biology, 11*, 649–659.

Lee, T. S., Krupa, R. A., Zhang, F., Hajimorad, M., Holtz, W. J., Prasad, N., et al. (2011). BglBrick vectors and datasheets: A synthetic biology platform for gene expression. *Journal of Biological Engineering, 5*, 12. http://dx.doi.org/10.1186/1754-1611-5-12.

Lee, J. W., Na, D., Park, J. M., Lee, J., Choi, S., & Lee, S. Y. (2012). Systems metabolic engineering of microorganisms for natural and non-natural chemicals. *Nature Chemical Biology, 8*, 536–546. http://dx.doi.org/10.1038/nchembio.970.

Leonard, E., Yan, Y., Lim, K. H., & Koffas, M. A. G. (2005). Investigation of two distinct flavone synthases for plant-specific flavone biosynthesis in Saccharomyces cerevisiae. *Applied and Environmental Microbiology, 71*, 8241–8248. http://dx.doi.org/10.1128/AEM.71.12.8241-8248.2005.

Lim, C. G., Wong, L., Bhan, N., Dvora, H., Xu, P., Venkiteswaran, S., et al. (2015). Development of a recombinant Escherichia coli strain for overproduction of plant pigment, anthocyanin. *Applied and Environmental Microbiology, 81*, 6276–6284. http://dx.doi.org/10.1128/AEM.01448-15.

Mairhofer, J., Scharl, T., Marisch, K., Cserjan-Puschmann, M., & Striedner, G. (2013). Comparative transcription profiling and in-depth characterization of plasmid-based and plasmid-free Escherichia coli expression systems under production conditions. *Applied and Environmental Microbiology, 79*, 3802–3812. http://dx.doi.org/10.1128/AEM.00365-13.

Moon, T. S., Dueber, J. E., Shiue, E., & Prather, K. L. J. (2010). Use of modular, synthetic scaffolds for improved production of glucaric acid in engineered E. coli. *Metabolic Engineering, 12*, 298–305. http://dx.doi.org/10.1016/j.ymben.2010.01.003.

Mutalik, V. K., Guimaraes, J. C., Cambray, G., Lam, C., Christoffersen, M. J., Mai, Q.-A., et al. (2013). Precise and reliable gene expression via standard transcription and translation initiation elements. *Nature Methods*, *10*, 354–360. http://dx.doi.org/10.1038/nmeth.2404.

Na, D., & Lee, D. (2010). RBSDesigner: Software for designing synthetic ribosome binding sites that yields a desired level of protein expression. *Bioinformatics*, *26*, 2633–2634. http://dx.doi.org/10.1093/bioinformatics/btq458.

Neidhardt, F. C., Bloch, P. L., & Smith, D. F. (1974). Culture medium for Enterobacteria. *Journal of Bacteriology*, *119*, 736–747.

Pham, V. D., Lee, S. H., Park, S. J., & Hong, S. H. (2015). Production of gamma-aminobutyric acid from glucose by introduction of synthetic scaffolds between isocitrate dehydrogenase, glutamate synthase and glutamate decarboxylase in recombinant Escherichia coli. *Journal of Biotechnology*, *207*, 52–57. http://dx.doi.org/10.1016/j.jbiotec.2015.04.028.

Reizman, I. M. B., Stenger, A. R., Reisch, C. R., Gupta, A., Connors, N. C., & Prather, K. L. J. (2015). Improvement of glucaric acid production in E. coli via dynamic control of metabolic fluxes. *Metabolic Engineering Communications*, *2*, 109–116. http://dx.doi.org/10.1016/j.meteno.2015.09.002.

Saini, M., Hong Chen, M., Chiang, C.-J., & Chao, Y.-P. (2015). Potential production platform of n-butanol in Escherichia coli. *Metabolic Engineering*, *27*, 76–82. http://dx.doi.org/10.1016/j.ymben.2014.11.001.

Salis, H. M., Mirsky, E. A., & Voigt, C. A. (2009). Automated design of synthetic ribosome binding sites to control protein expression. *Nature Biotechnology*, *27*, 946–950. http://dx.doi.org/10.1038/nbt.1568.

Santos, C. N. S., Koffas, M., & Stephanopoulos, G. (2011). Optimization of a heterologous pathway for the production of flavonoids from glucose. *Metabolic Engineering*, *13*, 392–400. http://dx.doi.org/10.1016/j.ymben.2011.02.002.

Seo, S. W., Yang, J.-S., Kim, I., Yang, J., Min, B. E., Kim, S., et al. (2013). Predictive design of mRNA translation initiation region to control prokaryotic translation efficiency. *Metabolic Engineering*, *15*, 67–74. http://dx.doi.org/10.1016/j.ymben.2012.10.006.

Shao, Z., Zhao, H., & Zhao, H. (2009). DNA assembler, an in vivo genetic method for rapid construction of biochemical pathways. *Nucleic Acids Research*, *37*, e16. http://dx.doi.org/10.1093/nar/gkn991.

Shetty, R. P., Endy, D., & Knight, T. F. (2008). Engineering BioBrick vectors from BioBrick parts. *Journal of Biological Engineering*, *2*, 5. http://dx.doi.org/10.1186/1754-1611-2-5.

Shiloach, J., & Fass, R. (2005). Growing E. coli to high cell density—A historical perspective on method development. *Biotechnology Advances*, *23*, 345–357. http://dx.doi.org/10.1016/j.biotechadv.2005.04.004.

Siedler, S., Stahlhut, S. G., Malla, S., Maury, J., & Neves, A. R. (2014). Novel biosensors based on flavonoid-responsive transcriptional regulators introduced into Escherichia coli. *Metabolic Engineering*, *21*, 2–8. http://dx.doi.org/10.1016/j.ymben.2013.10.011.

Solomon, K. V., Sanders, T. M., & Prather, K. L. J. (2012). A dynamic metabolite valve for the control of central carbon metabolism. *Metabolic Engineering*, *14*, 661–671. http://dx.doi.org/10.1016/j.ymben.2012.08.006.

Temme, K., Hill, R., Segall-Shapiro, T. H., Moser, F., & Voigt, C. A. (2012). Modular control of multiple pathways using engineered orthogonal T7 polymerases. *Nucleic Acids Research*, *40*, 8773–8781. http://dx.doi.org/10.1093/nar/gks597.

Thodey, K., Galanie, S., & Smolke, C. D. (2014). A microbial biomanufacturing platform for natural and semisynthetic opioids. *Nature Chemical Biology*, *10*, 837–844. http://dx.doi.org/10.1038/nchembio.1613.

Tyo, K. E. J., Ajikumar, P. K., & Stephanopoulos, G. (2009). Stabilized gene duplication enables long-term selection-free heterologous pathway expression. *Nature Biotechnology*, *27*, 760–765. http://dx.doi.org/10.1038/nbt.1555.

Wang, J., Guleria, S., Koffas, M. A., & Yan, Y. (2015). Microbial production of value-added nutraceuticals. *Current Opinion in Biotechnology*, *37*, 97–104. http://dx.doi.org/10.1016/j.copbio.2015.11.003.

Wu, J., Du, G., Chen, J., & Zhou, J. (2015). Enhancing flavonoid production by systematically tuning the central metabolic pathways based on a CRISPR interference system in Escherichia coli. *Scientific Reports*, *5*, 13477. http://dx.doi.org/10.1038/srep13477.

Xu, P., Gu, Q., Wang, W., Wong, L., Bower, A. G. W., Collins, C. H., et al. (2013). Modular optimization of multi-gene pathways for fatty acids production in E. coli. *Nature Communications*, *4*, 1409. http://dx.doi.org/10.1038/ncomms2425.

Xu, P., Li, L., Zhang, F., Stephanopoulos, G., & Koffas, M. (2014). Improving fatty acids production by engineering dynamic pathway regulation and metabolic control. *Proceedings of the National Academy of Sciences of the United States of America*, *111*, 11299–11304. http://dx.doi.org/10.1073/pnas.1406401111.

Xu, P., Vansiri, A., Bhan, N., & Koffas, M. A. G. (2012). ePathBrick: A synthetic biology platform for engineering metabolic pathways in E. coli. *ACS Synthetic Biology*, *1*, 256–266. http://dx.doi.org/10.1021/sb300016b.

Yan, Y., Huang, L., & Koffas, M. A. G. (2007). Biosynthesis of 5-deoxyflavanones in microorganisms. *Biotechnology Journal*, *2*, 1250–1262. http://dx.doi.org/10.1002/biot.200700119.

Yan, Y., Li, Z., & Koffas, M. A. G. (2008). High-yield anthocyanin biosynthesis in engineered Escherichia coli. *Biotechnology and Bioengineering*, *100*, 126–140. http://dx.doi.org/10.1002/bit.21721.

Zhang, F., Carothers, J. M., & Keasling, J. D. (2012). Design of a dynamic sensor-regulator system for production of chemicals and fuels derived from fatty acids. *Nature Biotechnology*, *30*, 354–359. http://dx.doi.org/10.1038/nbt.2149.

Zhang, J., Jensen, M. K., & Keasling, J. D. (2015). Development of biosensors and their application in metabolic engineering. *Current Opinion in Chemical Biology*, *28*, 1–8. http://dx.doi.org/10.1016/j.cbpa.2015.05.013.

Zhang, H., Li, Z., Pereira, B., & Stephanopoulos, G. (2015). Engineering E. coli-E. coli cocultures for production of muconic acid from glycerol. *Microbial Cell Factories*, *14*, 134. http://dx.doi.org/10.1186/s12934-015-0319-0.

Zhang, H., Pereira, B., Li, Z., & Stephanopoulos, G. (2015). Engineering Escherichia coli coculture systems for the production of biochemical products. *Proceedings of the National Academy of Sciences of the United States of America*, *112*, 8266–8271. http://dx.doi.org/10.1073/pnas.1506781112.

Zhao, S., Jones, J. A., Lachance, D. M., Bhan, N., Khalidi, O., Venkataraman, S., et al. (2015). Improvement of catechin production in Escherichia coli through combinatorial metabolic engineering. *Metabolic Engineering*, *28*, 43–53. http://dx.doi.org/10.1016/j.ymben.2014.12.002.

Zhou, K., Qiao, K., Edgar, S., & Stephanopoulos, G. (2015). Distributing a metabolic pathway among a microbial consortium enhances production of natural products. *Nature Biotechnology*, *33*, 377–383. http://dx.doi.org/10.1038/nbt.3095.

CHAPTER NINE

Reconstituting Plant Secondary Metabolism in *Saccharomyces cerevisiae* for Production of High-Value Benzylisoquinoline Alkaloids

M.E. Pyne, L. Narcross, E. Fossati, L. Bourgeois, E. Burton, N.D. Gold, V.J.J. Martin[1]

Centre for Structural and Functional Genomics, Concordia University, Montréal, QC, Canada
[1]Corresponding author: e-mail address: vincent.martin@concordia.ca

Contents

1. Introduction	196
2. BIA Biosynthetic Gene Sources and Selection of Candidate Genes	199
3. Yeast Functional Expression Strategies	200
3.1 Genetic Considerations	202
3.2 Enzyme Expression and Localization	203
3.3 Enzyme Activity	210
4. Pathway Reconstitution and Optimization	215
4.1 Pathway Assembly	216
4.2 Debottlenecking Strategies	216
5. Challenges and Future Directions	219
References	221

Abstract

Benzylisoquinoline alkaloids (BIAs) constitute a diverse class of plant secondary metabolites that includes the opiate analgesics morphine and codeine. Collectively, BIAs exhibit a myriad of pharmacological activities, including antimicrobial, antitussive, antispasmodic, and anticancer properties. Despite 2500 known BIA products, only a small proportion are currently produced though traditional crop-based manufacturing, as complex stereochemistry renders chemical synthesis of BIAs largely unfeasible. The advent of synthetic biology and sophisticated microbial engineering coupled with recent advances in the elucidation of plant BIA metabolic networks has provided growing motivation for producing high-value BIAs in microbial hosts. Here, we provide a technical basis for reconstituting BIA biosynthetic pathways in the common yeast *Saccharomyces cerevisiae*. Methodologies outlined in this chapter include fundamental techniques for expressing and assaying BIA biosynthetic enzymes, bioprospecting large libraries of BIA enzyme variants, and reconstituting and optimizing complete BIA formation pathways in yeast. To expedite construction of superior BIA-producing yeast strains, we emphasize high-throughput techniques. Finally, we identify fundamental challenges impeding deployment of yeast-based BIA production platforms and briefly outline future prospects to overcome such barriers.

1. INTRODUCTION

It has been estimated that 20% of plant species collectively produce 12,000 alkaloids, many of which have been exploited for centuries by traditional and modern medicine (Ziegler & Facchini, 2008). Approximately 2500 of these compounds constitute benzylisoquinoline alkaloids (BIAs), a structurally diverse family of metabolites that includes the opiates morphine and codeine. Plant species from the Berberidaceae, Menispermaceae, Papaveraceae, and Ranunculaceae families, including the opium poppy, *Papaver somniferum*, produce BIAs through complex secondary metabolic pathways (Ziegler & Facchini, 2008). BIAs exhibit a wide array of pharmacological activities, acting as analgesics (morphine and codeine), antibacterials (sanguinarine and berberine), antispasmodics (papaverine), and antitussives (codeine and noscapine) (Hagel & Facchini, 2013; Kutchan, 1995; Liscombe & Facchini, 2008; Minami et al., 2008; Narcross, Fossati, Bourgeois, Dueber, & Martin, 2016). Some BIAs (sanguinarine, berberine, and noscapine) have also shown promising anticancer properties (Barken, Geller, & Rogosnitzky, 2008; Bessi et al., 2012). BIA biosynthesis begins through condensation of dopamine and 4-hydroxyphenylacetaldehyde (4-HPAA), both derived from the aromatic amino acid pathway. The product of this reaction, (S)-norcoclaurine, exemplifies the characteristic scaffold common to all BIA backbones (Fig. 1). (S)-reticuline, the key branch point from which most BIAs are produced, is generated from (S)-norcoclaurine through one hydroxylation and three methylation reactions (Beaudoin & Facchini, 2014). Whereas the BIAs noscapine and sanguinarine, among others, are derived directly from (S)-reticuline, the morphinan class of BIAs are produced from (R)-reticuline via stereochemical inversion of the (S)-enantiomer (Fig. 1). In both cases, downstream BIAs are diversified through a complex network of chemical rearrangements and functional group transformations involving methyltransferases (O- and N-), FAD-dependent oxidoreductases, 2-oxoglutarate-dependent dioxygenases, acyl-CoA-dependent acyltransferases, NADPH-dependent reductases, and

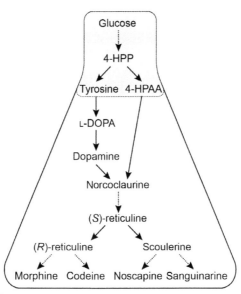

Fig. 1 BIA biosynthesis in yeast. Plant BIA secondary metabolic pathways (*unshaded*) can be linked to native yeast primary metabolism (*shaded*) via tyrosine and 4-hydroxyphenylacetaldehyde (4-HPAA), both metabolites from the aromatic amino acid pathway. Select BIA end products are shown derived from (*R*)- or (*S*)-reticuline. *Dashed arrows* encompass multiple successive enzymatic conversions. Other abbreviations: *4-HPP*, 4-hydroxyphenylpyruvate; L-*DOPA*, L-3,4-dihydroxyphenylalanine.

cytochrome P450s (Facchini et al., 2012). Notably, a preponderance of downstream BIA transformations is mediated by membrane-bound cytochrome P450s, underscoring the complexity of BIA biosynthesis and providing rationale for reconstituting BIA metabolism in yeast, as prokaryotes lack membrane-bound organelles required to support cytochrome P450 functionality (Fossati et al., 2014; Pompon, Louerat, Bronine, & Urban, 1996). BIA pathway complexity is compounded by the occurrence of a number of spontaneous chemical reactions in downstream pathways. For example, spontaneous chemical rearrangement drives formation of the BIA intermediates thebaine and codeinone (Fisinger, Grobe, & Zenk, 2007; Fossati, Narcross, Ekins, Falgueyret, & Martin, 2015; Theuns et al., 1984). Oxidation of dihydrosanguinarine, yielding the BIA end product sanguinarine, has also been shown to occur spontaneously despite the existence of a sanguinarine biosynthetic enzyme (Fossati et al., 2014; Trenchard, Siddiqui, Thodey, & Smolke, 2015).

Only a few select BIAs, including thebaine, noscapine, morphine, and codeine accumulate to appreciable levels *in planta* (Allen et al., 2004;

Frick, Kramell, Schmidt, Fist, & Kutchan, 2005), due in part to plant mutagenesis and selective breeding (Millgate et al., 2004; Nyman, 1980). Given that roughly 2500 structures are known, the exceptional pharmacological potential of plant BIAs remains largely untapped (Narcross et al., 2016). Currently, worldwide demand for opiates and other select BIAs is met through crop-based manufacturing, as total chemical synthesis is challenging and economically uncompetitive. Extracted BIAs, particularly thebaine, can be chemically derivatized, yielding the semisynthetic opioids oxycodone and hydrocodone, among many others (Rinner & Hudlicky, 2012). While it is possible to engineer native BIA-producing plants to generate higher quantities of target metabolites through implementation of genetic and metabolic engineering strategies (Facchini, 2001; Kutchan, 1995), plant secondary metabolism is highly complex and tools for sophisticated genetic manipulation of plants are lacking (Glenn, Runguphan, & O'Connor, 2013). Furthermore, crop-based manufacturing is dependent on slow plant growth, requires dedicated land for cultivation, and is susceptible to disease and drought. Instead, rapid advancements in interdisciplinary biological fields, such as synthetic biology, functional genomics, and systems biology, have prompted motivation for engineering microbes to produce high-value plant BIAs from simple sugars (Facchini et al., 2012). Such an accomplishment would alleviate our reliance on BIA-producing crops, as well as provide a framework for developing and characterizing novel BIA pharmaceuticals. In this context, a flurry of activity has been stimulated in recent years, including discovery of key enzymes in the morphinan, sanguinarine, and noscapine pathways, as well reconstitution of partial and complete BIA pathways in microbial hosts (Narcross et al., 2016). Although the model bacterium *Escherichia coli* has been engineered for de novo production of reticuline (Nakagawa et al., 2011), the common yeast *Saccharomyces cerevisiae* embodies the ideal cell factory for heterologous production of complex downstream BIA end products (Narcross et al., 2016). Yeast has recently been modified to produce an impressive array of high-value BIAs, including morphine, codeine, oxycodone, salutaridine, and sanguinarine (Fossati et al., 2014, 2015; Hawkins & Smolke, 2008; Thodey, Galanie, & Smolke, 2014). A challenge of these initial efforts involves the use of expensive BIA intermediates as feeding substrates for proof-of-principle formation of downstream BIAs. Because the BIA structural backbone is produced through condensation of two aromatic amino acid derivatives, the opportunity exists for linking plant BIA biosynthesis with yeast primary metabolism (Fig. 1). Accordingly, de novo synthesis of reticuline (DeLoache et al., 2015), and later thebaine and hydrocodone (Galanie, Thodey, Trenchard, Interrante, & Smolke, 2015), has been reported in yeast using simple substrates, eradicating the need for costly BIA supplementation. Despite these extraordinary feats of synthetic biology and metabolic engineering, however, yeast-based BIA production remains a proof of principle.

This chapter is intended to serve as a foundational resource for the heterologous production of plant BIAs in *S. cerevisiae*. We provide a roadmap for building yeast strains expressing nonnative BIA biosynthetic capabilities, starting from selection of candidate BIA genes through to full pathway reconstitution, including elucidation and resolution of metabolic bottlenecks. We identify technical challenges currently impeding deployment of yeast-based BIA production platforms, such as enhancing product yields and harnessing inherent enzyme promiscuity. Finally, we highlight the ethical and societal challenges arising from the construction of BIA-producing yeasts. Readers are directed to recent reviews for broader discussions of plant alkaloid biosynthesis (Hagel & Facchini, 2013; Ziegler & Facchini, 2008), metabolic engineering of plant systems for alkaloid production (Glenn et al., 2013), and microbial BIA production strategies (Narcross et al., 2016), as well as work detailing the use of bacterial hosts for BIA production (Minami et al., 2008; Nakagawa et al., 2011).

2. BIA BIOSYNTHETIC GENE SOURCES AND SELECTION OF CANDIDATE GENES

In addition to traditional methods of molecular cloning and virus-induced gene silencing, the advent of -omics technologies, namely transcriptomics, proteomics, and metabolomics, has accelerated the elucidation of BIA biosynthetic pathways and the discovery of cDNAs encoding BIA enzymatic activities (Hagel & Facchini, 2013). Metabolic pathways involved in the production of most BIAs have only been elucidated in recent years, including the sanguinarine, morphine, noscapine, and papaverine branches (Beaudoin & Facchini, 2013; Dang & Facchini, 2012; Desgagné Penix & Facchini, 2012; Hagel & Facchini, 2010; Hagel et al., 2012; Winzer et al., 2012). A notable example involves discovery of the elusive cDNA encoding the enzyme responsible for epimerization of (S)-reticuline in the committed step of the morphinan pathway, an achievement shared by three groups in 2015 (Farrow, Hagel, Beaudoin, Burns, & Facchini, 2015; Galanie et al., 2015; Winzer et al., 2015). To aid these efforts, a number of large-scale initiatives have recently been established, supplying an exceptional amount of plant DNA sequence and transcriptome data to the public domain. The 1000 Plants Project (http://www.onekp.com) was established in 2008 with the aim of delivering transcriptomes for 1000 distinct plant species possessing notable biosynthetic capabilities. In a more focused initiative, the PhytoMetaSyn Project (http://www.phytometasyn.ca) has identified 75 nonmodel plant species that produce high-value terpenoids, alkaloids, and polyketides (Facchini et al., 2012; Xiao et al., 2013). BIAs represent one of six subgroups of metabolites targeted by the PhytoMetaSyn Project. Collectively, these resources provide a basis for assembling complex BIA biosynthetic pathways in yeast by delivering publicly accessible transcriptome databases rich in BIA biosynthetic diversity.

Since enzymes from every step of BIA biosynthetic networks have been elucidated and characterized in recent years, canonical cDNAs can be used as queries to assemble vast libraries comprised of orthologous enzyme variants. Candidate cDNAs are first identified and selected on the basis of DNA or protein sequence similarity. Plant-specific transcriptome archives, such as The 1000 Plants Project and the PhytoMetaSyn Project, or traditional databases, such as GenBank, can be queried using full-length cDNA or protein sequences. Alternatively, if detailed structural or mechanistic knowledge is available for target BIA biosynthetic enzymes, specific protein domains or conserved active site motifs can be used as queries with the aim of identifying variants employing similar reaction mechanisms. Full-length orthologous enzyme variants can then be compared using sequence alignment software. Candidate enzyme variants identified in this manner are then utilized in additional rounds of bioprospecting to enhance library diversity. Once a library of promising BIA biosynthetic variants has been compiled, a phylogenetic tree can be constructed to provide a visual representation of diversity. cDNA corresponding to each enzyme variant is then codon-optimized, synthesized by a DNA synthesis vendor, and reconstituted in yeast for expression and functional characterization using the methods described in the ensuing sections (Sections 3 and 4).

3. YEAST FUNCTIONAL EXPRESSION STRATEGIES

Owing to the rapidly declining cost of DNA synthesis, expression of heterologous genes and pathways in yeast has become an increasingly high-throughput endeavor. Extensive libraries of gene variants can be purchased from DNA synthesis vendors and assayed for functionality in parallel, a strategy referred to as bioprospecting. Different haploid strains have been employed for reconstituting BIA pathways in *S. cerevisiae*. These strains include CEN.PK2-1D (*MATα ura3-52 trp1-289 leu2-3_112 his3-Δ1 MAL2-8C SUC2*) (Fossati et al., 2014, 2015; Galanie et al., 2015), W303α (*MATα leu2-3,112 trp1-1 can1-100 ura3-1 ade2-1 his3-11,15*) (Galanie et al., 2015; Thodey et al., 2014), and BY4741 (*MATa his3Δ1 leu2Δ0 met15Δ0 ura3Δ0*) (DeLoache et al., 2015). In this section, we outline fundamental considerations associated with functional expression of heterologous enzymes and pathways in *S. cerevisiae*, including procedures for assessing BIA enzyme expression, localization, and activity (Fig. 2).

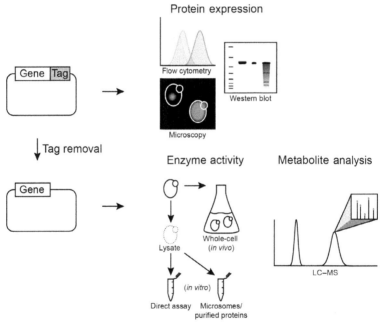

Fig. 2 Overview of technical protocols outlined in this chapter. Target genes cloned as translational fusion proteins with fluorescent tags that can be used to monitor protein expression with techniques which include flow cytometry, fluorescence microscopy, and immunoblotting (Western blotting). Fluorescent tags are removed from target gene constructs for analysis of enzyme activity. Strains expressing target biosynthetic genes can be supplemented with BIA intermediates for in vivo analysis of pathway functionality. Alternatively, following cell lysis, crude cell lysates, partially purified yeast microsomes, or highly purified enzyme preparations can be analyzed for enzyme functionality in vitro. LC–MS is the preferred technique for monitoring reaction products of both in vivo and in vitro assays. (See the color plate.)

3.1 Genetic Considerations

Although many downstream BIA biosynthetic pathways are found exclusively in plants (eg, pathways from (S)-norcoclaurine to sanguinarine, noscapine, and morphine), genes involved in upstream pathways and BIA derivatization have been imported from a diverse array of organisms, including bacteria and mammals (Galanie et al., 2015). To ensure high levels of gene expression, it is important that coding sequences are codon-optimized for yeast during gene synthesis (Sharp, Tuohy, & Mosurski, 1986). Genes can be synthesized as modular expression cassettes, containing a promoter and transcriptional terminator, or coding sequences can be inserted into common plasmids to generate gene expression cassettes (Jansen, Wu, Schade, Thomas, & Whiteway, 2005). Alternatively, the powerful homologous recombination capability of yeast can be harnessed for in vivo assembly of promoters, terminators, and target BIA biosynthetic genes (Shao, Zhao, & Zhao, 2009). To promote high-level expression of heterologous genes, a large collection of constitutive yeast promoters and expression-enhancing terminators have been characterized (Curran, Karim, Gupta, & Alper, 2013; Sun et al., 2012). Common promoters utilized specifically for expression of BIA biosynthetic enzymes in yeast include ones from the yeast cytochrome C (P_{CYC1}), pyruvate decarboxylase (P_{PDC1}), glyceraldehyde-3-phosphate dehydrogenase (P_{TDH3}), fructose 1,6-bisphosphate aldolase (P_{FBA1}), alcohol dehydrogenase (P_{ADH1}), phosphoglucose isomerase (P_{PGI1}), and 3-phosphoglycerate kinase (P_{PGK1}) genes (Fossati et al., 2014, 2015; Galanie & Smolke, 2015).

BIA gene expression cassettes and full biosynthetic pathways can be expressed from autonomous plasmids or integrated into the yeast genome. While plasmid-borne expression embodies the traditional approach, plasmid-free expression systems have become the favored platform for synthetic biology, largely due to major advancements in genome editing. Presently, a number of yeast BIA production studies have incorporated a combination of plasmid and chromosomal expression approaches (DeLoache et al., 2015; Fossati et al., 2014; Galanie & Smolke, 2015). Nevertheless, the discovery and implementation of robust genome engineering technologies, especially the bacterial CRISPR–Cas9 system, have revolutionized genome editing in a plethora of eukaryotic organisms (Hsu, Lander, & Zhang, 2014) and vastly simplified chromosomal integration of complex multigene parts. The CRISPR–Cas9 system has proven to be especially powerful for genome editing and metabolic engineering of *S. cerevisiae*

(DiCarlo et al., 2013; Horwitz et al., 2015; Jakočiūnas, Bonde, et al., 2015). As CRISPR–Cas9 editing continues to be expanded and refined, the construction of chromosomal expression platforms will become increasingly user friendly (Doudna & Charpentier, 2014; Jakočiūnas, Jensen, & Keasling, 2015).

3.2 Enzyme Expression and Localization

Following introduction of a heterologous gene into yeast cells, it is recommended that expression of the gene be confirmed through detection of the corresponding protein product. While enzyme activity is sometimes assayed directly following expression of a target gene using the methods outlined in Section 3.3, it is preferred to confirm enzyme expression prior to assaying activity. An array of techniques are available for analyzing expression of heterologous proteins in yeast. In Sections 3.2.1–3.2.3, we detail three common techniques, namely flow cytometry, fluorescence microscopy, and immunoblotting (Fig. 2). We also provide representative data obtained in our laboratory using each of these techniques in the context of yeast BIA production (Fig. 3).

Analysis of gene expression using flow cytometry and microscopy involve the use of fluorescent fusion proteins. Here, we briefly outline the construction of translational fusions of target BIA biosynthetic proteins with common fluorescent protein tags. Subsequent evaluation of culture fluorescence facilitates simple screening of protein expression in yeast, as an absence of fluorescence signals poor protein expression, thus precluding further analysis. In this regard, it is generally assumed that addition of a small fluorescent protein tag does not affect expression of the target protein (Snapp, 2005). We focus later on the green fluorescent protein (GFP), which has found widespread utility as a gene reporter and protein fusion tag. For use of alternative fusion tags, we refer readers to a comprehensive study by Lee, Lim, and Thorn (2013) detailing the utilization of 19 distinct fluorescent tags in *S. cerevisiae*. Tagged proteins are employed to first assess enzyme expression and, if successful, untagged variants are constructed for analysis of functionality and assembly of BIA biosynthetic pathways in yeast. A procedure for excision of fusion tags using our in-house pBOT expression vector system (discussed later) is included in Section 3.2.2 and is employed following confirmation of enzyme expression using flow cytometry and/or fluorescence microscopy.

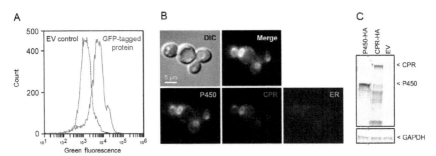

Fig. 3 Representative data validating heterologous protein expression and localization in *S. cerevisiae*. (A) Detection of a heterologous C-terminal GFP-tagged cytochrome P450 reductase (CPR) by flow cytometry. Yeast cells were grown in synthetic complete medium at 30°C and 200 rpm for 4 h and analyzed for GFP fluorescence using a BD Biosciences Accuri C6 flow cytometer. Expression of the GFP-tagged CPR from a pBOT vector was observed through a shift toward higher green fluorescence (excitation 488 nm; emission 510/15 nm), compared to the empty vector (EV) control. (B) Colocalization of a heterologous plant cytochrome P450 and its cognate CPR partner to the endoplasmic reticulum (ER). The C-terminal GFP-tagged cytochrome P450 and the C-terminal mCherry-tagged CPR were coexpressed using two compatible pBOT plasmids. The ER-Tracker Blue-White DPX dye was used to visualize the ER. Superposition of fluorescence images was used to visualize cytochrome P450 and CPR colocalization. Images were visualized at 100× magnification and 700 ms exposure using a Nikon Eclipse Ti epifluorescence inverted microscope and captured using a Nikon high-resolution (4908 × 3264 pixels) color camera DsRi2 CMOS. *Red* mCherry signals were excited with LED 555 nm and filtered with the TxRed filter cube; *green* GFP signals were excited with LED 480 nm and filtered with the CFP/YFP filter cube; *blue* ER-Tracker Blue-White DPX dye signal was excited with LED 405 nm and filtered with the DAPI filter cube. Differential interference contrast (DIC) is shown in the *top left panel*. (C) Immunoblot of an HA-tagged plant cytochrome P450 (P450-HA) and its cognate HA-tagged CPR (CPR-HA) expressed independently in *S. cerevisiae* using compatible pBOT plasmids. Samples were collected from cultures grown in synthetic complete medium at 30°C and 200 rpm for 24 h. Approximately 2×10^7 cells were pelleted and suspended in 50 μL of lysis buffer (50 mM Tris–HCl, pH 7.4, containing 150 mM NaCl, 2 mM MgCl$_2$, and 0.1% Nonidet P-40) plus 50 μL of 2× Laemmli buffer. Samples were boiled for 5 min, resolved by SDS-PAGE (30 μL/lane), and transferred to a nitrocellulose membrane for detection of the HA-epitope using the mouse anti-HA antibody HA-C5 (Abcam). The EV control was added to exclude nonspecific binding of the anti-HA antibody. Glyceraldehyde 3-phosphate dehydrogenase (GAPDH) was used as loading control and probed using rabbit anti-GAPDH (Rockland Immunochemicals). Imaging was performed using an Odyssey imager and IRDye® secondary antibodies (LI-COR Biosciences). (See the color plate.)

Our custom pBOT vector series, derived from pGREG *E. coli–S. cerevisiae* shuttle vectors (Jansen et al., 2005), is designed to facilitate both expression and functional analysis of target enzymes (Fig. 2). Any gene can be inserted into the pBOT *Sap*I restriction enzyme recognition sites using standard PCR and restriction endonuclease cloning, resulting in a translational fusion with a common fluorescent protein (ie, GFP) (Fig. 4). A 36-nucleotide linker fuses the target coding sequence in-frame with the fluorescent tag. The use of the Type IIS *Sap*I restriction endonuclease enables scarless cloning of target genes, as the enzyme cleaves DNA within user-defined nucleotide sequences. The pBOT vectors have been designed such that ligation reconstitutes a functional Kozak sequence (5′-AAACA-3′) at the 5′ end of the target gene, which is followed by the 5′-AUG-3′ start codon. Hence, to insert a target gene into the pBOT vector series, the following DNA product must be generated via PCR or gene synthesis (the *Sap*I core recognition sites are underlined):

5′-GCTCTTCTACA-[target gene coding sequence without stop codon]-GGCTGAAGAGC-3′

Following validation of protein expression, the linker and fluorescent marker can be removed via *Kas*I digestion and intramolecular ligation (Fig. 4). The *Kas*I sites flanking the GFP gene can also be utilized for inserting alternative fluorescent tags. Each pBOT vector contains a unique auxotrophic marker (*LEU2*, *URA3*, *HIS3*, or *TRP1*) and promoter–terminator pair to facilitate plasmid cotransformation and construction of multigene assemblies via homology-mediated cloning techniques. Finally, if many gene variants are purchased in parallel, the construction of fluorescent tagged variants can be expedited by supplying DNA synthesis vendors with the target pBOT plasmid. Target proteins are then supplied within the desired destination vector, eliminating the need for further manipulations prior to yeast transformation.

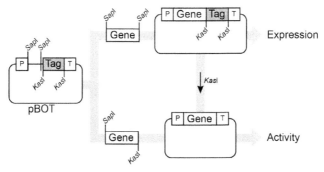

Fig. 4 Custom pBOT series of *S. cerevisiae* expression vectors. Target genes are cloned into the pBOT series of vectors for analysis of expression and enzyme functionality. Genes can be cloned as fluorescent fusion proteins using *Sap*I. Following validation of protein expression, fluorescent tags are removed by *Kas*I digestion and plasmid recircularization. Alternatively, the gene of interest can be cloned directly without a fluorescent tag using *Sap*I and *Kas*I. Abbreviations: *P*, promoter; *T*, terminator. (See the color plate.)

3.2.1 Flow Cytometry

Flow cytometry is particularly advantageous for high-throughput analysis of gene expression, as diverse enzyme libraries comprised of dozens or even hundreds of variants can be screened in a single day. In this section, we provide two procedures for assessing fluorescence of BIA biosynthetic fusion proteins using flow cytometry. The first protocol exemplifies a typical workflow for quantitative analysis of culture fluorescence. Cells are inoculated into growth medium, cultivated to exponential phase, and washed in sterile water prior to flow cytometry analysis. This procedure facilitates accurate determination and comparison of mean culture fluorescence (Fig. 3A). Conversely, we provide a condensed protocol for rapidly assessing fluorescence of single colonies using flow cytometry. We have found that it is unnecessary to cultivate and wash cells to distinguish GFP-expressing colonies from nonfluorescent ones. Hence, the modified protocol provides a rapid means of profiling colonies for GFP expression by picking colonies directly from selective agar plates and suspending in sterile water. Although this method does not provide an accurate determination of mean culture fluorescence, it is sufficient for identifying GFP-positive colonies. The following protocols can be performed using microcentrifuge tubes, 12-strip PCR tubes, or 96-well microtiter plates depending on the number of enzymes to be screened.

3.2.1.1 Protocol: Quantitative Determination of Mean Culture Fluorescence Using GFP-Tagged Proteins and Flow Cytometry

1. Inoculate yeast colonies in triplicate into 150 µL of selective Yeast Nitrogen Base (YNB) medium with 2% glucose. Include a negative control strain lacking a GFP tag and a positive control strain previously validated for GFP fluorescence using flow cytometry.
2. Grow strains at 30°C and 400 rpm for 16–20 h.
3. The following day add 900 µL of selective YNB to the cultures.
4. Continue growing strains at 30°C and 400 rpm for approximately 4 h.
5. Centrifuge cultures at $3200 \times g$ for 5 min.
6. Carefully aspirate medium from each tube or well using a Pasteur pipette.
7. Suspend cells in 1 mL of sterile water.
8. Transfer 100 µL of cells to sterile tubes, PCR strips, or microtiter plates.
9. Incubate at room temperature for 1 h.
10. Vortex tubes to ensure even mixing immediately prior to performing flow cytometry analysis.

11. Flow cytometry analysis is performed using a BD Biosciences Accuri C6 flow cytometer. Control strains are utilized to set side and forward scatter gates, as well as autofluorescence gates. The Green signal for GFP is monitored at an excitation and emission of 488 and 533 nm, respectively. Mean fluorescence from 50,000 cell events is measured to compare sample fluorescence.

3.2.1.2 Protocol: Rapid Identification of Fluorescent Colonies Expressing GFP-Tagged Proteins Using Flow Cytometry

1. Streak out one transformant colony on selective YNB agar plates. Ensure streaking sufficiently dilutes cells to give distinct, isolated colonies. Include a negative control strain lacking a GFP tag and a positive control strain previously validated for GFP fluorescence using flow cytometry.
2. Incubate plates for 2–3 days at 30°C (colonies should be large enough to pick). We recommend using only fresh colonies for the rapid flow cytometry protocol.
3. Pick three colonies (ie, three replicates) from each streak plate and suspend in 100 μL of sterile water. Use a sterile P200 micropipette tip to gently touch colonies (do not scoop full colonies). Cell suspensions should be cloudy.
4. Vortex tubes to ensure even mixing immediately prior to performing the flow cytometry analysis.
5. Perform flow cytometry analysis as outlined earlier. Colony fluorescence is evaluated based on measurements from 10,000 cell events.

3.2.2 Fluorescence Microscopy

Fluorescence microscopy is advantageous for analyzing expression of target proteins, as insights can also be gained into enzyme localization within yeast cells. For example, fluorescence provides a means of assessing localization of cytochrome P450s to the endoplasmic reticulum. Such a strategy was employed to analyze expression and localization of a novel cytochrome P450 tyrosine hydroxylase utilized for de novo biosynthesis of (S)-reticuline through C-terminal fusion to a yellow fluorescent protein (DeLoache et al., 2015). If a cytochrome P450 and its cytochrome P450 reductase (CPR) partner are tagged with different fluorescent tags, microscopy can be utilized to determine colocalization of the protein partners to the endoplasmic reticulum (Fig. 3B). In these instances, organelle-specific yeast tracker dyes are used to assess protein localization to various subcellular compartments (Wolinski & Kohlwein, 2014). For an in-depth discussion of protein colocalization by fluorescence microscopy, we suggest the review by Dunn, Kamocka, and McDonald (2011). As with flow cytometry, target proteins are first tagged through fusion with fluorescent proteins. Plasmid-borne expression is commonly employed to assess protein localization. However, chromosomal integration of tagged proteins tends to yield more uniform expression across a population of cells (Ryan et al., 2014). While microscopy is generally a low-throughput and time-intensive technique, strategies for adapting microscopy for high-throughput library screening have recently been developed (Chen et al., 2015). Here, we provide a general protocol for analysis of BIA enzyme expression and localization using fluorescence microscopy of GFP protein fusion constructs.

3.2.2.1 Protocol: Analysis of BIA Enzyme Expression and Localization in Yeast Using Fluorescence Microscopy

1. Inoculate yeast colonies in triplicate into 3 mL of the appropriate selective synthetic complete medium with 2% glucose. Include a negative control strain lacking a GFP tag and a positive control strain previously validated for GFP fluorescence using microscopy.
2. Grow strains at 30°C and 200 rpm for 16–20 h.
3. Inoculate 300 μL of overnight cultures into 3 mL of the appropriate selective synthetic complete medium with 2% glucose.
4. Grow cultures for 4 h at 30°C and 200 rpm (approximately two cell doublings).
5. Centrifuge 500 μL of cultures at $4000 \times g$ for 5 min.
6. Pour off supernatant and wash cells once with water or phosphate-buffered saline (PBS) and suspend in 100 μL of water for visualization.
7. Spot 9 μL of cultures (unfixed) onto plain glass slides for visualization.
8. A range of dyes have been developed for assessing subcellular localization to specific organelles (Johnson & Spence, 2010). For example, the dye ER-Tracker Blue-White DPX blue is employed to evaluate localization to the endoplasmic reticulum. This dye is added to the growth medium at 1 μM after 4 h of growth, followed by incubation for 1 h at 30°C and 200 rpm. Cells are then washed twice with PBS to remove residual dye and suspended in water for visualization. Visualization is performed using a Nikon Eclipse Ti epifluorescence inverted research microscope. Appropriate excitation LED/emission filters are chosen for each fluorophore. For example, the green GFP signal is excited with LED 480 nm and filtered with the fluorescence filter cube for CFP/YFP, while ER-staining using the ER-Tracker Blue-White DPX dye signal

is excited with LED 405 nm and visualized with the DAPI filter cube. Images are taken at 100× magnification and are captured with a Nikon high-resolution (4908 × 3264 pixels) color camera DsRi2 CMOS.
Following confirmation of target protein expression and localization using flow cytometry and/or fluorescence microscopy, it is routine to remove the fluorescent tag prior to assaying enzyme activity. Fluorescent tags can be excised from our custom pBOT vector constructs using traditional restriction enzyme cloning. Proteins with validated GFP fluorescence can then be tested for activity as described in Section 3.3 (Fig. 2). The GFP tag is excised from the vector by *Kas*I digestion, which is followed by circularization of the resulting vector backbone via intramolecular ligation. Alternatively, untagged constructs can be PCR-amplified and cloned directly into the *Kas*I and *Sap*I recognition sites of the appropriate pBOT vector.

3.2.3 Immunoblotting

Immunoblots, commonly referred to as Western blots, are widely employed to assess expression of heterologous proteins. In the context of yeast BIA production, immunoblots have been utilized to confirm expression of a number of key BIA biosynthetic enzymes in yeast, including salutaridine synthase and a CPR (Fig. 3C; Fossati et al., 2015). A common approach entails the creation of a translational protein fusion consisting of the target protein expressed with a small epitope tag (Kolodziej & Young, 1991). The protein of interest is then detected through recognition by an antibody specific to the tag. One of the most widely employed epitopes is the human influenza hemagglutinin (HA) tag comprised of the amino acid sequence YPYDVPDYA, required for recognition by the anti-HA antibody (Wilson et al., 1984). Given the longstanding history and popularity of immunoblotting in protein biology, laboratory procedures are highly standardized (Johnson, Gautsch, Sportsman, & Elder, 1984) and, therefore, are not detailed in this chapter. Instead, readers are directed to useful resources describing immunoblotting for confirmation of protein expression in yeast systems (Antebi & Fink, 1992; Fossati et al., 2015; Shen, Selvakumar, Stanford, & Hopper, 1993; Sherwood, Tsang, & Osley, 1993). While immunoblotting provides insights into protein molecular weight and potential protein degradation issues, the technique is low throughput and not amenable to screening large enzyme libraries. Dot-blot immunodetection, on the other hand, is an efficient alternative to traditional Western blots for high-throughput evaluation of protein expression using antibodies specific to common epitope tags (Zeder-Lutz, Cherouati, Reinhart, Pattus, & Wagner, 2006).

3.3 Enzyme Activity

Several techniques are available for evaluating activity of heterologous enzymes. With the exception of a robust biosensor for assaying production of the BIA precursor L-3,4-dihydroxyphenylalanine (L-DOPA) (DeLoache et al., 2015) (refer to Section 4.2.2), there are no examples of truly high-throughput screens for evaluating BIA biosynthesis in microbes. Presently, activity of BIA biosynthetic enzymes is assessed using several low- or medium-throughput techniques. In this section, we present methods for evaluating activity of BIA biosynthetic enzymes using in vitro or in vivo assays (Fig. 2). Whereas in vitro enzyme assays employ purified, partially purified, or crude enzyme preparations, in vivo analysis involves whole cells as biocatalysts. In both instances, assays are performed using BIA pathway intermediates as substrates and enzyme activity is evaluated based on liquid chromatography–mass spectrometry (LC–MS) analysis of assay products (Section 3.3.3). BIA enzymatic assays may be performed to evaluate a single enzyme, multiple enzymes expressed in the same yeast strain, or multiple enzyme preparations incubated simultaneously or sequentially to simulate a reconstituted pathway. BIA biosynthetic enzymes validated for expression, localization, and functionality are then utilized for reconstituting enzyme blocks or full BIA biosynthetic pathways in yeast (Section 4).

3.3.1 In Vitro Enzyme Assays (Purified Proteins, Microsomal Preparations, or Crude Cell Lysates)

Purified enzyme preparations offer the most accurate determination of enzymatic activity and provide the basis for the measurement of enzyme kinetics. Enzyme purification is most commonly accomplished through the addition of a small epitope tag to the target protein followed by isolation via solid-state purification using an appropriate column. Conversely, enzymes localized to the endoplasmic reticulum, such as cytochrome P450s and CPRs, can be selectively isolated via microsomal purification. Owing to the abundance of cytochrome P450s in plant secondary metabolism, microsomal assays are often a critical facet of BIA production strategies. Accordingly, we refer the reader to the seminal paper by Pompon et al. (1996) for an in-depth overview of yeast microsome preparation. The simplest and perhaps most attractive alternative to purified enzymes for assaying activity of heterologous enzymes is crude cell lysates. In this regard, cultures expressing target proteins can be lysed and added directly to in vitro enzyme assays without additional treatments or purifications. This method substantially simplifies the screening of heterologous proteins for functional activity, but suffers from an inability to control specific assay variables, such as concentration of the target protein and the presence of interfering proteins and metabolites. Regardless of the source of the target protein, in vitro enzymatic assays necessitate efficient lysis of host cells. Numerous techniques are available for breaking yeast cell walls, including enzymatic digestion, bead beating, the French press, and homogenization. Due to its low cost and overall simplicity, we outline a procedure below for bead beating. Additionally, we provide a brief overview of assaying cytochrome P450s in vitro using yeast microsomal preparations.

3.3.1.1 Protocol: Preparation of Yeast Crude Cell Lysates, Microsomal Preparations, and Purified Proteins

1. Inoculate target yeast colonies into 3 mL of the appropriate selective synthetic complete medium with 2% glucose. Include a negative control strain lacking the target protein to be assayed and, if available, a positive control strain previously validated for functional activity of the target protein or metabolic pathway.
2. Grow strains at 30°C and 200 rpm for 16–20 h.
3. Inoculate overnight cultures into the appropriate selective synthetic complete medium with 2% glucose to achieve an initial OD_{600} of approximately 0.1.
4. Grow cultures to early logarithmic phase (typically 4–6 h) at 30°C and 200 rpm. Incubation time can vary and is dependent on the target protein. For example, prior time-course analyses of target protein expression and stability using immunoblot or flow cytometry (Section 3.2) can provide insights into growth and cultivation conditions for optimization of target protein production.
5. Harvest cells by centrifugation at $3200 \times g$ for 5 min.
6. Wash cells once with a chosen buffer (eg, 100 mM Tris–HCl, pH 7.5). The wash buffer should be the same buffer as the enzymatic reaction buffer.
7. Add 100 μL of acid-washed glass beads (400 μm).
8. Suspend cells in 900 μL of chosen buffer and 100 μL of $10 \times$ complete protease inhibitor cocktail (Roche). The ratio of cells:lysis buffer dictates the final concentration of proteins in the crude cell lysate. If the target protein is expressed at low levels, the total volume of buffer and protease inhibitor cocktail can be reduced. A higher concentration of protein is also desired in assays involving costly substrates in small reaction volumes.

3.3.1.2 Protocol: In Vitro Assay of BIA Biosynthetic Enzymes

Enzymes involved in BIA biosynthesis have been characterized in a variety of buffers in the range of pH 7–9, including glycine (Rueffer, Zumstein, & Zenk, 1990), HEPES (Ikezawa, Iwasa, & Sato, 2008), phosphate buffer (Chávez, Rolf, Gesell, & Kutchan, 2011), and Tris–HCl (Minami, Dubouzet, Iwasa, & Sato, 2007). A characteristic protocol for assaying cytochrome P450 activity using microsomal preparations follows:

1. In a final volume of 50 μL, add the following reaction components:
   - 20 μg microsomal preparation containing a cytochrome P450 and CPR
   - 50 mM HEPES–NaOH, pH 7.6
   - 0.5 mM NADPH
   - 125 μM substrate
   - H$_2$O to 50 μL
2. Incubate for 90 min at 30°C.
3. Stop the reaction by precipitating proteins with 100 μL of methanol.
4. Centrifuge reactions at 15,000 × g for 5 min.
5. Analyze supernatants for BIA products using LC–MS.

3.3.2 In Vivo Cell-Feeding Assays

Cell-feeding assays represent the most powerful tool for assessing BIA pathway functionality in yeast. Since only a few select BIA pathway intermediates can be obtained commercially (eg, norlaudanosoline, reticuline, salutaridine, scoulerine, stylopine, thebaine, morphine, codeine, and oripavine), cell-feeding assays provide a means of monitoring flux through various pathway checkpoints dictated by availability of BIA intermediates. Cell-feeding assays are particularly advantageous for assaying many strains in parallel, as throughput can be enhanced by performing assays in 96-well microtiter plate format. However, in vivo assays involving whole-cell biocatalysts produce challenges associated with the uptake of supplemented BIA substrates and fractioning of substrate and products into cell and supernatant fractions. Later, we provide a procedure for BIA substrate feeding assays based on an optimized two-stage strategy whereby cell growth and BIA production are spatially separated. BIA-producing strains are first grown in glucose medium devoid of BIA pathway intermediates. Once cultures reach a sufficient density, cells are centrifuged and suspended in BIA assay buffer at pH 8 (eg, 10 mM Tris–HCl, pH 8, containing 1 mM EDTA). While a pH of 8 contrasts the pH of growing yeast cultures (typically pH 3–5), pH 8 does not affect yeast viability (Fossati et al., 2015) (refer to

Section 4.2.4 for additional discussion). Furthermore, an increased proportion of many supplemented BIAs is associated with the cellular fraction with increasing pH, indicating more efficient substrate uptake. Percent conversion of BIA substrates to products is also enhanced at elevated pH.

The protocol below describes cell-feeding assays in 96-well plate format using 2 mL deep-well plates. In addition, three 300 µL 96-well polypropylene plates are required for storage of extracted BIA samples per 96-sample deep-well plate employed. BIA reaction products are subsequently characterized using LC–MS analysis (Section 3.3.3). In total, the protocol spans approximately 4 days from inoculation of yeast cultures through to LC–MS analysis. For lower throughput analysis, the procedure can be modified using culture tubes for cell growth (1–3 mL working volume) and microcentrifuge tubes for BIA extraction.

3.3.2.1 Protocol: In Vivo Cell-Feeding Assays

1. Inoculate yeast colonies in triplicate into 150 µL of selective YNB medium. Include a negative control strain containing the same host genotype without BIA biosynthetic genes and a positive control strain previously characterized or one expressing the target BIA biosynthetic genes from a plasmid (for chromosomally expressed pathways).
2. Grow strains at 30°C and 400 rpm for 16–20 h.
3. The next day add 900 µL of selective YNB to the cultures.
4. Incubate at 30°C and 400 rpm for roughly 5 h.
5. Pellet cells in the 96-well plate by centrifugation at $3200 \times g$ for 10 min.
6. Aspirate medium from each well without disturbing the cell pellet using a Pasteur pipette.
7. Suspend cell pellets in 400 µL of TE (10 mM Tris–HCl, pH 8, and 1 mM EDTA) supplemented with 5 µM of the appropriate BIA feeding substrate. Stock solutions of BIA substrates are prepared at 10–30 mM in an appropriate solvent.
8. Incubate cultures at 30°C and 400 rpm for approximately 16 h.
9. The next day pellet cells in the 96-well plate by centrifugation at $3200 \times g$ for 10 min.
10. Using a multichannel pipette, remove 200 µL of the supernatant and reserve in a 300 µL polypropylene 96-well plate (this constitutes the supernatant fraction).
11. Centrifuge the 96-well plate containing the cultures again at $3200 \times g$ for 10 min.
12. Aspirate medium from each well without disturbing the cell pellet.

13. Suspend cells in 400 µL of methanol and cover the 96-well plate with an aluminum seal to reduce evaporation.
14. Incubate the 96-well plate at 4°C and 950 rpm for 30 min.
15. Centrifuge at $3200 \times g$ for 10 min.
16. Remove 200 µL of the supernatant and reserve in a 300 µL polypropylene 96-well plate (this constitutes the cell pellet fraction).
17. Centrifuge the supernatant fraction at $3200 \times g$ for 10 min (a pellet will not be visible).
18. Meanwhile, aliquot 50 µL of methanol to each well of a new 300 µL polypropylene 96-well plate.
19. Transfer 50 µL of supernatant from the supernatant fraction plate to the new 96-well plate containing 50 µL of methanol (this constitutes a 1:1 supernatant fraction:methanol mix).

3.3.3 LC–MS Analysis of In Vitro and In Vivo Assay Products

LC–MS is employed to analyze BIA pathway intermediates and end products generated through both in vitro and in vivo assays. While alternative techniques such as nuclear magnetic resonance spectroscopy can be used, MS has the advantage of a low limit of detection (generally <1 nM). A low limit of detection also ensures that costly BIA substrates can be used in low concentrations in supplementation assays.

Prior to MS analysis, compounds are first separated on a reverse-phase liquid chromatography column, such as a C18 column, which serves two purposes. First, chromatography ensures that salts and other impurities are removed prior to MS analysis. Second, compounds elute at characteristic times depending on factors that include elution gradient and flow rate, which can be exploited for identification. Reverse-phase C18 columns separate a variety of BIAs based on polarity, as polar compounds elute prior to nonpolar ones. Since BIA enantiomers coelute from a C18 column, chiral chromatography is required in applications demanding separation of enantiomer mixtures. However, the ability of a given chiral chromatography column to separate enantiomers varies dramatically and must be determined empirically. Furthermore, the ability of a column to resolve enantiomers of one BIA does not guarantee that it can be employed for chiral analysis of even closely related BIAs. Nevertheless, two chiral chromatography columns have proven effective for separation of BIA enantiomers: the CHIRALCEL OD-H (4.6×250 mm; Daicel Chemical Industries) reliably separated (R)- and (S)-reticuline (Nakagawa et al., 2011), while the Shodex ORpak CDBS-453 column (FUTECS Chromatography) has proven effective for separation of (R)- and (S)-norcoclaurine (DeLoache et al., 2015). Following elution from the column, samples are injected directly into a mass spectrometer. If the mass spectrometer is accurate enough to determine the exact mass of the ion, the chemical formula of the ion can be elucidated directly. In some instances, this analysis is sufficient for determining the identity of a particular compound. If exact mass and elution times do not permit accurate determination of ions, MS/MS analysis can be employed. Helium bombardment generates characteristic ion transitions that can be detected using single reaction monitoring. This technique necessitates prior determination of ion transitions using pure standards, which can be difficult to obtain. Later, we provide experiment details validated for LC–MS detection of key BIA biosynthetic pathway intermediates and products.

3.3.3.1 Protocol: Conditions for Separation of BIA Assay Intermediates and Products Using Liquid Chromatography

BIAs can be separated on a Perkin Elmer Series 200 Micropump with an Agilent Zorbax Rapid Resolution HT C18 column (2.1×30 mm, 1.8 μm) using the following solvents and elution profile: solvent A, 0.1% formic acid; solvent B, 100% acetonitrile, 0.1% formic acid; gradient: 0–2 min at 5% B, 2–10 min at 5–100% B (linear gradient), 10–11 min at 100% B, 11–11.1 min at 100–5% B, 11.1–13 min at 5% B; flow rate of 0.25 mL/min.

3.3.3.2 Protocol: Conditions for Detection of BIA Assay Intermediates and Products Using Mass Spectrometry

Samples eluted from liquid chromatography can be analyzed with a 7T-LTQ FT ICR from Thermo Scientific, which allows for the analysis of exact mass of positively charged ionized BIAs with an accuracy of <2 ppm, under the following conditions: source temperature, 350°C; capillary voltage, 5 kV; scanning range, 250–450 AMU; resolution, 100,000.

4. PATHWAY RECONSTITUTION AND OPTIMIZATION

Reconstitution of a full BIA biosynthetic pathway embodies the coordinated expression and concerted activity of multiple heterologous enzymatic activities within a single yeast host. Given the complexity of plant secondary metabolism, yeast-based BIA production strategies to date have incorporated up to 23 nonnative activities from a wide range of organisms, including plants, bacteria, and mammals (Galanie et al., 2015). In this section, we describe strategies for building complex heterologous BIA biosynthetic pathways using a stepwise enzyme block strategy (Fossati et al., 2014, 2015). We also provide common genetic and bioprocessing strategies for alleviating pathway bottlenecks and optimizing flux through the target BIA production pathway.

4.1 Pathway Assembly

Rather than reconstituting an extensive target BIA biosynthesis pathway in a single host strain, it has proven advantageous to first partition pathways into smaller tractable blocks of sequential enzymatic activities (Fossati et al., 2014, 2015). Given that many BIA pathway intermediates are not available commercially, pathway blocks should be separated on the basis of availability of feeding substrates. For example, within the morphinan biosynthesis branch from (R)-reticuline, commercially available feeding substrates include (R)-reticuline, salutaridine, thebaine, oripavine, and codeine (Fossati et al., 2014; Thodey et al., 2014). By employing this strategy, functionality of pathway blocks expressing multiple enzymes can be easily assessed using in vivo cell-feeding assays (Fossati et al., 2014). Blocks of enzymes validated for functionality are then employed for stepwise reconstitution of the full target pathway in a single host strain. This approach has proven effective in the production of sanguinarine from (R,S)-norlaudanosoline, in which commercially available (R,S)-reticuline and (S)-stylopine served as bridging intermediates using a three-block design (Fossati et al., 2014). Enzyme blocks were first assayed using plasmid-borne expression and subsequently incorporated into the yeast genome for full pathway reconstruction. Once reconstituted, pathways can be enhanced by identifying and resolving weak points in the pathway commonly referred to as metabolic bottlenecks.

4.2 Debottlenecking Strategies

Metabolic flux through a heterologous pathway is limited by the enzymatic step possessing the lowest catalytic efficiency. Such bottlenecks are easily identified and are manifested through a build-up of pathway intermediates, signaling poor substrate conversion. In this regard, cell-feeding assays have proven paramount for unveiling BIA pathway bottlenecks. Product yield can be calculated for each enzymatic step or enzyme block to identify putative bottlenecks as targets for improvement. Later, we discuss four key strategies that have proven indispensable in heterologous BIA production schemes. These strategies encompass fundamental genetic approaches, enzyme bioprospecting, enzyme evolution, and culture optimization.

4.2.1 Genetic Tuning

Genetic tuning represents the most fundamental strategy for improving pathway flux and product yield. Tuning can be achieved through manipulation of a range of genetic factors outlined in Section 3.1, including gene dosage and promoter and RBS strength. For plasmid-based expression approaches, gene dosage can be altered through swapping of plasmid replication origins. While the classical yeast 2 μ replicon affords high copy number, low copy alternatives include yeast centromere plasmids and yeast artificial chromosomes. It is important to note that high copy number does not necessarily correlate with optimal enzyme activity, as excessive gene dosing can have deleterious effects on host cell metabolism and plasmid stability. Hence, it is often advantageous to assay multiple expression strategies when employing plasmid-based strategies. In chromosome-encoded expression platforms, genes are first introduced as single copies and, if necessary, gene dosing can be enhanced through subsequent rounds of integration (Galanie & Smolke, 2015). Copy number is then titrated until the target bottleneck is alleviated or enhancement in product yield or conversion is no longer observed. Currently, techniques for enhancing gene copy number in chromosomal expression platforms are laborious, generally necessitating iterative rounds of gene integration. Consequently, robust techniques facilitating simultaneous integration of a gene cassette into multiple loci are paramount for resolving pathway bottlenecks by enhancing gene dosage. Multiplexed genome editing strategies have recently been reported for simultaneous integration of multiple genetic parts in *S. cerevisiae* using CRISPR–Cas9 (Horwitz et al., 2015; Jakočiūnas, Bonde, et al., 2015; Ryan et al., 2014). The application of these techniques in the context of yeast-based BIA production should expedite the construction of highly productive BIA-producing strains. An alternative strategy for mitigating metabolic bottlenecks embodies promoter selection, as up to a sixfold difference in enzyme activity can be achieved through promoter swapping (Sun et al., 2012).

4.2.2 Directed Enzyme Evolution

Directed enzyme evolution provides an empirical approach to alleviating pathway bottlenecks through protein engineering. Using this method, a vast library of gene variants is generated through random mutagenesis of the gene sequence encoding the target protein. Screening the resulting library is an arduous task often due to vast sequence diversity, necessitating high-throughput screening methods to identify variants exhibiting the greatest catalytic efficiency. Thus, debottlenecking strategies based on directed enzyme evolution hinge on the availability of highly sensitive screening assays. An example of such a screen in the context of yeast BIA production involves the use of an enzyme-coupled biosensor to monitor intracellular levels of L-DOPA, an early BIA precursor derived from tyrosine (DeLoache et al., 2015). It was observed that L-DOPA concentration could be precisely monitored through expression of a plant DOPA dioxygenase, whereby L-DOPA concentration was linked with production of betaxanthin, a yellow fluorescent pigment. Use of the resulting biosensor to screen both natural and mutagenized enzyme variants enabled increases in L-DOPA yield and dopamine production by 2.8- and 7.4-fold, respectively, demonstrating the power of highly sensitive screening assays for identifying improved BIA biosynthetic enzymes. At present, the L-DOPA-specific biosensor represents the only high-throughput screening tool for assaying large libraries of BIA enzyme variants. The development of analogous assays for use in screening downstream BIA metabolites represents a critical step toward enhancing BIA product yields. Since several end product BIAs, such as berberine (yellow) and sanguinarine (orange-red), are intensely colored compounds, promising opportunities exist for establishing high-throughput assays based on colorimetric screens. Unfortunately, current BIA end product titers preclude the implementation of such assays.

4.2.3 Bioprospecting of Enzyme Variants

Bioprospecting of orthologous genes from a wide array of plant species offers a valuable alternative to protein engineering and directed evolution. While enzyme variants corresponding to any enzymatic step within a target pathway can be screened to enhance catalytic efficiency, pathway bottlenecks present the most promising opportunities for enzyme bioprospecting. BIA biosynthetic gene variants can be easily identified by querying genome and transcriptome databases of BIA-producing plants, including the 1000 Plants Project and the PhytoMetaSyn Project described briefly in Section 2. Due to the mounting availability of DNA sequence and transcriptome data available within the public domain, enzyme bioprospecting offers vast opportunities for enhancing overall BIA pathway efficiency by tapping into nature's metabolic diversity.

4.2.4 Cultivation Conditions

A number of bioprocessing conditions have been investigated in BIA production schemes involving both precursor supplementation and de novo approaches (Fossati et al., 2015; Galanie & Smolke, 2015; Trenchard et al., 2015). These parameters include temperature, selection of growth substrate, and pH. Temperature has been shown to be a key factor in the production of some downstream BIAs, as activity of many plant cytochrome P450s is enhanced through cultivation at suboptimal growth temperatures. For example, stylopine production increased 3.4-fold through cultivation at 25°C compared to the typical temperature of 30°C, likely through enhanced enzyme stability at lower temperatures (Trenchard & Smolke, 2015). On the other hand, temperature had no effect on production of canadine and berberine (Galanie & Smolke, 2015). Selection of the growth substrate also affects BIA biosynthesis, as galactose outcompeted other common carbon sources during cofeeding with norlaudanosoline for conversion to stylopine, canadine, or berberine (Galanie & Smolke, 2015; Trenchard & Smolke, 2015). Presently, little is known about the effect of carbon source on BIA biosynthesis, highlighting the importance of characterizing BIA-producing strains under a range of bioprocessing conditions. However, it is clear that culture pH is one of the most important determinants of BIA production in yeast hosts. Optimal growth of yeast occurs at pH 4.5–5.5, although many strains exhibit an impressive tolerance to acidic environments spanning pH 3–7. This phenotype conflicts with conclusions drawn from BIA production studies (Fossati et al., 2014, 2015), as BIA conversion and derivatization are enhanced by increasing pH over a range of 3–8 (Narcross et al., 2016). Such a discrepancy in pH optima between yeast culture growth and plant BIA biosynthesis points to a two-stage production strategy, in which biomass accumulation and BIA production exemplify spatially distinct processes (Narcross et al., 2016). Indeed, procedures for cell-feeding assays described in Section 3.3.2 employ such two-stage processes, whereby cell growth occurs under standard conditions and is succeeded by BIA biosynthesis at an elevated pH (pH 8) (Fossati et al., 2014). Hence, yeast BIA bioprocesses should be characterized at a variety of pHs. We recommend pH in the range of 6–8 to maximize BIA conversion and limit loss of BIA intermediates and products to secretion.

5. CHALLENGES AND FUTURE DIRECTIONS

As heterologous BIA production strategies to date are confined to proof of concept, immense efforts will be required to boost BIA production titers first to laboratory scale and eventually to industrial scale. While scaling represents an important engineering challenge in itself, enhancing BIA production yields and harnessing enzyme promiscuity are more immediate production challenges. Despite major achievements in recent years, BIA product titers remain very low (DeLoache et al., 2015; Galanie et al., 2015; Trenchard et al., 2015). Present de novo BIA production titers in yeast include 0.3 and 6.4 µg/L for the downstream BIAs hydrocodone and thebaine, respectively (Galanie et al., 2015), and 19.2 (Trenchard et al., 2015) and 80.6 µg/L (DeLoache et al., 2015) for the upstream intermediate (S)-reticuline, highlighting poor catalytic efficiency of many downstream BIA enzymes. It has been estimated that titers of approximately 5 g/L, an increase by a factor of 10^6 based on current BIA levels, would be required for microbial BIA production to compete with traditional methods of BIA extraction (Galanie et al., 2015). Accordingly, low yield is arguably the most salient challenge facing microbial BIA production strategies. Improving BIA yields will involve dramatic enhancement of overall pathway efficiency through iterative rounds of systematic debottlenecking to improve catalytic efficiency at multiple enzymatic steps throughout the BIA biosynthetic pathway. It is evident that an integrated approach combining genetic, biochemical, and bioprocessing strategies will be required to elevate BIA yields to competitive levels. On the other hand, the inherent substrate promiscuity exhibited by many BIA biosynthetic enzymes exemplifies a much more insidious obstacle (Narcross et al., 2016), while spontaneous chemical rearrangements pose additional pathway challenges to heterologous BIA biosynthesis. Harnessing enzyme promiscuity will require a myriad of sophisticated interventions, including genetic tuning strategies to enhance activity of trunk pathway enzymes, protein engineering to suppress off-target activities, and spatial or temporal separation of promiscuous enzymes from pathway intermediates (Narcross et al., 2016). Dual spatiotemporal strategies have also been envisioned in the context of yeast BIA production (Thodey et al., 2014). Alternatively, natural enzyme variants exhibiting reduced off-pathway activities can be bioprospected and screened using methods described in this chapter. The inherent promiscuity exhibited by many BIA biosynthetic enzymes is testament to the complexity of plant secondary metabolism and, therefore, successful deployment of yeast-based BIA production platforms will incorporate innovative strategies for reinforcing productive pathways and limiting or entirely abolishing off-target reactions.

It is noteworthy that the BIA class of plant alkaloids is comprised of a large number of compounds that are controlled under international law. Several of these substances can be further modified to yield an array of highly addictive opioids, including heroin and oxycodone. Since opioid abuse is epidemic in many countries (Manchikanti et al., 2012), microbial BIA production platforms bring potential for illicit exploitation. Theoretically, BIA-producing yeast strains could be modified to convert simple sugars into recreational drugs using preexisting "home-brew" technology. Although present microbial BIA yields thwart the prospect of abuse, it is paramount that the ethical and societal implications of microbial BIA production be weighed well in advance of the deployment of industrial BIA bioprocesses.

REFERENCES

Allen, R. S., Millgate, A. G., Chitty, J. A., Thisleton, J., Miller, J. A., Fist, A. J., et al. (2004). RNAi-mediated replacement of morphine with the nonnarcotic alkaloid reticuline in opium poppy. *Nature Biotechnology*, *22*, 1559–1566.

Antebi, A., & Fink, G. R. (1992). The yeast Ca^{2+}-ATPase homologue, PMR1, is required for normal Golgi function and localizes in a novel Golgi-like distribution. *Molecular Biology of the Cell*, *3*, 633–654.

Barken, I., Geller, J., & Rogosnitzky, M. (2008). Noscapine inhibits human prostate cancer progression and metastasis in a mouse model. *Anticancer Research*, *28*, 3701–3704.

Beaudoin, G. A., & Facchini, P. J. (2013). Isolation and characterization of a cDNA encoding (S)-cis-N-methylstylopine 14-hydroxylase from opium poppy, a key enzyme in sanguinarine biosynthesis. *Biochemical and Biophysical Research Communications*, *431*, 597–603.

Beaudoin, G. A., & Facchini, P. J. (2014). Benzylisoquinoline alkaloid biosynthesis in opium poppy. *Planta*, *240*, 19–32.

Bessi, I., Bazzicalupi, C., Richter, C., Jonker, H. R., Saxena, K., Sissi, C., et al. (2012). Spectroscopic, molecular modeling, and NMR-spectroscopic investigation of the binding mode of the natural alkaloids berberine and sanguinarine to human telomeric G-quadruplex DNA. *ACS Chemical Biology*, *7*, 1109–1119.

Chávez, M. L. D., Rolf, M., Gesell, A., & Kutchan, T. M. (2011). Characterization of two methylenedioxy bridge-forming cytochrome P450-dependent enzymes of alkaloid formation in the Mexican prickly poppy *Argemone mexicana*. *Archives of Biochemistry and Biophysics*, *507*, 186–193.

Chen, R., Rishi, H. S., Potapov, V., Yamada, M. R., Yeh, V. J., Chow, T., et al. (2015). A barcoding strategy enabling higher-throughput library screening by microscopy. *ACS Synthetic Biology*, *4*, 1205–1216.

Curran, K. A., Karim, A. S., Gupta, A., & Alper, H. S. (2013). Use of expression-enhancing terminators in *Saccharomyces cerevisiae* to increase mRNA half-life and improve gene expression control for metabolic engineering applications. *Metabolic Engineering*, *19*, 88–97.

Dang, T.-T. T., & Facchini, P. J. (2012). Characterization of three O-methyltransferases involved in noscapine biosynthesis in opium poppy. *Plant Physiology*, *159*, 618–631.

DeLoache, W. C., Russ, Z. N., Narcross, L., Gonzales, A. M., Martin, V. J., & Dueber, J. E. (2015). An enzyme-coupled biosensor enables (S)-reticuline production in yeast from glucose. *Nature Chemical Biology*, *11*, 465–471.

Desgagné Penix, I., & Facchini, P. J. (2012). Systematic silencing of benzylisoquinoline alkaloid biosynthetic genes reveals the major route to papaverine in opium poppy. *The Plant Journal, 72*, 331–344.

DiCarlo, J. E., Norville, J. E., Mali, P., Rios, X., Aach, J., & Church, G. M. (2013). Genome engineering in *Saccharomyces cerevisiae* using CRISPR-Cas systems. *Nucleic Acids Research, 41*, 4336–4343.

Doudna, J. A., & Charpentier, E. (2014). The new frontier of genome engineering with CRISPR-Cas9. *Science, 346*, 1258096.

Dunn, K. W., Kamocka, M. M., & McDonald, J. H. (2011). A practical guide to evaluating colocalization in biological microscopy. *American Journal of Physiology. Cell Physiology, 300*, C723–C742.

Facchini, P. J. (2001). Alkaloid biosynthesis in plants: Biochemistry, cell biology, molecular regulation, and metabolic engineering applications. *Annual Review of Plant Biology, 52*, 29–66.

Facchini, P. J., Bohlmann, J., Covello, P. S., De Luca, V., Mahadevan, R., Page, J. E., et al. (2012). Synthetic biosystems for the production of high-value plant metabolites. *Trends in Biotechnology, 30*, 127–131.

Farrow, S. C., Hagel, J. M., Beaudoin, G. A., Burns, D. C., & Facchini, P. J. (2015). Stereochemical inversion of (S)-reticuline by a cytochrome P450 fusion in opium poppy. *Nature Chemical Biology, 11*, 728–732.

Fisinger, U., Grobe, N., & Zenk, M. H. (2007). Thebaine synthase: A new enzyme in the morphine pathway in *Papaver somniferum*. *Natural Product Communications, 2*, 249–253.

Fossati, E., Ekins, A., Narcross, L., Zhu, Y., Falgueyret, J.-P., Beaudoin, G. A., et al. (2014). Reconstitution of a 10-gene pathway for synthesis of the plant alkaloid dihydrosanguinarine in *Saccharomyces cerevisiae*. *Nature Communications, 5*, 3283.

Fossati, E., Narcross, L., Ekins, A., Falgueyret, J.-P., & Martin, V. J. (2015). Synthesis of morphinan alkaloids in *Saccharomyces cerevisiae*. *PLoS One, 10*, e0124459.

Frick, S., Kramell, R., Schmidt, J., Fist, A. J., & Kutchan, T. M. (2005). Comparative qualitative and quantitative determination of alkaloids in narcotic and condiment *Papaver somniferum* cultivars. *Journal of Natural Products, 68*, 666–673.

Galanie, S., & Smolke, C. D. (2015). Optimization of yeast-based production of medicinal protoberberine alkaloids. *Microbial Cell Factories, 14*, 1–13.

Galanie, S., Thodey, K., Trenchard, I. J., Interrante, M. F., & Smolke, C. D. (2015). Complete biosynthesis of opioids in yeast. *Science, 349*, 1095–1100.

Glenn, W. S., Runguphan, W., & O'Connor, S. E. (2013). Recent progress in the metabolic engineering of alkaloids in plant systems. *Current Opinion in Biotechnology, 24*, 354–365.

Hagel, J. M., Beaudoin, G. A., Fossati, E., Ekins, A., Martin, V. J., & Facchini, P. J. (2012). Characterization of a flavoprotein oxidase from opium poppy catalyzing the final steps in sanguinarine and papaverine biosynthesis. *Journal of Biological Chemistry, 287*, 42972–42983.

Hagel, J. M., & Facchini, P. J. (2010). Dioxygenases catalyze the O-demethylation steps of morphine biosynthesis in opium poppy. *Nature Chemical Biology, 6*, 273–275.

Hagel, J. M., & Facchini, P. J. (2013). Benzylisoquinoline alkaloid metabolism—A century of discovery and a brave new world. *Plant and Cell Physiology, 54*, 647–672.

Hawkins, K. M., & Smolke, C. D. (2008). Production of benzylisoquinoline alkaloids in *Saccharomyces cerevisiae*. *Nature Chemical Biology, 4*, 564–573.

Horwitz, A. A., Walter, J. M., Schubert, M. G., Kung, S. H., Hawkins, K., Platt, D. M., et al. (2015). Efficient multiplexed integration of synergistic alleles and metabolic pathways in yeasts via CRISPR-Cas. *Cell Systems, 1*, 88–96.

Hsu, P. D., Lander, E. S., & Zhang, F. (2014). Development and applications of CRISPR-Cas9 for genome engineering. *Cell, 157*, 1262–1278.

Ikezawa, N., Iwasa, K., & Sato, F. (2008). Molecular cloning and characterization of CYP80G2, a cytochrome P450 that catalyzes an intramolecular C–C phenol coupling of (S)-reticuline in magnoflorine biosynthesis, from cultured *Coptis japonica* cells. *Journal of Biological Chemistry*, *283*, 8810–8821.

Jakočiūnas, T., Bonde, I., Herrgård, M., Harrison, S. J., Kristensen, M., Pedersen, L. E., et al. (2015). Multiplex metabolic pathway engineering using CRISPR/Cas9 in *Saccharomyces cerevisiae*. *Metabolic Engineering*, *28*, 213–222.

Jakočiūnas, T., Jensen, M. K., & Keasling, J. D. (2015). CRISPR/Cas9 advances engineering of microbial cell factories. *Metabolic Engineering*, *34*, 44–59.

Jansen, G., Wu, C., Schade, B., Thomas, D. Y., & Whiteway, M. (2005). Drag&Drop cloning in yeast. *Gene*, *344*, 43–51.

Johnson, D. A., Gautsch, J. W., Sportsman, J. R., & Elder, J. H. (1984). Improved technique utilizing nonfat dry milk for analysis of proteins and nucleic acids transferred to nitrocellulose. *Gene Analysis Techniques*, *1*, 3–8.

Johnson, I., & Spence, M. (2010). *A guide to fluorescent probes and labeling technologies* (11th ed.). Eugene, OR: Life Technologies Corporation.

Kolodziej, P. A., & Young, R. A. (1991). [35] Epitope tagging and protein surveillance. *Methods in Enzymology*, *194*, 508–519.

Kutchan, T. M. (1995). Alkaloid biosynthesis—The basis for metabolic engineering of medicinal plants. *The Plant Cell*, *7*, 1059.

Lee, S., Lim, W. A., & Thorn, K. S. (2013). Improved blue, green, and red fluorescent protein tagging vectors for *S. cerevisiae*. *PLoS One*, *8*, e67902.

Liscombe, D. K., & Facchini, P. J. (2008). Evolutionary and cellular webs in benzylisoquinoline alkaloid biosynthesis. *Current Opinion in Biotechnology*, *19*, 173–180.

Manchikanti, L., Helm, S., II, Fellows, B., Janata, J. W., Pampati, V., Grider, J. S., et al. (2012). Opioid epidemic in the United States. *Pain Physician*, *15*, ES9–ES38.

Millgate, A. G., Pogson, B. J., Wilson, I. W., Kutchan, T. M., Zenk, M. H., Gerlach, W. L., et al. (2004). Analgesia: Morphine-pathway block in *top1* poppies. *Nature*, *431*, 413–414.

Minami, H., Dubouzet, E., Iwasa, K., & Sato, F. (2007). Functional analysis of norcoclaurine synthase in *Coptis japonica*. *Journal of Biological Chemistry*, *282*, 6274–6282.

Minami, H., Kim, J.-S., Ikezawa, N., Takemura, T., Katayama, T., Kumagai, H., et al. (2008). Microbial production of plant benzylisoquinoline alkaloids. *Proceedings of the National Academy of Sciences*, *105*, 7393–7398.

Nakagawa, A., Minami, H., Kim, J.-S., Koyanagi, T., Katayama, T., Sato, F., et al. (2011). A bacterial platform for fermentative production of plant alkaloids. *Nature Communications*, *2*, 326.

Narcross, L., Fossati, E., Bourgeois, L., Dueber, J. E., & Martin, V. J. (2016). Microbial factories for the production of benzylisoquinoline alkaloids. *Trends in Biotechnology*, *34*(3), 228–241.

Nyman, U. (1980). Selection for high thebaine/low morphine content in *Papaver somniferum* L. *Hereditas*, *93*, 121–124.

Pompon, D., Louerat, B., Bronine, A., & Urban, P. (1996). Yeast expression of animal and plant P450s in optimized redox environments. *Methods in Enzymology*, *272*, 51–64.

Rinner, U., & Hudlicky, T. (2012). *Synthesis of morphine alkaloids and derivatives* (pp. 33–66). Springer. Alkaloid synthesis.

Rueffer, M., Zumstein, G., & Zenk, M. H. (1990). Partial purification and properties of S-adenosyl-l-methionine:(S)-tetrahydroprotoberberinecis-N-methyltransferase from suspension-cultured cells of *Eschscholtzia* and *Corydalis*. *Phytochemistry*, *29*, 3727–3733.

Ryan, O. W., Skerker, J. M., Maurer, M. J., Li, X., Tsai, J. C., Poddar, S., et al. (2014). Selection of chromosomal DNA libraries using a multiplex CRISPR system. *eLife*, *3*, e03703.

Shao, Z., Zhao, H., & Zhao, H. (2009). DNA assembler, an in vivo genetic method for rapid construction of biochemical pathways. *Nucleic Acids Research, 37*, e16.

Sharp, P. M., Tuohy, T. M., & Mosurski, K. R. (1986). Codon usage in yeast: Cluster analysis clearly differentiates highly and lowly expressed genes. *Nucleic Acids Research, 14*, 5125–5143.

Shen, W.-C., Selvakumar, D., Stanford, D. R., & Hopper, A. K. (1993). The *Saccharomyces cerevisiae LOS1* gene involved in pre-tRNA splicing encodes a nuclear protein that behaves as a component of the nuclear matrix. *Journal of Biological Chemistry, 268*, 19436–19444.

Sherwood, P., Tsang, S., & Osley, M. (1993). Characterization of HIR1 and HIR2, two genes required for regulation of histone gene transcription in *Saccharomyces cerevisiae*. *Molecular and Cellular Biology, 13*, 28–38.

Snapp, E. (2005). Design and use of fluorescent fusion proteins in cell biology. *Current Protocols in Cell Biology, 27*. 21.4:21.4.1–21.4.13.

Sun, J., Shao, Z., Zhao, H., Nair, N., Wen, F., Xu, J. H., et al. (2012). Cloning and characterization of a panel of constitutive promoters for applications in pathway engineering in *Saccharomyces cerevisiae*. *Biotechnology and Bioengineering, 109*, 2082–2092.

Theuns, H. G., Lenting, H. B., Salemink, C. A., Tanaka, H., Shibata, M., Ito, K., et al. (1984). Neodihydrothebaine and bractazonine, two dibenz[d, f]azonine alkaloids of Papaver bracteatum. *Phytochemistry, 23*, 1157–1166.

Thodey, K., Galanie, S., & Smolke, C. D. (2014). A microbial biomanufacturing platform for natural and semisynthetic opioids. *Nature Chemical Biology, 10*, 837–844.

Trenchard, I. J., Siddiqui, M. S., Thodey, K., & Smolke, C. D. (2015). De novo production of the key branch point benzylisoquinoline alkaloid reticuline in yeast. *Metabolic Engineering, 31*, 74–83.

Trenchard, I. J., & Smolke, C. D. (2015). Engineering strategies for the fermentative production of plant alkaloids in yeast. *Metabolic Engineering, 30*, 96–104.

Wilson, I. A., Niman, H. L., Houghten, R. A., Cherenson, A. R., Connolly, M. L., & Lerner, R. A. (1984). The structure of an antigenic determinant in a protein. *Cell, 37*, 767–778.

Winzer, T., Gazda, V., He, Z., Kaminski, F., Kern, M., Larson, T. R., et al. (2012). A Papaver somniferum 10 gene cluster for synthesis of the anticancer alkaloid noscapine. *Science, 336*, 1704–1708.

Winzer, T., Kern, M., King, A. J., Larson, T. R., Teodor, R. I., Donninger, S. L., et al. (2015). Morphinan biosynthesis in opium poppy requires a P450-oxidoreductase fusion protein. *Science, 349*, 309–312.

Wolinski, H., & Kohlwein, S. D. (2014). *Single yeast cell imaging* (pp. 91–109). New York City, NY: Springer. Yeast genetics.

Xiao, M., Zhang, Y., Chen, X., Lee, E.-J., Barber, C. J., Chakrabarty, R., et al. (2013). Transcriptome analysis based on next-generation sequencing of non-model plants producing specialized metabolites of biotechnological interest. *Journal of Biotechnology, 166*, 122–134.

Zeder-Lutz, G., Cherouati, N., Reinhart, C., Pattus, F., & Wagner, R. (2006). Dot-blot immunodetection as a versatile and high-throughput assay to evaluate recombinant GPCRs produced in the yeast *Pichia pastoris*. *Protein Expression and Purification, 50*, 118–127.

Ziegler, J., & Facchini, P. J. (2008). Alkaloid biosynthesis: Metabolism and trafficking. *Annual Review of Plant Biology, 59*, 735–769.

CHAPTER TEN

Engineering Microbes to Synthesize Plant Isoprenoids

K. Zhou*,1, S. Edgar†, G. Stephanopoulos†,1
*National University of Singapore, Singapore, Singapore
†Massachusetts Institute of Technology, Cambridge, MA, United States
1Corresponding authors: e-mail address: kang.zhou@nus.edu.sg; gregstep@mit.edu

Contents

1. Introduction — 226
2. Improving Production of C20 ISMs by Modular Genetic Engineering of *Escherichia coli* — 228
   2.1 Constructing Expression Vectors for IsoSs — 230
   2.2 Testing the Expression Vectors — 235
3. Improving Production of C20 ISMs by Optimizing Bioreactor Operation — 236
   3.1 Bioreactor Specifications — 237
   3.2 Protocol for Operating the Bioreactor — 237
4. Oxygenating C20 ISMs by Using *E. coli–Saccharomyces cerevisiae* Coculture — 238
   4.1 Expression of a CYP in *S. cerevisiae* — 239
   4.2 Characterization of the *S. cerevisiae* — 243
   4.3 Coculture of the *S. cerevisiae* with an Isoprenoid-Overproducing *E. coli* — 243
5. Discussion — 244
References — 244

Abstract

Humans constantly look for faster, more economical, and more sustainable ways to produce chemicals that originally harvested from nature. Over the past two decades, substantial progress has been made toward this goal by harnessing enzymes and cells as biocatalysts. For example, enzymes of slow-growing plants can be reconstituted in microbes, which empower them with the ability to produce useful plant metabolic compounds from sugars faster than plants. In this chapter, we provide protocols for producing isoprenoids – a large group of useful natural products – in microbes. It has been found that expression of genes encoding plant enzymes and selected endogenous genes must be delicately adjusted in microbes, otherwise isoprenoid production is negatively affected. Therefore, we focus on how to balance gene expression in *Escherichia coli* and use process engineering to increase its isoprenoid production. We also introduce our recent work on the use of microbial consortia and provide protocols for using yeast to help *E. coli* functionalize its isoprenoid product. Together, the methods and protocols provided here should be useful to researchers who aim to use microbes to synthesize novel isoprenoids.

1. INTRODUCTION

Isoprenoids are a large group of natural products, which contain tens of thousands of chemicals with diverse functional groups (Cuellar & van der Wielen, 2015; Krivoruchko & Nielsen, 2015). Presently, many isoprenoids can be found on the global market as pharmaceuticals, nutrients, fragrances, flavoring substances, bulk chemicals, and fuels (Ajikumar et al., 2008). Most of these compounds were originally discovered in plants and – for many of them – extraction of plant materials remains a major production route (Ajikumar et al., 2010). However, it becomes increasingly difficult to meet the growing demand of many isoprenoids from plant extraction, due to the slow growth rate and low isoprenoid content of plants. For some relatively simple isoprenoids such as carotenoids, total chemical synthesis methods have been developed (Shen et al., 2011), but it is challenging to similarly produce more complex compounds, especially those with cyclic structures and multiple oxygenated functional groups. To tackle this challenge, different approaches are under investigation, including genetic modification of plants to increase isoprenoid content (Wilson & Roberts, 2014), development of plant cell culture in which plant biomass accumulates faster (Wilson, Cummings, & Roberts, 2014), and heterologous production by using engineered microbes (Biggs, De Paepe, Santos, De Mey, & Kumaran Ajikumar, 2014). We discuss the last approach in this chapter.

Microbes are ideal hosts for producing isoprenoids, because they grow fast, require little land/water resources, and naturally produce the building blocks of all isoprenoids (Fig. 1), isopentenyl pyrophosphate (IPP) and dimethylallyl pyrophosphate (DMAPP). There are also abundant genetic tools available for engineering microbes. Theoretically, any plant-produced isoprenoid can be produced in microbes – using its endogenous IPP/DMAPP as substrates – by expressing plant isoprenoid synthases (IsoSs) and functionalizing enzymes (Fig. 1).

Intensive research has been done to harness the biosynthetic capacity of bacteria for isoprenoid production. In bacteria, glycolytic intermediates are converted into IPP and DMAPP via methylerythritol pyrophosphate (MEP) pathway (Fig. 1), which consumes energy (in terms of ATP) and reducing equivalents (in terms of NADPH and NADH). MEP pathway flux is weak in nonengineered bacteria due to their low cellular demand for isoprenoids, so various strategies have been developed to increase flux of the pathway (Fig. 2), including overexpression of the rate-limiting enzymes

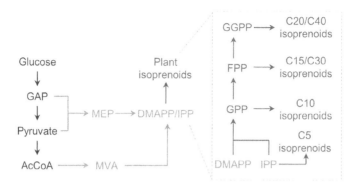

Fig. 1 Schematic illustration of metabolic pathways for production of plant isoprenoids in microbes. Both MEP and MVA pathways can convert intermediates of central metabolism into IPP and DMAPP, two building blocks for all isoprenoids. IPP and DMAPP can be converted into various plant isoprenoids by plant isoprenoid synthases and functionalizing enzymes. *GAP*, glyceraldehyde 3-phosphate; *AcCoA*, acetyl CoA; *MEP*, methylerythritol 4-phosphate; *MVA*, mevalonate; *DMAPP*, dimethylallyl pyrophosphate; *IPP*, isopentenyl pyrophosphate; *GPP*, geranyl pyrophosphate; *FPP*, farnesyl pyrophosphate; *GGPP*, geranylgeranyl pyrophosphate.

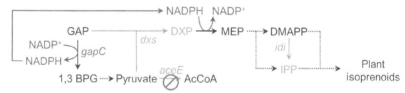

Fig. 2 Three rational strategies to increase flux through the MEP pathway – overexpression of rate-limiting genes (*dxs* and *idi*), inactivation of competing pathway (△*aceE*), and enrichment of essential cofactors (NADPH). *GAP*, glyceraldehyde 3-phosphate; *1,3 BPG*, 1,3-bisphosphoglyceric acid; *DXP*, 1-deoxy-D-xylulose 5-phosphate; *MEP*, methylerythritol 4-phosphate; *DMAPP*, dimethylallyl pyrophosphate; *IPP*, isopentenyl pyrophosphate; *Dxs*, the gene encoding 1-deoxy-D-xylulose 5-phosphate synthase; *aceE*, the gene encoding subunit of E1p component of pyruvate dehydrogenase complex; *idi*, isopentenyl diphosphate isomerase.

(Yuan, Rouviere, Larossa, & Suh, 2006), inactivation of competing pathways by gene knockout (Alper, Jin, Moxley, & Stephanopoulos, 2005), enrichment of the cofactors by engineering central metabolism (Martínez, Zhu, Lin, Bennett, & San, 2008), and combinatorial manipulations coupled with high-throughput screening (Alper & Stephanopoulos, 2007; Tyo, Ajikumar, & Stephanopoulos, 2009). Among them, the first strategy was direct and effective – expressing DXS (the first enzyme in the pathway) pushed carbon flux from central metabolism into the pathway, and having more IDI (the last enzyme in the pathway) efficiently isomerized DMAPP into IPP, which is needed in larger quantity than DMAPP in synthesis of most (C10 and heavier) isoprenoids (Yuan et al., 2006). However, one complication associated with this strategy is that there is an optimal expression status for the MEP enzymes: when their levels are too low the overall pathway flux is limited; when expression levels were too high, isoprenoid production could be inhibited, possibly due to accumulation and active efflux of an MEP intermediate (MEC), as revealed in a metabolite profiling study (Zhou, Zou, Stephanopoulos, & Too, 2012a, 2012b). This optimal expression status is also heavily influenced by downstream IsoSs so it must be experimentally determined each time when new IsoSs are expressed or IsoSs' expression is altered (Ajikumar et al., 2010).

In 2010, we developed a method that can rapidly locate optimal expression level of both MEP enzymes and downstream IsoSs (Ajikumar et al., 2010) and have applied the method to improve production of taxadiene, the scaffold molecule of anticancer drug paclitaxel (a C20 isoprenoid). According to this method, important MEP enzymes are lumped together and regarded as a module, and downstream IsoSs (geranylgeranyl pyrophosphate synthase and taxadiene synthase in the case) are treated as another module. Expression level of each module can be systematically altered by using various promoters and replication origins, and a large number of combinations can be subsequently generated by combining variants of each module. Quantifying isoprenoid production for all combinations of modular gene expression then reveals the optimal expression levels for both modules. In Section 2 of this chapter, we will describe detailed protocols for implementing a simplified version of this method for producing C20 isoprenoid scaffold molecules (ISMs). In the following sections, we will provide detailed protocols for scaling up isoprenoid production in laboratory bioreactor and oxygenating C20 ISMs by using microbial consortia. We conclude with a discussion of how to adapt protocols described in this chapter for producing isoprenoids other than the C20 ones.

2. IMPROVING PRODUCTION OF C20 ISMs BY MODULAR GENETIC ENGINEERING OF *ESCHERICHIA COLI*

E. coli is a model Gram-negative bacterium that has been widely used for both academic study and industrial application within the biotech industry. There are many techniques that can alter activity of an enzyme in *E. coli* – they can be classified into two groups, those altering quantity of the enzyme and those affecting specific enzymatic activity. The latter group includes protein engineering (eg, mutagenesis, domain shuffling, fusion, and truncation), expression of chaperones that help folding of the target enzyme, and altering local concentration of its substrates (Zhou et al., 2012a, 2012b). The former group primarily focuses on altering the copy number of the gene that encodes the enzyme, the promoter that drives transcription of the gene, and the ribosomal binding site that influences translation efficiency of the gene's transcript (Ajikumar et al., 2010). Because the former group is less dependent on the sequence and function of the target enzyme, they have been widely used to genetically engineer *E. coli*. Here, we illustrate how to locate optimal expression levels of MEP genes and IsoSs for isoprenoid production by using different promoters and replication origins.

Our prior study on taxadiene production indicated that MEP genes need to be expressed at relatively low level – while IsoSs should be strongly overexpressed (Ajikumar et al., 2010). This study allows simplification of the procedure for locating optimal gene expression level (Fig. 3): MEP genes are all integrated into genome of *E. coli*, under three different promoters (P_{trc}, P_{T5}, and P_{T7}); IsoSs are expressed on three multicopy plasmids (p5, p10, and p20), all with a strong promoter (P_{T7}). Each multicopy plasmid is transformed into an engineered *E. coli*, creating nine strains in total. Analyzing their isoprenoid production could reveal the optimal expression levels of both MEP genes and IsoSs.

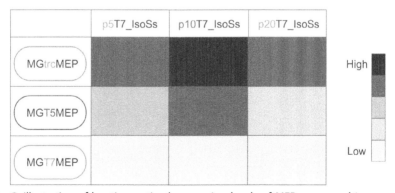

Fig. 3 Illustration of locating optimal expression levels of MEP genes and isoprenoid synthases by modular genetic engineering of *E. coli*. Color indicates level of isoprenoid production. *MGtrcMEP*, *E. coli* MG1655 ($\Delta recA$, $\Delta endA$, DE3, $\Delta araA$::P_{trc}_dxs-idi-ispDF); *MGT5MEP*, *E. coli* MG1655 ($\Delta recA$, $\Delta endA$, DE3, $\Delta araA$::P_{T5}_dxs-idi-ispDF); *MGT7MEP*, *E. coli* MG1655 ($\Delta recA$, $\Delta endA$, DE3, $\Delta araA$::P_{T7}_dxs-idi-ispDF). (See the color plate.)

2.1 Constructing Expression Vectors for IsoSs
2.1.1 IsoS-Encoding Genes
Our focus here is on C20 isoprenoids because their scaffold molecules are easy to handle – they are not volatile, yet still can cross cell membrane, thus simplifying purification. The example used in the following protocols is miltiradiene, the scaffold molecule of tanshinone, a C20 therapeutic isoprenoid found in *Salvia miltiorrhiza* (Zhou, Qiao, Edgar, & Stephanopoulos, 2015). As all C20 ISMs, miltiradiene is derived from geranylgeranyl pyrophosphate (GGPP) (Fig. 1). Cyclization of GGPP into miltiradiene can be catalyzed by a fusion enzyme in *E. coli*, encoded by *ksl_cps*, which can be synthesized through commercial gene synthesis service. We have cloned a GGPP synthase from *Taxus canadensis* into the three expression vectors, which can be used to supply GGPP for production of any C20 isoprenoids (Fig. 4).

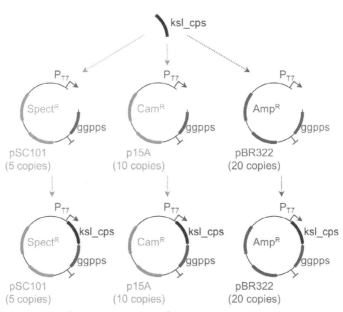

Fig. 4 Construction of expression vectors for C20 isoprenoid synthases by using CLIVA method. *Ksl_cps*, the gene encoding miltiradiene synthase; *Ggpps*, the gene encoding geranylgeranyl pyrophosphate synthase; P_{T7}, T7 promoter; $Spect^R$, spectinomycin resistance gene; Cam^R, chloramphenicol resistance gene; Amp^R, ampicillin resistance gene; *pSC101*, pSC101 replication origin, whose copy number is *E. coli* is estimated to be 5; *p15A*, p15A replication origin, whose copy number in *E. coli* is estimated to be 10; *pBR322*, pBR322 replication origin, whose copy number in *E. coli* is estimated to be 20. (See the color plate.)

2.1.2 Primer Design

Ksl_cps is inserted into the three vectors by using CLIVA method (Zou, Zhou, Stephanopoulos, & Too, 2013). DNA fragments used in CLIVA method need to be generated by using PCR, where modified oligos are used to introduce phosphorothioate bonds to DNA fragments (Fig. 5). The phosphorothioate bonds can be cleaved by using iodine and ethanol at basic condition, generating long, specific sticky ends, which ensure accurate annealing of the DNA fragments. The fragments are covalently linked after the annealed fragments are transformed into *E. coli* (Fig. 5). A common forward primer is designed for amplifying all three plasmids – it is pXT7_F with sequence "GATCCAAT*CCAGGTCT*AA," where "*" denotes a phosphorothioate bond – and the common reverse primer for all three plasmids is pXT7-R with sequence "CATCATGC*CCATGGTA*TA." Forward primer for *ksl_cps* (Sequence 1) is ksl_cps_F with sequence "TACCATGG*GCATGATG*AGCCTGGCATTTAAT," and reverse primer for ksl_cps is ksl_cps_R with sequence "AGACCTGG*ATTGGATC*TTAGGCAACCGGCTC."

2.1.3 PCR Reaction

For each PCR reaction, the following components are mixed in a PCR tube: 25 μL of Q5 master mixture (NEB), 2.5 μL of forward primer (10 μM), 2.5 μL of reverse primer (10 μM), 5 μL of plasmid or IsoSC gene (1 ng/μL), and 20 μL of ultrapure water. The capped PCR tube is incubated in a thermocycler according to the following program: 98°C for 30 s, 35 cycles of (98°C for 10 s, 55°C for 30 s, and 72°C for X s), 72°C for

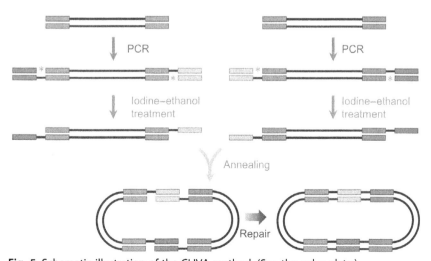

Fig. 5 Schematic illustration of the CLIVA method. (See the color plate.)

SEQUENCE 1 Sequence of *ksl_cps*

ATGAGCCTGGCATTTAATCCGGCAGCAACCGCATTTAGCGGTAATGGTGCACGTAGC
CGTCGTGAAAACTTTCCGGTTAAACATGTTACCGTTCGTGGTTTTCCGATGATTACCA
ATAAAAGCAGCTTTGCCGTTAAATGCAATCTGACCACCACCGATCTGATGGGCAAAA
TTGCAGAAAAATTCAAAGGCGAGGATAGCAATTTTCCGGCAGCCGCAGCAGTTCAGC
CTGCAGCAGATATGCCGAGCAATCTGTGTATTATTGATACCCTGCAGCGTCTGGGTG
TTGATCGTTATTTTCGTAGCGAAATTGATACCATCCTGGAAGATACCTATCGTCTGTGG
CAGCGTAAAGAACGTGCAATTTTTAGCGATACCGCAATTCATGCAATGGCATTTCGTC
TGCTGCGTGTTAAAGGTTATGAAGTTAGCAGCGAAGAACTGGCACCGTATGCAGATC
AAGAACATGTGGATCTGCAGACCATTGAAGTTGCAACCGTTATTGAACTGTATCGTGC
AGCACAAGAACGTACCGGTGAAGATGAAAGCAGCCTGAAAAAACTGCATGCATGGA
CCACCACATTTCTGAAACAGAAACTGCTGACCAATAGCATCCCGGATAAAAAACTGC
ACAAACTGGTGGAATACTATCTGAAAAACTTTCACGGCATTCTGGATCGTATGGGTGT
TCGTCAGAATCTGGATCTGTACGATATTAGTTATTACCGTACCAGCAAAGCAGCCAAT
CGTTTTAGTAATCTGTGCTCCGAAGATTTTCTGGCATTTGCACGTCAGGATTTTAACAT
TTGTCAGGCACAGCATCAGAAAGAACTGCAGCAGCTGCAACGTTGGTATGCCGATTG
TAAACTGGATACCCTGAAATATGGTCGTGATGTTGTTCGTGTTGCAAATTTTCTGACCA
GCGCAATTATTGGTGATCCGGAACTGAGTGATGTTCGTATTGTTTTTGCACAGCATATT
GTTCTGGTGACCCGCATCGATGATTTTTTTGATCATCGTGGTAGCCGTGAAGAGAGCT
ACAAAATTCTGGAACTGATCAAAGAATGGAAAGAAAAACCGGCAGCAGAATATGGT
AGCGAAGAAGTTGAAATTCTGTTCACCGCAGTGTATAATACCGTGAATGAACTGGCA
GAACGTGCCCATGTTGAACAGGGTCGTAGCGTTAAAGATTTCCTGATTAAACTGTGG
GTGCAGATCCTGAGCATCTTTAAACGTGAGCTGGATACCTGGTCAGATGATACCGCA
CTGACCCTGGATGATTATCTGAGCGCAAGCTGGGTTAGCATTGGTTGTCGTATTTGTA
TTCTGATGTCCATGCAGTTCATTGGCATCAAACTGTCAGATGAAATGCTGCTGAGCGA
AGAATGTATTGATCTGTGTCGTCATGTTAGCATGGTGGATCGCCTGCTGAATGATGTT
CAGACCTTTGAAAAAGAACGCAAAGAGAATACCGGTAATAGCGTTACCCTGCTGCTG
GCAGCAAATAAAGATGATAGCAGTTTTACCGAAGAAGAGGCAATTCGTATTGCAAAA
GAAATGGCCGAATGTAATCGTCGTCAGCTGATGCAGATTGTGTATAAAACCGGTACA
ATTTTTCCGCGTCAGTGCAAAGATATGTTTCTGAAAGTTTGCCGCATTGGGTGTTATCT
GTATGCAAGCGGTGATGAATTTACCAGTCCGCAGCAGATGATGGAAGATATGAAAAG
CCTGGTTTATGAACCGCTGACCATTCATCCGCTGGTTGCAAATAATGTTCGCGGTAAA
GGTGGTGGTTCTGCAAGCCTGAGCAGCACCATTCTGAGCCGTAGTCCGGCAGCACGT
CGTCGTATTACACCGGCAAGCGCAAAACTGCATCGTCCGGAATGTTTTGCAACCAGC
GCATGGATGGGTAGCAGCAGCAAAAATCTGAGCCTGAGCTATCAGCTGAACCACAA
AAAAATCAGCGTTGCAACCGTTGATGCACCGCAGGTTCATGATCACGATGGCACCAC
CGTTCATCAGGGTCATGATGCAGTTAAAAACATTGAAGATCCGATCGAATATATCCGT
ACCCTGCTGCGTACCACCGGTGATGGTCGTATTAGCGTTAGCCCGTATGATACCGCA
TGGGTTGCAATGATTAAAGATGTTGAAGGTCGTGATGGTCCGCAGTTTCCGAGCAGC
CTGGAATGGATTGTTCAGAATCAGCTGGAAGATGGTAGCTGGGGTGATCAGAAACTG
TTTTGTGTTTATGATCGTCTGGTGAATACCATTGCATGTGTTGTTGCACTGCGTAGCTG

SEQUENCE 1 Sequence of *ksl_cps*—cont'd

GAATGTTCATGCACATAAAGTTAAACGTGGCGTGACCTATATCAAAGAAAACGTGGA
TAAACTGATGGAAGGCAACGAAGAACATATGACCTGTGGTTTTGAAGTTGTTTTTCCG
GCACTGCTGCAGAAAGCAAAAAGCCTGGGTATCGAAGATCTGCCGTATGATTCACCG
GCAGTTCAAGAGGTTTATCATGTTCGTGAACAAAAACTGAAACGCATTCCGCTGGAAA
TCATGCATAAAATTCCGACCAGTCTGCTGTTTAGCCTGGAAGGTCTGGAAAATCTGGA
TTGGGACAAACTGCTGAAACTGCAGAGCGCAGATGGTAGTTTTCTGACCAGCCCGAG
CAGTACCGCATTTGCATTTATGCAGACCAAAGATGAGAAATGCTATCAGTTCATCAAA
AACACCATCGACACCTTTAATGGTGGTGCACCGCATACCTATCCGGTTGATGTTTTTGG
TCGTCTGTGGGCAATTGATCGCCTGCAGCGTCTGGGTATTAGCCGTTTTTTTGAACCG
GAAATTGCAGATTGTCTGAGCCACATTCACAAATTCTGGACCGATAAAGGTGTTTTTA
GCGGTCGTGAAAGCGAATTTTGCGATATTGATGATACCAGTATGGGTATGCGTCTGA
TGCGTATGCATGGTTATGATGTTGATCCGAATGTTCTGCGCAACTTTAAACAGAAAG
ATGGCAAATTTAGCTGCTATGGTGGTCAGATGATTGAAAGCCCGAGCCCGATTTATAA
CCTGTATCGTGCAAGCCAGCTGCGTTTTCCGGGTGAAGAAATTCTGGAAGATGCCAA
ACGTTTTGCCTATGATTTCCTGAAAGAAAAACTGGCCAATAACCAGATCCTGGATAAA
TGGGTTATTAGCAAACATCTGCCGGATGAAATTAAACTGGGCCTGGAAATGCCGTGG
CTGGCAACCCTGCCTCGTGTTGAAGCAAAATACTATATTCAGTATTATGCCGGTAGCG
GTGATGTGTGGATTGGTAAAACCCTGTATCGCATGCCGGAAATTAGCAATGATACCTA
TCATGATCTGGCCAAAACCGATTTTAAACGTTGTCAGGCAAAACACCAGTTTGAGTGG
CTGTATATGCAAGAATGGTATGAAAGCTGCGGCATTGAAGAATTTGGCATTAGCCGTA
AAGATCTGCTGCTGAGCTATTTTCTGGCAACCGCAAGCATCTTTGAACTGGAACGTAC
CAATGAACGTATTGCATGGGCCAAAAGCCAGATTATTGCAAAAATGATCACCAGCTTT
TTTAACAAAGAAACCACGAGCGAAGAAGATAAACGCGCACTGCTGAATGAACTGGG
TAACATTAATGGTCTGAATGATACCAATGGTGCAGGTCGTGAAGGTGGTGCGGGTA
GCATTGCACTGGCCACCCTGACCCAGTTTCTGGAAGGTTTTGATCGTTATACCCGTCA
CCAGCTGAAAAATGCATGGTCAGTTTGGCTGACCCAGCTGCAGCATGGTGAAGCAG
ATGATGCAGAACTGCTGACCAATACCCTGAATATTTGTGCCGGTCATATTGCCTTTCGC
GAAGAAATCCTGGCACACAATGAATATAAAGCACTGAGCAATCTGACGAGCAAAATT
TGTCGCCAGCTGAGCTTTATTCAGAGCGAAAAAGAAATGGGTGTGGAAGGTGAAATT
GCCGCAAAAAGCAGCATTAAAAACAAAGAACTGGAAGAGGATATGCAGATGCTGGT
TAAACTGGTTCTGGAGAAATATGGTGGTATTGACCGCAATATCAAAAAAGCATTTCTG
GCAGTTGCCAAAACCTATTATTACCGTGCATATCATGCAGCCGATACCATTGATACCC
ATATGTTCAAAGTTCTGTTTGAGCCGGTTGCCTAA

120 s, and held at 16°C, where X is 20 times length (kb) of the gene or vector (eg, X is 50 for a 2.5 kb gene and 140 for a 7 kb vector).

2.1.4 DNA Electrophoresis and Recovery

Ten microliters of $6 \times$ DNA loading dye are added to each PCR reaction mixture, which is briefly vortexed and loaded into a well in a 0.8%

agarose-TAE gel that is prestained with SYBR-Safe DNA Gel Stain. The electrophoresis is carried out at 120 V for 40 min, and the DNA bands are carefully excised on Blue Light Transilluminator (pearl biotech). DNA is recovered from the gel by using DNA Gel Extraction Kit by following manufacture's instruction (at the last step of DNA extraction, DNA is eluted with 40 μL of ultrapure water).

2.1.5 Iodine Treatment and Ethanol Precipitation

The following components are mixed in a 1.7-mL Axygen centrifuge tube: 5.5 μL of 1 M Tris solution (pH 9), 10 μL of 30 g/L iodine solution (solvent is ethanol), and 40 μL of the purified PCR product from the last step. The solution is incubated at 70°C for 5 min by using a water bath, and standard sodium acetate ethanol precipitation method is used to purify the treated DNA (at the last step, 15 μL of ultrapure water is used to resuspend the DNA).

2.1.6 DNA Assembly

By this point, there are four purified, digested DNA samples – ksl_cps, p5T7 fragment, p10T7 fragment, and p20T7 fragment. *Ksl_cps* is assembled with each of the three vector fragments (Fig. 4). In each case, 3 μL of ksl_cps and 3 μL of a vector fragment are mixed in a PCR tube, which is then incubated in a thermocycler according to the following program: 80°C for 60 s, 50°C for 60 s, and gradually decreased to 16°C at the rate of 0.1°C/s.

2.1.7 Cell Transformation

Two microliters of each DNA assembly reaction mixture are added into 50 μL of DH5α electrocompetent *E. coli* (commercially available), which is electroporated in a 0.1-cm cuvette at 1.8 kV. Two hundred microliters of SOC medium are added into the cuvette to dilute the cell suspension, and the diluted solution is transferred into a 14-mL round-bottom Falcon tube, which is incubated at 37°C/250 rpm for 30 min. Two hundred microliters of the cell suspension are plated onto a LB agar plate with proper antibiotics (50 mg/L spectinomycin for vector p5T7-ksl_cps-ggpps, 34 mg/L chloramphenicol for p10T7-ksl_cps-ggpps, and 50 mg/L carbenicillin for vector p20T7-ksl_cps-ggpps), which is incubated at 37°C for 8–24 h. One colony is picked from each plate, and inoculated into 5 mL of LB medium with proper antibiotics in a 50-mL Falcon tube, which is then incubated at 37°C/250 rpm for 8–24 h. Plasmid is extracted from each cell culture by using a commercial Plasmid Miniprep Kit, and verified via Sanger sequencing.

2.2 Testing the Expression Vectors

2.2.1 Competent Cell Preparation and Cell Transformation

Strains MGtrcMEP, MGT5MEP, and MGT7MEP (Fig. 3) are first made into electrocompetent cells: a colony is inoculated into 10 mL of LB medium in 50 mL of Falcon tube, and it is incubated at 37°C/250 rpm until cell density reaches 0.5 (OD600), after which the cell pellet is washed three times by using 10% glycerol and resuspended in 200 µL of 10% glycerol. Each strain is transformed with the three IsoSC-expressing vector individually by using electroporation as described in Section 2.1.7 (1 µL of the plasmid is used in each transformation). One colony of each resulting strain is inoculated into 5 mL of LB medium with proper antibiotics in a 50-mL Falcon tube, which is then incubated at 37°C/250 rpm for 8–24 h. Five hundred microliters of the cell culture are mixed with 500 µL of 80% glycerol solution (solvent is water) to make *E. coli* glycerol stock, which can be stored at $-80°C$ freezer for up to 5 years.

2.2.2 Preparation of E. coli Seed Culture

Small amounts of *E. coli* glycerol stock are inoculated into 1 mL of LB medium with proper antibiotics – held in 14 mL of Falcon tube – by using disposable 1 µL inoculation loop. The tube is incubated at 37°C/250 rpm for 10–16 h.

2.2.3 Preparation of 1000× K3 Trace Elements Stock Solution

The following components are dissolved in 1 L of deionized water: 5 g $CaCl_2 \cdot 2H_2O$, 1.6 g $MnCl_2 \cdot 4H_2O$, 0.38 g $CuCl_2 \cdot 2H_2O$, 0.5 g $CoCl_2 \cdot 6H_2O$, 0.94 g $ZnCl_2$, 0.0311 g H_3BO_3, and 0.4 g $Na_2EDTA \cdot 2H_2O$. The mixture is autoclaved at 121°C for 15 min.

2.2.4 Preparation of K3 Basal Medium

Four grams of $(NH_4)_2HPO_4$ and 13.3 g of KH_2PO_4 are dissolved in 0.9 L of deionized water, and the solution is autoclaved at 121°C for 15 min.

2.2.5 Preparation of K3 Master mix

Two point five milliliters of 0.1 M ferric citrate solution (autoclaved), 1 mL of 4.5 g/L thiamine solution (sterilized by using 0.2 µm filter), 3 mL of 4 mM Na_2MoO_4 (autoclaved), 1 mL of 1000× K3 trace elements stock solution, and 1 mL of 1 M $MgSO_4$ solution (autoclaved) are mixed in a sterile 50-mL Falcon tube. The resulting solution can be stored at room temperature for no more than 3 weeks.

2.2.6 Testing C20 Isoprenoid Production in Test Tube

Ninety milliliters of K3 basal medium, 0.17 mL of K3 master mix, 10 mL of 200 g/L xylose solution (autoclaved), and 100 μL of 50% (v/v) Sigma antifoam B (autoclaved) are mixed aseptically in a sterile container. One milliliter of the mixture is transferred to a sterile VWR 16 × 100 mm (diameter and height) round-bottom culture tube, and 10 μL of fully grown *E. coli* seed culture is added. The tube is incubated at 30°C/250 rpm until cell density reaches 0.5–1.0 (OD600), and 1 μL of 100 mM IPTG solution is added to induce isoprenoid production. The tube is then incubated at 22°C/250 rpm for 72 h, and 100 μL of the cell culture is withdrawn at the end for isoprenoid analysis.

2.2.7 Extraction of C20 Isoprenoids for GCMS Analysis

One hundred microliters of cell suspension, 100 μL of ethyl acetate, and 100 μL of 0.5 mm (diameter) glass beads are added into a 1.5-mL Eppendorf safe-lock tube. The tube is vortexed at 2000 rpm for 20 min by using VWR heavy-duty vortex, and then centrifuged at 18,000 × g for 2 min. Twenty microliters of the ethyl acetate phase are carefully withdrawn (without touching the aqueous phase) for GCMS analysis.

2.2.8 GCMS Analysis

One microliter of the ethyl acetate phase is injected into GCMS injector at 250°C – C20 ISMs can be vaporized at the temperature. Agilent HP-5ms column is generally suitable for analyzing C20 ISMs, but oven temperature rising program needs to be customized to achieve good separation of target compounds in each case. The program for miltiradiene is 100°C for 1 min, ramping up to 175°C at 15°C/min, ramping up to 220°C at 4°C/min, ramping up to 290°C at 50°C/min, and 290°C for 1 min. Helium is used as carrier gas at 1 mL/min. Mass spectrometry is operated at scan mode (40–400 m/z). Total ion counts are used for quantification of C20 isoprenoids with calibration curves constructed by using authentic standards.

3. IMPROVING PRODUCTION OF C20 ISMs BY OPTIMIZING BIOREACTOR OPERATION

Once gene expression levels of the MEP pathway and those of IsoSs are optimized, the resulting strain can be cultured in well-controlled bioreactor to further increase the isoprenoid titer. The higher product titer and large volume of bioreactor culture facilitate purification of desired isoprenoids in laboratory scale. Usually, following bioreactor optimization, milligrams of C20 isoprenoids can easily be obtained, which could be used in NMR analysis to confirm compound structures, or used as substrate to identify downstream functionalizing enzymes in enzymatic reactions. In this section, we provide detailed protocols for improving titer of C20 ISMs via bioreactor optimization.

3.1 Bioreactor Specifications

A lab scale bioreactor (1–10 L) can be used in this experiment and should include (1) a dissolved oxygen (DO) probe and a stirrer that can be controlled according to online DO reading, (2) a pH probe and two pump heads that can be used to add acid and alkaline to maintain pH, (3) a temperature probe and a heating/cooling system, (4) two pump heads that are used to feed xylose solution (carbon source) and diammonium phosphate solution (nitrogen and phosphorus source), (5) a sparger that is connected to compressed air source via a flow rate controller, (6) a chilled condenser that reduces loss of volume via evaporation, (7) a sampling port, and (8) an inoculation port.

3.2 Protocol for Operating the Bioreactor

3.2.1 Preparation of the Bioreactor

One liter bioreactor is used here as an example with working volume of 500 mL. The bioreactor is assembled according to manufacturer's instruction, and 450 mL of K3 basal medium (Section 2.2.4) is added. Its pH probe is calibrated by using pH 7 and pH 4 standard solution before the reactor is autoclaved at 121°C for 15 min, after which if DO probe is a polargraphic one it needs to be polarized by connecting it to DO meter overnight before calibration. DO probe is calibrated by using nitrogen (0) and air (100%).

3.2.2 Initiating Bioreactor Run and Bioreactor Setting

The following components are added aseptically into the bioreactor through the inoculation port: 850 μL of sterile K3 master mix (Section 2.2.5), 50 mL of 200 g/L xylose solution (autoclaved), 0.5 mL of 50% Sigma antifoam B solution (autoclaved), and 5 mL of fully grown seed culture (Section 2.2.2). Agitation speed is set at 280 rpm, temperature is set at 30°C, and air flow is set at 0.5 L/min. pH is controlled at 7 by using 240 g/L NaOH solution and 0.5 M HCl solution. Temperature is programmed to drop to 22°C when DO decreases to 40%. When temperature reaches 22°C, 0.5 mL of 100 mM IPTG stock solution is added. Agitation speed is increased (up to 800 rpm, this threshold may vary depending on the bioreactor) when

DO drops below 30%. One milliliter of 50% Sigma antifoam B solution (autoclaved) is added in case of foaming.

3.2.3 Feeding the Bioreactor

E. coli needs large quantity of carbon, nitrogen, and phosphorus to achieve high cell density during bioreactor cultivation. These elements are supplied in the form of xylose and diammonium phosphate over the course of fermentation. It is recommended to pump 500 g/L xylose solution (autoclaved) and 400 g/L diammonium phosphate solution (autoclaved) into the bioreactor at a constant rate. Such linear feeding strategy is easy to implement and can be used to control cell growth under carbon-limited condition. Xylose feeding is started when xylose concentration drops below 10 g/L. Optimal rate of xylose feeding needs to be determined empirically, and often lies in the range between 5 and 15 g/day. Feed rate of the diammonium phosphate solution is always set to be 10% of that of the xylose solution. The bioreactor is stopped at 120 h after IPTG induction.

3.2.4 Sampling Bioreactor and HPLC Analysis

Samples are taken every 24 h for cell density (OD600), GCMS, and HPLC analysis. Samples for GCMS analysis are processed as described in Sections 2.2.7 and 2.2.8. One milliliter of cell suspension is taken for HPLC analysis: it is clarified by centrifugation at $16{,}000 \times g$ for 1 min, and the supernatant is sterilized by using 13 mm 0.2 μm Nylon filter. Ten microliters of the filtrate is injected into Aminex HPX-87H column. The mobile phase is 5 mM sulfuric acid, column temperature is 50°C, and each run lasts for 20 min. The HPLC analysis reveals concentration of xylose and fermentation by-products such as acetate.

3.2.5 Purification of Isoprenoids

C20 ISMs can be extracted by ethyl acetate and purified by using HPLC with C18 column and acetonitrile as mobile phase. Because this is a standard procedure in organic chemistry, detailed protocols are not described here.

4. OXYGENATING C20 ISMs BY USING *E. COLI*–*SACCHAROMYCES CEREVISIAE* COCULTURE

The last two steps of the MEP pathway are catalyzed by iron–sulfur-containing proteins, which are hypersensitive to oxidizing agents – they can be inactivated even by molecular oxygen (Artsatbanov et al., 2012).

Yet functionalization of C20 ISMs often requires oxygenation reactions that are catalyzed by cytochrome P450 (CYP), which often generate large amounts of reactive oxygen species (Zhou et al., 2015). As a result, previous attempts revealed that introduction of CYP into *E. coli* disturbed its MEP pathway and substantially reduced isoprenoid titer (Ajikumar et al., 2010). To address this incompatibility, we spatially separated the CYPs for oxygenating C20 ISMs from the upstream pathways by distributing them into two microbes (Fig. 6) – the first microbe produced an isoprenoid scaffold that can diffuse into the second species of the consortium where it is oxygenated by CYPs (Zhou et al., 2015). This way, reactive oxygen species generated by CYPs would not destroy the iron–sulfur clusters that are located in a different cell.

Here we provide detailed protocols for expression of a CYP in *S. cerevisiae*, characterization of the *S. cerevisiae*, and coculture of the *S. cerevisiae* with an *E. coli* that produces isoprenoid scaffold.

4.1 Expression of a CYP in *S. cerevisiae*

4.1.1 Constructing Yeast Expression Vector for CYP

We have constructed integrative expression vector pYPRC15-URA-UGPD, which can be used to express a CYP and a CYP reductase (CPR) in *S. cerevisiae* (Fig. 7). It is based on pUC19 that can be replicated in *E. coli* and contains two 500 base-pair homologous arms that direct integration into YPRC15 locus on yeast genome. The vector also contains a CYC terminator, an enhanced GPD promoter, and a URA3 marker. Forward primer for amplifying this plasmid is pYPRC15_F with sequence "CATCATCA*TCACCATC*AC,", and the reverse primer is

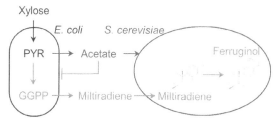

Fig. 6 Oxygenating C20 isoprenoid scaffold molecule by using mutualistic microbial consortium. Miltiradiene is used here as an example – it is produced by *E. coli* and oxidized into ferruginol in *S. cerevisiae*. PYR, pyruvate; GGPP, geranylgeranyl pyrophosphate; *E. coli* produces acetate as a by-product that inhibits its own growth, which is used here to build a mutualistic relationship between the two species – acetate serves as carbon source of the yeast, which in return removes the inhibitor (acetate) for *E. coli*. (See the color plate.)

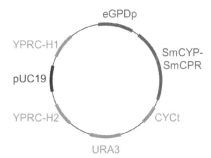

Fig. 7 Plasmid map of pYPRC15-URA-UGPD-SmCYP_SmCPR. *pUC19*, pUC19 plasmid backbone; *YPRC-H1*, 500 bp homologous region upstream YPRC15 locus of *S. cerevisiae* genome; *YPRC-H2*, 500 bp homologous region downstream YPRC15 locus of *S. cerevisiae*; *eGPDp*, enhanced GPD promoter; *CYCt*, CYC terminator; *URA3*, URA3 marker; *SmCYP-SmCPR*, chimeric CYP-CPR gene, which is used as an example in this chapter and involved in oxygenation of miltiradiene. (See the color plate.)

pYPRC15_R with sequence "CATTTTGT*TTGTTTAT*GTG." CYP and CPR are fused in this design (linker's sequence is "GGCAGCA CCGGATCC"), and SmCYP and SmCPR are used as examples here (the sequence is shown in Sequence 2). Forward primer for SmCYP is SmCYP_F with sequence "ATAAACAA*ACAAAATG*GACTCTTTTCCATT ATTG"; reverse primer for SmCYP is SmCYP_R with sequence "GATCCGG*TGCTGCC*GGACTTGACGATTGG." Forward primer for SmCPR is SmCPR_F with sequence "GGCAGCA*CCGGATC* CGAACCATCCTCTAAAAA," and reverse primer for SmCPR is SmCPR_R with sequence "GATGGTGA*TGATGATG*CCAGACAT CTCTCAAGTA." pYPRC15-URA-UGPD, SmCYP, and SmCPR are amplified with corresponding primers and assembled according to procedure described in Sections 2.1.3–2.1.7.

4.1.2 Yeast Transformation

The constructed yeast expression vector is integrated into YPRC15 locus of *S. cerevisiae* BY4700 (MATa ura3Δ0) by using commercial kit 'S. c. EasyComp Transformation Kit' (Invitrogen) and following the manufacturer's instruction. Transformant is selected on YNB CSM-URA glucose agar plate (Section 4.1.3) – colonies usually appear after being incubated at 30°C for 2 days. A colony is inoculated into 10 mL of YNB CSM-URA glucose medium (Section 4.1.4) in a 50-mL Falcon tube, and incubated at 30°C/250 rpm for 2 days. Then 0.5 mL of the cell suspension is mixed with 0.5 mL of 80% glycerol solution (autoclaved) to make yeast glycerol stock, which can be stored at −80°C for up to 5 years,

SEQUENCE 2 Sequence of SmCYP_SmCPR
ATGGACTCTTTTCCATTATTGGCTGCCTTGTTTTTCATTGCTGCTACTATTACCTTCTTGT
CCTTCCGTAGAAGAAGAAATTTGCCACCAGGTCCATTTCCATATCCAATCGTTGGTAA
TATGTTGCAATTGGGTGCTAACCCACATCAAGTTTTTGCTAAGTTGTCTAAGAGATAC
GGTCCATTGATGTCCATTCATTTGGGTTCCTTGTACACCGTTATAGTCTCTTCACCAGA
AATGGCCAAAGAAATCTTGCATAGACATGGTCAAGTTTTCTCCGGTAGAACTATTGCT
CAAGCTGTTCATGCTTGTGATCACGATAAGATTTCTATGGGTTTTTTGCCAGTTGCCTC
TGAATGGAGAGATATGAGAAAGATCTGCAAAGAACAAATGTTCTCCAATCAATCCAT
GGAAGCTTCTCAAGGTTTGAGAAGACAAAAGTTGCAACAATTATTGGACCACGTCCA
AAAGTGTTCTGATTCTGGTAGAGCTGTTGATATTAGAGAAGCTGCTTTCATTACCACC
TTGAATTTGATGTCTGCTACCTTGTTTTCTTCACAAGCTACCGAATTTGATTCCAAGGC
TACCATGGAATTCAAAGAAATTATTGAAGGTGTTGCCACCATCGTTGGTGTTCCAAAT
TTTGCTGATTACTTCCCAATCTTAAGACCATTCGATCCACAAGGTGTTAAGAGAAGAG
CTGATGTTTTTTTCGGTAAGTTGTTGGCCAAGATCGAAGGTTATTTGAACGAAAGATT
GGAATCCAAGAGAGCTAATCCAAACGCTCCAAAGAAGGATGATTTCTTGGAAATCGT
TGTCGATATCATCCAAGCCAACGAATTCAAGTTGAAAACCCATCATTTCACCCACTTG
ATGTTGGATTTGTTTGTTGGTGGTTCTGATACCAACACCACTTCTATTGAATGGGCTAT
GTCTGAATTGGTTATGAACCCAGATAAGATGGCTAGATTGAAGGCTGAATTGAAATC
TGTTGCTGGTGACGAAAAGATCGTTGATGAATCTGCTATGCCAAAGTTGCCATACTTG
CAAGCTGTTATCAAAGAAGTCATGAGAATTCATCCACCTGGTCCTTTGTTGTTACCAA
GAAAAGCTGAATCCGATCAAGAAGTCAACGGTTACTTAATTCCAAAGGGTACTCAAA
TCTTGATTAACGCTTACGCCATTGGTAGAGATCCATCTATTTGGACTGATCCAGAAAC
TTTTGACCCAGAAAGATTCTTGGACAACAAGATCGATTTCAAGGGTCAAGACTACGA
ATTATTGCCATTTGGTTCAGGTAGAAGAGTTTGTCCAGGTATGCCATTGGCTACTAGA
ATATTGCATATGGCTACTGCTACTTTGGTTCACAATTTCGATTGGAAGTTGGAAGATG
ATTCTACTGCTGCTGCTGATCATGCTGGTGAATTATTTGGTGTTGCTGTTAGAAGAGC
AGTCCCATTGAGAATTATTCCAATCGTCAAGTCCGGCAGCACCGGATCCGAACCATC
CTCTAAAAAGTTGTCCCCATTGGATTTCATTACCGCCATTTTGAAGGGTGATATTGAA
GGTGTTGCTCCAAGAGGTGTTGCAGCTATGTTGATGGAAAACAGAGATTTGGCTATG
GTTTTGACTACCTCTGTTGCTGTTTTGATTGGTTGCGTTGTTGTTTTGGCTTGGAGAAG
AACTGCTGGTTCTGCTGGTAAAAAACAATTGCAACCACCAAAGTTGGTTGTTCCAAA
ACCAGCTGCTGAACCTGAAGAAGCTGAAGACGAAAAAACTAAGGTCAGTGTTTTCT
TCGGTACTCAAACTGGTACTGCTGAAGGTTTTGCTAAAGCTTTTGCCGAAGAAGCTA
AAGCTAGATATCCACAAGCTAAGTTCAAGGTTATCGATTTGGATGATTACGCTGCCG
ATGATGATGAATACGAAGAAAAGTTGAAGAAAGAATCCTTGGCCTTCTTCTTCTTGG
CTTCTTATGGTGATGGTGAACCTACTGATAATGCTGCTAGATTTTACAAGTGGTTCAC
CGAAGGTAAGGATAGAGAAGATTGGTTGAAGAACTTGCAATACGGTGTTTTTGGTTT
GGGTAACAGACAATACGAACACTTCAACAAGATTGCCATCGTTGTCGATGATTTGAT
TACTGAACAAGGTGGTAAGAAGTTGGTTCCAGTTGGTTTAGGTGATGATGATCAATG
CATCGAAGATGATTTTTCCGCTTGGAGAGAATTGGTTTGGCCAGAATTGGATAAGTT
GTTGAGAAATGAAGATGATGCTACTGTTGCTACTCCATATACCGCTGTTGTTTTACAA
TACAGAGTTGTCTTGCACGATCAAACTGATGGTTTGATCACAGAAAATGGTTCTCCAA
ATGGTCATGCTAACGGTAACACTATCTATGATGCTCAACATCCATGTAGAGCTAACG

Continued

SEQUENCE 2 Sequence of SmCYP_SmCPR—cont'd

TTGCTGTTAGAAGAGAATTGCATACTCCAGCTTCAGATAGATCTTGTACCCATTTGGA
ATTCGATACTTCAGGTACTGGTTTGGTTTACGAAACTGGTGATCATGTTGGTGTTTAC
TGCGAAAACTTGTTGGAAAATGTCGAAGAAGCCGAAAAGTTATTGAACTTGTCTCCA
CAAACCTACTTCTCCGTTCATACTGATAACGAAGATGGTACTCCATTGTCTGGTTCTT
CATTGCCACCACCATTTCCACCATGTACTTTGAGAACTGCTTTGACTAAGTACGCCGA
TTTGATTTCTATGCCAAAGAAGTCTGTTTTGGTTGCTTTGGCTGAATACGCCTCTAATC
AATCAGAAGCTGATAGATTGAGATACTTGGCTTCACCAGATGGTAAAGAAGAATACG
CCCAATATATCGTTGCCTCCCAAAGATCATTATTGGAAGTTATGGCTGAATTCCCATC
TGCTAAACCACCATTGGGTGTTTTTTTGCTGCTATTGCTCCTAGATTGCAACCTAGAT
TCTACTCCATTTCTTCCTCCCCAAAAATTGCTCCAACTAGAGTTCATGTTACCTGTGCT
TTGGTTTATGATAAGACTCCAACTGGTAGAATCCATAAGGGTATTTGTTCTACCTGGA
TTAAGAACGCTGTTCCATTGGAAGAATCTTCAGATTGCTCTTGGGCTCCAATTTTCATC
AGAAACTCTAACTTTAAGTTGCCAGCCGATCCAAAGGTTCCAATTATCATGGTTGGTC
CAGGTACAGGTTTAGCTCCTTTTAGAGGTTTCTTACAAGAAAGATTGGCCTTGAAAGA
ATCTGGTGCTGAATTGGGTCCAGCTATTTTGTTTTTTGGTTGTAGAAACAGAAAGATGG
ACTTCATATACGAAGATGAATTGAACTCCTTCGTTAAGGTTGGTGCCATTTCTGAATTG
ATCGTTGCTTTTTCTAGAGAAGGTCCAGCCAAAGAATACGTTCAACATAAGATGTCTC
AAAGAGCCTCCGATATTTGGAAGATGATATCTGATGGTGGTTACATGTACGTTTGTGG
TGATGCTAAAGGTATGGCTAGAGATGTTCATAGAACCTTGCATACCATTGCTCAAGAA
CAAGGTTCTTTGTCATCTTCTGAAGCAGAAGGTATGGTCAAAAACTTGCAAACTACTG
GTAGATACTTGAGAGATGTCTGGCATCATCATCACCATCACTAA

4.1.3 YNB CSM-URA Glucose Agar Plate

Seven-point 5 g of agar is mixed with 350 mL of deionized water and autoclaved. When it is cooled down to 60°C, 50 mL of $10 \times$ YNB solution (with ammonium sulfate, filtration sterilized), 50 mL of $10 \times$ CSM-URA solution (complete synthetic medium with uracil dropped off, autoclaved), and 50 mL of 200 g/L glucose (autoclaved) are added. Twenty milliliters of the mixture are poured into a 90-mm petri dish to make an YNB CSM-URA glucose agar plate.

4.1.4 YNB CSM-URA Glucose Medium

Fifty milliliters of $10 \times$ YNB solution (with ammonium sulfate, filtration sterilized), 50 mL of $10 \times$ CSM-URA solution (complete synthetic medium with uracil dropped off, autoclaved), 50 mL of 200 g/L glucose (autoclaved), and 350 mL of deionized water (autoclaved) are mixed to make 500 mL of YNB CSM-URA glucose medium.

4.2 Characterization of the *S. cerevisiae*

Activity of the CYP expressed in *S. cerevisiae* can be quantified by feeding its substrate (ISM) into its culture and monitoring substrate oxygenation.

4.2.1 Preparation of Isoprenoid-DMSO Stock Solution

C20 isoprenoid scaffold is highly soluble in DMSO, a solvent that does not affect yeast growth when its concentration is less than 1% (v/v) in yeast culture. Two milligrams of the HPLC-purified isoprenoid (Section 3.2.5) are dissolved in 0.4 mL of DMSO (5 g/L of isoprenoid), and this solution is sterilized by using 13 mm 0.2 µm Nylon filter.

4.2.2 Preparation of Seed Culture

Small quantity of yeast glycerol stock is inoculated into 1 mL of YNB CSM-URA glucose medium – held in 14 mL Falcon tube – by using disposable 1 µL inoculation loop. The tube is incubated at 30°C/250 rpm for 2 days.

4.2.3 Feeding Isoprenoid into Yeast Culture

Fifteen microliters of fully grown yeast culture is inoculated into 1.5 mL of YPD medium in a sterile VWR 16 × 100 mm (diameter and height) round-bottom culture tube and incubated at 30°C/250 rpm. When cell density reached 1 (OD600), 15 µL of the isoprenoid-DMSO stock solution is added. The incubation continued for 96 h, and 100 µL of cell suspension is taken every 24 h for GCMS analysis, which is carried out as described in Sections 2.2.7–2.2.8.

4.3 Coculture of the *S. cerevisiae* with an Isoprenoid-Overproducing *E. coli*

4.3.1 Preparation of Yeast Preculture

The *S. cerevisiae* strain constructed in Section 4.1.2 and the best *E. coli* strain obtained in Section 2.2.6 are cocultured in the same bioreactor. To ensure their coexistence, they are cultured in a medium which contains xylose as the sole carbon source. The two microbes thus have a mutualistic relationship in this case: *S. cerevisiae* cannot utilize xylose and relies on *E. coli* to access carbon; *E. coli* naturally produces acetate – a by-product inhibitory to its own growth – and depends on *S. cerevisiae* to remove it. To ensure no acetate accumulation in bioreactor, an excess of yeast relative to *E. coli* needs to be inoculated and the needed preculture is prepared by addition of 1 mL of fully grown seed culture into 50 mL of YPD, followed by incubation at 30°C/250 rpm for 24 h.

4.3.2 Bioreactor Run

Bioreactor used is the same to that described in Section 3.1, and it is prepared according to protocol in Section 3.2.1. The bioreactor is initiated and operated as described in Section 3.2.2 except that pellet of 50 mL of fully grown yeast preculture is inoculated together with the *E. coli* seed culture – the yeast pellet is resuspended in the *E. coli* seed culture prior to inoculation. Feeding strategy applied here is the same to that in Section 3.2.3. Sample is taken every 24 h for cell density, GCMS, and HPLC analysis. GCMS samples are processed as described in Sections 2.2.7–2.2.8, and HPLC samples are processed according to Section 3.2.4. For cell density analysis, 500 µL of coculture are loaded onto 1 mL of 45% sucrose solution in a 14-mL Falcon tube, and the tube is centrifuged at $2100 \times g$ for 2 min. After centrifugation, all yeast cells are pelleted and most *E. coli* cells remain in supernatant. Then cell density of each species can be quantified by measuring absorbance at 600 nm (OD600).

5. DISCUSSION

Protocols provided in this chapter need to be revised for producing other types of isoprenoids. Plasmids for expressing IsoSs in this chapter all contain GGPP synthase (Fig. 4), which needs to be replaced if target isoprenoids are derived from FPP, GPP, DMAPP, or IPP. Isoprenoids with larger backbones (C30, C40, etc.) generally cannot diffuse through cell membranes, thus their biosynthetic pathway cannot be distributed between two cells. The small isoprenoids (C5 and C10) are very volatile, so they should be recovered from off-gas exit of the bioreactor or trapped in the bioreactor by using a nontoxic organic solvent such as dodecane. C15 isoprenoids have similar properties to the C20 ones; hence, the same bioreactor configuration and operating conditions can be applied to this type of isoprenoids.

REFERENCES

Ajikumar, P. K., Tyo, K., Carlsen, S., Mucha, O., Too, H. P., & Stephanopoulos, G. (2008). Terpenoids: Opportunities for biosynthesis of natural product drugs using engineered microorganisms. *Molecular Pharmaceutics, 5,* 167–190.

Ajikumar, P. K., Xiao, W. H., Tyo, K. E., Wang, Y., Simeon, F., Leonard, E., et al. (2010). Isoprenoid pathway optimization for Taxol precursor overproduction in Escherichia coli. *Science, 330,* 70–74.

Alper, H., Jin, Y. S., Moxley, J. F., & Stephanopoulos, G. (2005). Identifying gene targets for the metabolic engineering of lycopene biosynthesis in Escherichia coli. *Metabolic Engineering, 7,* 155–164.

Alper, H., & Stephanopoulos, G. (2007). Global transcription machinery engineering: A new approach for improving cellular phenotype. *Metabolic Engineering, 9*, 258–267.

Artsatbanov, V. Y., Vostroknutova, G. N., Shleeva, M. O., Goncharenko, A. V., Zinin, A. I., Ostrovsky, D. N., et al. (2012). Influence of oxidative and nitrosative stress on accumulation of diphosphate intermediates of the non-mevalonate pathway of isoprenoid biosynthesis in corynebacteria and mycobacteria. *Biochemistry (Moscow), 77*, 362–371.

Biggs, B. W., De Paepe, B., Santos, C. N., De Mey, M., & Kumaran Ajikumar, P. (2014). Multivariate modular metabolic engineering for pathway and strain optimization. *Current Opinion in Biotechnology, 29*, 156–162.

Cuellar, M. C., & van der Wielen, L. A. (2015). Recent advances in the microbial production and recovery of apolar molecules. *Current Opinion in Biotechnology, 33*, 39–45.

Krivoruchko, A., & Nielsen, J. (2015). Production of natural products through metabolic engineering of Saccharomyces cerevisiae. *Current Opinion in Biotechnology, 35*, 7–15.

Martínez, I., Zhu, J., Lin, H., Bennett, G. N., & San, K.-Y. (2008). Replacing Escherichia coli NAD-dependent glyceraldehyde 3-phosphate dehydrogenase (GAPDH) with a NADP-dependent enzyme from Clostridium acetobutylicum facilitates NADPH dependent pathways. *Metabolic Engineering, 10*, 352–359.

Shen, R., Jiang, X., Ye, W., Song, X., Liu, L., Lao, X., et al. (2011). A novel and practical synthetic route for the total synthesis of lycopene. *Tetrahedron, 67*, 5610–5614.

Tyo, K. E., Ajikumar, P. K., & Stephanopoulos, G. (2009). Stabilized gene duplication enables long-term selection-free heterologous pathway expression. *Nature Biotechnology, 27*, 760–765.

Wilson, S. A., Cummings, E. M., & Roberts, S. C. (2014). Multi-scale engineering of plant cell cultures for promotion of specialized metabolism. *Current Opinion in Biotechnology, 29*, 163–170.

Wilson, S. A., & Roberts, S. C. (2014). Metabolic engineering approaches for production of biochemicals in food and medicinal plants. *Current Opinion in Biotechnology, 26*, 174–182.

Yuan, L. Z., Rouviere, P. E., Larossa, R. A., & Suh, W. (2006). Chromosomal promoter replacement of the isoprenoid pathway for enhancing carotenoid production in E. coli. *Metabolic Engineering, 8*, 79–90.

Zhou, K., Qiao, K., Edgar, S., & Stephanopoulos, G. (2015). Distributing a metabolic pathway among a microbial consortium enhances production of natural products. *Nature Biotechnology, 33*, 377–383.

Zhou, K., Zou, R., Stephanopoulos, G., & Too, H. P. (2012a). Enhancing solubility of deoxyxylulose phosphate pathway enzymes for microbial isoprenoid production. *Microbial Cell Factories, 11*, 148.

Zhou, K., Zou, R., Stephanopoulos, G., & Too, H. P. (2012b). Metabolite profiling identified methylerythritol cyclodiphosphate efflux as a limiting step in microbial isoprenoid production. *PloS One, 7*, e47513.

Zou, R., Zhou, K., Stephanopoulos, G., & Too, H. P. (2013). Combinatorial engineering of 1-deoxy-D-xylulose 5-phosphate pathway using cross-lapping in vitro assembly (CLIVA) method. *PloS One, 8*, e79557.

CHAPTER ELEVEN

Natural Product Biosynthesis in *Escherichia coli*: *Mentha* Monoterpenoids

H.S. Toogood*, S. Tait*, A. Jervis†, A. Ní Cheallaigh‡, L. Humphreys§, E. Takano*, J.M. Gardiner‡, N.S. Scrutton*,[1]

*Manchester Institute of Biotechnology, Faculty of Life Sciences, University of Manchester, Manchester, United Kingdom
†Manchester Institute of Biotechnology, SYNBIOCHEM, University of Manchester, Manchester, United Kingdom
‡School of Chemistry, Manchester Institute of Biotechnology, University of Manchester, Manchester, United Kingdom
§GlaxoSmithKline, Medicines Research Centre, Stevenage, United Kingdom
[1]Corresponding author: e-mail address: Nigel.Scrutton@manchester.ac.uk

Contents

1. Introduction — 248
2. Operon Construction — 251
   2.1 Vector and Gene Selections — 251
   2.2 Assembly of *Mentha*-Like Biosynthetic Pathways in *E. coli* — 253
3. Multigene Protein Expression and Validation — 257
4. Biotransformations — 258
   4.1 General Monoterpenoid Reaction and Detection — 259
   4.2 Alternative Protocols — 260
5. Further Studies — 261
   5.1 Expression Constructs — 261
   5.2 Strain and Expression Optimization — 262
   5.3 High-Throughput Optimization Studies — 263
6. Conclusions — 266
Acknowledgments — 267
References — 267

Abstract

The era of synthetic biology heralds in a new, more "green" approach to fine chemical and pharmaceutical drug production. It takes the knowledge of natural metabolic pathways and builds new routes to chemicals, enables nonnatural chemical production, and/or allows the rapid production of chemicals in alternative, highly performing organisms. This route is particularly useful in the production of monoterpenoids in microorganisms, which are naturally sourced from plant essential oils. Successful pathways are constructed by taking into consideration factors such as gene selection, regulatory elements, host selection and optimization, and metabolic considerations of the host organism. Seamless pathway construction techniques enable a "plug-and-play" switching of genes and regulatory parts to optimize the metabolic functioning in vivo. Ultimately, synthetic biology approaches to microbial monoterpenoid production may revolutionize "natural" compound formation.

1. INTRODUCTION

Natural products are organic compounds produced enzymatically by living organisms, often through pathways of secondary metabolism (Yonekura-Sakakibara & Saito, 2009). While secondary metabolites are not essential for survival, they often give the organism advantages for survival. For example, cytotoxic compounds produced by some plants can act as a form of chemical warfare against predators and competing organisms (Hunter, 2008). As natural products often have biological or pharmacological activity, they have traditionally been included, or are the active components of both traditional and modern medicines (Atanasov et al., 2015). Additional structural diversity is achievable through chemical, and more recently enzymatic biosynthesis, and synthetic analogues can be prepared with increased potency and purity. It is therefore not surprising that natural products have often been seen as the starting points in drug discovery screens (Hunter, 2008; Li & Vederas, 2009).

Many natural products have complex structures, often containing regiospecific functionalization and chiral centers. These features are largely responsible for their biological activity and potency. Traditionally, these compounds were obtained by extractions and extensive purifications from native plant species. This route is often problematic for several reasons: (i) plant natural products are often in low concentrations and require long growth periods; (ii) yields of compound(s) are subject to regional, seasonal, and climate variations; and (iii) purification procedures are often complex, costly, and challenging if pure enantiomeric compounds are needed. Consequently, there have been many studies focussed on standardizing and increasing the yields of primary and secondary natural products in both native plants and cell cultures (Hussain et al., 2012). Complete chemical synthesis of natural products provides an alternative route; however, the lengthy steps involved can result in an overall low yield of products, and the use of costly and/or toxic chemicals. For example, the total chemical synthesis of Taxol, a potent anticancer agent originally isolated from *Taxus brevifolia*

(Pacific yew tree), yielded only 0.02% of the final product (Nicolaou, Yang, Liu, Ueno, & Nantermet, 1994).

Limonene and derivatives are the most abundant monocyclic monoterpenoids in nature (Duetz, Bouwmeester, van Beilen, & Witholt, 2003). (S)-Limonene is naturally found in herbs such as *Mentha* species (peppermint and spearmint), while the more common (R)-enantiomer is a major oil constituent of orange and lemon peels. Limonene derivatives are often used in fragrances, perfumes, and flavors, such as menthol isomers and carvone giving the distinctive aroma of peppermint and spearmint, respectively (Fig. 1). Menthol isomers are commonly used as flavors in oral hygiene products, food, and beverages. Carveol is an additive of cosmetics, while pulegone is frequently found in perfumes and aromatherapy products. Within the pharmaceutical industry, carvone and carveol are known to have anticancer properties, while menthol has antibacterial properties against *Staphylococcus aureus* and *Escherichia coli* (Ajikumar et al., 2008).

There is a high demand for limonene and derivatives (eg, menthol oil ca. 31,000 t/$373–401 US million pa); however, the flavor and fragrance industries supply most of these compounds from natural sources (extract from *Mentha* spp.) to be compatible with use in food products. These sources are at the mercy of environmental conditions, volatile prices, and arable land competition for more profitable crops (eg, for biofuels production). Additionally, menthol oil requires expensive and low yielding steam distillation and filtration processes to produce a usable commodity (Lange et al., 2011; Lawrence, 2007). There are two major synthetic routes to menthol; the Haarmann–Reimer route from *m*-cresol and the Takasago synthesis starting with β-pinene (Sell, 2003). These processes generate annual yields of crude menthol to the value of around US$ 300 million (Croteau et al., 2005). However, given that the latter product is not from natural sources, alternative clean (bio)synthetic routes to these compounds are a commercially attractive prospect.

An alternative "natural" route to pure organic compounds is to utilize microorganisms as biological factories, using either existing or introduced de novo pathways (Chemler & Koffas, 2008; Mitchell, 2011). This synthetic biology approach generates fine chemicals via an enzymatic route utilizing rapidly growing, cost effective, and even food-compatible microorganisms grown on nonpetroleum-based renewable feedstock (Ajikumar et al., 2010; Friedman & Ellington, 2015; Winter & Tang, 2012). For example, *E. coli*, *Saccharomyces cerevisiae* (Baker's yeast), and cyanobacteria have been shown to produce plant terpenoids, such as limonene, by the inclusion of plant terpene

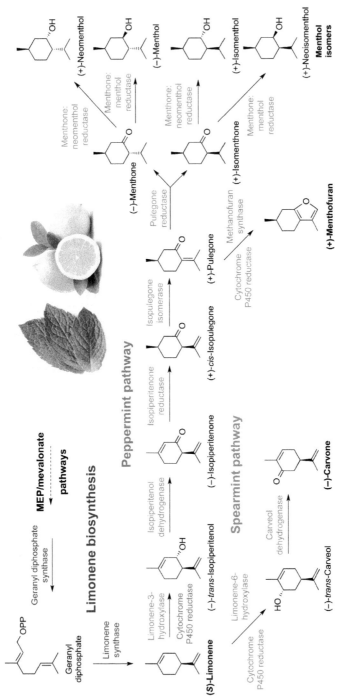

Fig. 1 Biosynthetic pathways of essential oil monoterpenoids by *Mentha piperita* (peppermint) and *M. spicata* (spearmint). The enzymes involved in each pathway are color coded for limonene (*blue*), peppermint (*red*), and spearmint (*magenta*) biosynthetic pathways (Croteau, Davis, Ringer, & Wildung, 2005; Toogood et al., 2015). (See the color plate.)

synthases, and the upregulation of enzymes involved in isoprene precursor production (Alonso-Gutierrez et al., 2013; Carter, Peters, & Croteau, 2003; Farmer & Liao, 2001; Jackson, Hart-Wells, & Matsuda, 2003; Kiyota, Okuda, Ito, Hirai, & Ikeuchi, 2014; Martin, Pitera, Withers, Newman, & Keasling, 2003). The semisynthetic industrial-scale production (ca. 35 t per annum) of the sesquiterpene lactone artemisinin by Sanofi, a major active ingredient in modern malarial treatments, has shown commercial success, in comparison to traditional extractions and purifications from natural sources (sweet wormwood) (Chang & Keasling, 2006). In the case of semisynthetic Taxol production, a multivariate-modular approach was successful, where the pathways were partitioned into (i) native upstream methylerythritol phosphate (MEP) pathway to form isopentenyl pyrophosphate and (ii) a heterologous downstream taxadiene (taxol precursor)-forming pathway (Ajikumar et al., 2010).

Recent work has shown progress in a synthetic biology approach to the formation of menthol isomers by *E. coli* using biosynthetic pathway precursors (Fig. 1) as the starting material (Toogood et al., 2015). In this approach, a de novo enzymatic route based on the *Mentha piperita* biosynthetic pathway was constructed in *E. coli*, and whole-cell extracts were successful in the synthesis of menthol from pulegone. Given the potential of synthetic biology techniques to revolutionize the production of "naturally sourced" organic compounds (Breitling & Takano, 2015), this paper will discuss general techniques applicable to this rapidly advancing technology, using the menthol production study protocols of "operon" construction and natural product synthesis/analysis (Toogood et al., 2015; Yadav, De Mey, Giaw Lim, Kumaran Ajikumar, & Stephanopoulos, 2012). The methodologies are suitable for generating constructs complete with reporter tags and multiple repeating sequences, the latter of which is often challenging to obtain.

2. OPERON CONSTRUCTION
2.1 Vector and Gene Selections

One advantage of constructing biosynthetic pathways in *E. coli*, as opposed to yeast, is the ability to create de novo synthetic polycistronic gene modules (operons) (Blount, Weenink, & Ellis, 2012). This reduces the DNA burden, as a single promoter/terminator pair can potentially control multiple genes. The eukaryotic nature of yeast makes it more challenging to predict gene regulation; however, the inclusion of synthetic regulatory networks in *S. cerevisiae* shows the potential of generating biofactories in yeast (Blount et al., 2012). Given that the *E. coli* molecular biology "toolkit" is considerably more extensive than in yeast and other organisms, it is not surprising that new synthetic biology pathways are often constructed and tested in bacterial systems. Successful constructs can then be imported into more publically acceptable microorganisms (eg, Baker's yeast) and grown on renewable waste feedstock. To date, terpenoid metabolic engineering in microorganisms has been focused mostly on the production of carotenoids, such as lycopene and β-carotene, precursors for approved drugs, such as artemisinin and paclitaxel, and other terpenoids (Ajikumar et al., 2008).

There are a variety of commercially available vectors for the expression of one or multiple genes in *E. coli*. Perhaps one of the most commonly used collections within research facilities is the pET system (Merck Millipore, Billerica, Massachusetts; http://www.merckmillipore.com), where protein expression is controlled by the presence of an IPTG-inducible bacteriophage T7 RNA polymerase within the host cell. The arabinose-inducible pBAD system is also popular (Invitrogen, ThermoFisher, Waltham, Massachusetts; https://www.thermofisher.com), but it typically yields lower protein expression levels. However, it has the advantage of a tighter regulation of protein expression over pET vectors. Inclusive within these vectors are a variety of antibiotic (selection) genes and a variety of protein tags and cleavage sites. The addition of these N- or C-terminal tags to enzymes can be adventitious for several reasons: (i) increase soluble protein expression, (ii) rapid affinity chromatography of the target protein(s), and (iii) detection of recombinant protein by Western blotting.

Transcription of multiple genes within a bacterial operon is possible due to the presence of Shine–Dalgarno (SD) sequences; ribosomal-binding sites within messenger RNA upstream of the start codon of successive genes. This enables the ribosome to continue to initiate protein synthesis from the next gene without the need for a second full promoter sequence. The SD sequences can vary in length and nucleotide composition, which ultimately affects the expression levels of the downstream gene. There are a variety of algorithms' available for both predicting SD sequences in bacterial genomes (Starmer, Stomp, Vouk, & Bitzer, 2006), as well as designing novel sequences to maximize expression of a specific gene (eg, http://parts.igem.org/Ribosome_Binding_Sites/Design). Other factors to consider when designing synthetic operons are the choice of the promoter/terminator pair and even the order of the genes in the construct (Starmer et al., 2006). With very long constructs (greater than five genes), secondary promoter elements may be incorporated to boost downstream gene expression.

The selection of the gene homologues is an important consideration in successful biofactory design. For example, simply choosing the exact gene sequences encoding enzymes responsible for a native plant secondary metabolite pathway may not be successful in *E. coli*. This could be due to factors including insoluble protein expression, absence of plant-specific posttranslational modifications, incorrect protein folding and/or cofactor incorporation, and poor *E. coli*-specific codon usage in the plant genes. Therefore, gene sequences may need to be altered, such as codon optimization for the chosen host and the removal of any plant-specific targeting sequences. Alternatively, selection of gene homologues from bacteria may increase the likelihood of soluble, functional protein expression, as may selecting genes with a higher predicted protein solubility (Wilkinson & Harrison, 1991).

A recent study describing the generation of menthol isomers from pulegone in *E. coli* (Toogood et al., 2015) highlights the importance of some of the considerations above when designing operons in *E. coli*. In this approach, the vector pET21b was chosen due to the high expression levels of the individual proteins and the presence of hexa-histidine tags. The latter was important to enable the positive identification of recombinant protein expression via Western blotting and to enable successful gene expression in one case (menthone: neomenthol reductase (MNMR); Fig. 1). Gene selections were taken from both the native *M. piperita* pathway (Croteau et al., 2005) and a previously published bacterial homologue (Mansell et al., 2013). Each gene was synthesized, employing codon-optimization for improved *E. coli* soluble expression. The following sections will describe techniques employed in the construction of partial mint biosynthetic pathways and discuss the general applicability of these techniques in operon generation.

2.2 Assembly of *Mentha*-Like Biosynthetic Pathways in *E. coli*

The ideal cloning strategy involves a seamless stitching of genes and regulatory parts together, without the limitations of restriction digests and the associated cloning "scars." A number of techniques and commercial kits are available that fulfill these requirements, such as In-Fusion cloning (Zhu, Cai, Hall, & Freeman, 2007), Gibson assembly (Gibson et al., 2009), ligase cycling reaction (de Kok et al., 2014), and megaprimer PCR (Unger, Jacobovitch, Dantes, Bernheim, & Peleg, 2010). The chosen assembly method for the *Mentha* pathways in *E. coli* was the In-Fusion system (Toogood et al., 2015), where linearized vectors and gene insert(s) are fused in a precise order and orientation by the presence of 15 bp overlaps at each end of the DNA fragments.

2.2.1 General Assembly Protocols

The success of seamless cloning DNA assembly is dependent on the presence of unique sequences at each end of the DNA parts, to enable DNA fragments to be joined in the correct order and orientation. Any intergenic regions or other DNA alterations (eg, SD sequences) are added to the parts by PCR amplification prior to construct assembly. The vector must also be linearized at the required insertion site.

The following are the general protocols utilized to enable the successful generation of the three-gene expression construct (Toogood et al., 2015). The vector (pET21b) already contained the first gene, namely, double bond reductase (NtDBR) from *Nicotiana tabacum*, containing a C-terminal His_6-tag. The two genes to be inserted were menthone: menthol reductase (MMR) and MNMR from *M. piperita* to form the DMN construct (Fig. 1). Both genes contained a C-terminal His_6-tag, however, MNMR also required an N-terminal His_6-tag to enable successful protein expression in *E. coli* (Toogood et al., 2015). Additionally, both MMR and MNMR required the addition of SD sequences at the 5′ end to enable coexpression of the three genes.

1. Inverse PCR of NtDBR-pET21b to generate a linearized single-gene construct. The forward and reverse primers anneal to the vector terminator region and 3′-end of NtDBR, respectively. This is followed by selective digestion of the template DNA by the restriction enzyme *Dpn*I.

    Note: *Dpn*I requires the presence of N6-methyladenine within the recognition sequence to cleave DNA. Therefore, the template DNA must have been produced by a Dam^+ *E. coli* strain for digestion to occur.

2. PCR amplification of the two additional genes, incorporating a SD sequence at the 5′ end of the genes. The 5′ and 3′ regions also contain a 15 bp overlap with the DNA sequences to be ligated to.

3. In-Fusion ligation reaction between the three DNA-linearized parts followed by transformation into Stella competent *E. coli* cells (Clontech, Mountain View, California; www.clontech.com). The transformation mixture is plated onto an antibiotic selection plate and incubated overnight at 37°C.

    Note: This strain of *E. coli* is specifically designed to survive the presence of toxic additives found in the In-Fusion reaction mixture. Using other competent cell strains may lead to no colony formation.

4. Colonies are screened for the presence of the two additional genes by colony pick PCR (Woodman, 2008). Confirmation of the correct sequence and gene orientations is performed by DNA sequencing.

2.2.2 In-Fusion Reactions with Repeating Sequences

Difficulties may arise when one or both termini of multiple pairs of DNA parts contain the same sequence with a high melting temperature. For example, with construct DMN, the junction between both NtDBR–MMR and MMR–MNMR contain identical C-terminal regions (CTCGAGCACCACC ACCACCACCACTGA) and 5′ SD sequence (GGAGGACAGCTAA), with a combined melting temperature of 73°C (Toogood et al., 2015). To avoid this, C-terminal His_6-tags can be eliminated, and/or nonidentical SD sequences may be used between different gene pairs. Where this is not optimal, primer design for gene amplification is critical to ensure there are no repeating units contained within the 15 bp overhangs.

One solution is to ensure that the repeating unit is wholly contained within either the 3′ end of one gene or 5′ end of the next gene, with no repeating sequence spanning both genes of the interface. For example, construct DMN could be constructed with nonidentical SD sequences (S1 and S2; Fig. 2A), while maintaining the C-terminal His_6-tags. In this case, NtDBR-pET21b could be linearized with the addition of the first SD sequence by PCR. The second gene MMR is amplified with both the first and second SD sequences at the 5′ and 3′ end, respectively. The final gene MNMR is amplified with the second SD sequence, and vector terminator region overhangs at the 5′ and 3′ ends, respectively. Therefore, the required 15 bp overlaps are contained wholly within either the nonidentical SD sequences or the terminator region of the vector (Fig. 2A). This would allow a one-step assembly reaction to be performed with all genes containing C-terminal His_6-tags.

In cases where the repeating sequence spans both genes within the fusion interface, a multistep approach can be employed (Fig. 2B). The published DMN construct method required both NtDBR-pET21b and MMR to be PCR amplified with identical SD sequences added as the 15 bp overhang at the 3′ and 5′ ends, respectively (Toogood et al., 2015). However, the 3′ end of MMR was amplified without its stop codon, and contained the 5′ end of the vector terminator region. This was followed by a two-gene construct In-Fusion ligation, then latter linearization by PCR at the junction between MMR and the terminator region. The final gene

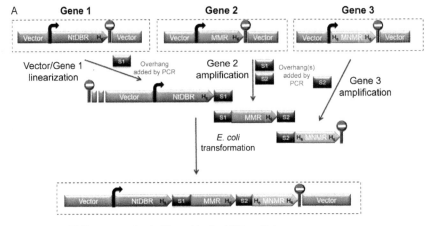
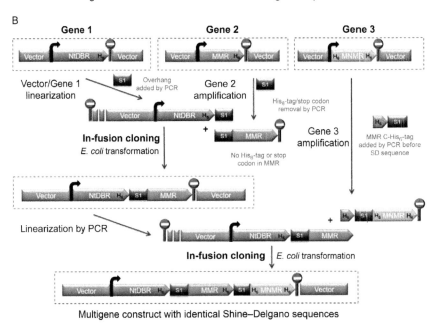
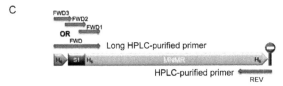

Fig. 2 PCR-based In-fusion cloning to create multigene expression constructs. (A) Single In-fusion cloning step protocol for constructs without significant repeating sequences. (B) Multiple In-fusion cloning steps protocol where repeating sequences (His_6-tag—Shine–Dalgarno sequence) occur at the cloning termini. (C) Construction of large sequence overhangs at the 5′ end of gene MNMR by PCR. *H6*, hexa-histidine tag; *S1/S2*, Shine–Dalgarno sequences. (See the color plate.)

MNMR was amplified with a very long 5′ overhang containing the His$_6$-tag/stop codon of MMR followed by the SD sequence (Fig. 2B). The 3′ end aligns to the vector terminator region, and the final three-gene In-Fusion assembly was conducted. Generation of lengthy overhangs (30–80 bp) can be readily achieved by PCR using either very long HPLC-purified primers, or a cascading series of smaller overlapping oligos (Fig. 2C). In the latter case, the ratio of the oligos is biased in favor of the outermost oligo (eg, oligos FWD1:FWD2:FWD3 with ratio 1:2:5) to maximize the amplification of full-length PCR product (Toogood et al., 2015). Therefore, the presence of repeating units at the junction between genes to be assembled does not necessarily limit the success of seamless cloning strategies.

3. MULTIGENE PROTEIN EXPRESSION AND VALIDATION

Successful production of monoterpenoids by microbial hosts is not solely dependent on sufficient expression levels of soluble, active recombinant enzymes. Other factors play important roles such as (i) impact on central metabolism, (ii) toxicity of the recombinant enzyme overexpression and monoterpenoid substrate(s)/product, (iii) colocalization of the enzymes with the substrate(s) of secondary metabolite production, (iv) competing pathways depleting monoterpenoid precursors or catalyzing side reactions, and (v) sufficient cofactor/coenzyme supply for both primary and secondary metabolism. These latter factors often require microbial strain engineering to generate robust, productive monoterpenoid biofactories. However, when designing and testing a new biosynthetic pathway in *E. coli*, it is important to demonstrate successful expression and catalytic functioning of the recombinant enzymes independent of other constraints of in vivo monoterpenoid production. Western blots are a useful technique to positively identify the presence of recombinant proteins separated by SDS-PAGE analysis.

1. Transformation of each operon constructs into *E. coli* expression strains. Grow small starter cultures in selective medium under noninducing conditions.

    Note: Constructs in pET vectors require *E. coli* (DE3) strains for expression of the genes. To minimize leaky expression in uninduced cultures, add 0.2% glucose to the medium (Novy & Morris, 2001).

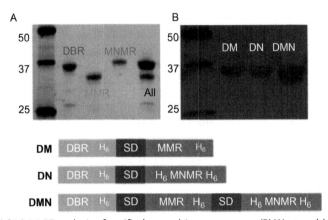

Fig. 3 (A) SDS-PAGE analysis of purified recombinant enzymes. (B) Western blot analysis of cell extracts of three operons of menthol biosynthesis. (See the color plate.)

2. Inoculate larger selective media with the starter cultures and express the recombinant proteins under conditions optimal for maximal soluble protein production.

    Note: Optimal conditions are determined by varying expression conditions such as inducer concentration, induction growth phase, postinduction temperature, and medium composition. In this case, constructs DMN, DM (NtDBR–MMR), and DN (NtDBR–MNMR) were successfully expressed in autoinduction medium (Fig. 3).

3. Generate cell-free extracts of expressed recombinant proteins. SDS-PAGE analysis will indicate the presence of overexpression bands (Fig. 3A).

    Note: Recombinant proteins containing affinity tags (eg, N- or C-His$_6$) can be partially purified prior to SDS-PAGE analysis to improve the detection of moderately expressed proteins.

4. Confirmation of expression is performed by a Western blot using anti-6X His tag antibodies (Fig. 3B).

    Note: Detection of recombinant proteins of similar migration distance by SDS-PAGE may be separated by native PAGE prior to Western blots.

4. BIOTRANSFORMATIONS

To avoid the complications of in vivo biotransformations (eg, cofactor supply and substrate transportation), initial monoterpenoid production tests are performed with either cell-free extracts or permeabilized cells (Weyler & Heinzle, 2015), with an external cofactor (recycling) supply. This also allows the detection of by-products catalyzed by native *E. coli* enzymes, giving insight into possible strain engineering requirements to knock out/down competing pathways. The volatility of these monoterpenoids enables their easy detection by GC-based methods.

4.1 General Monoterpenoid Reaction and Detection

1. Biotransformations (1 mL) are composed of buffer (50 mM Tris, pH 7.0) containing substrate (1 mM pulegone), external hydride source (NADP$^+$/glucose dehydrogenase/glucose cofactor recycling system), and enzyme(s) source (cell-free extract or permeabilized cells). Reactions are incubated at 25–30°C for 2–24 h.

    Note: Extract quantities within the reactions are dependent on the state of the samples. For example, permeabilized cells and crude cell extracts may be viscous and/or precipitate during the reaction, making later solvent extractions difficult. Cofactor recycling systems can be replaced with NADPH at three to five times the concentration of the initial substrate. Shorter reaction times are particularly suitable for substrates and/or products that rapidly degrade under the reaction conditions. Monoterpenoid losses due to high volatility can be minimized by the addition of a solvent overlay in the reaction, such as 10% dodecane.

2. Substrate and product(s) are extracted into ethyl acetate (0.9 mL) containing an internal standard (0.5% *sec*-butylbenzene), dried with anhydrous magnesium sulfate, and quantified by GC.

    Note: Quantitative analysis was performed by a comparison of peak areas to authentic standards of known concentrations. GC–MS can also be used as the compound analyzer, enabling positive identification of any side products.

3. Reaction extracts (1 μL) are analyzed by gas chromatography on a DB-WAX column (30 m; 0.32 mm; 0.25-μm film thickness) with an FID detector. This column successfully separates the two isomers of menthone and four isomers of menthol (Fig. 4). Unknown products are identified by GC analysis in combination with mass spectrometry (GC–MS; inert XL-EI/CI MSD with triple axis detector) using a Zebron ZB-Semi Volatiles column (15 m × 0.25 mm × 0.25-μm film thickness).

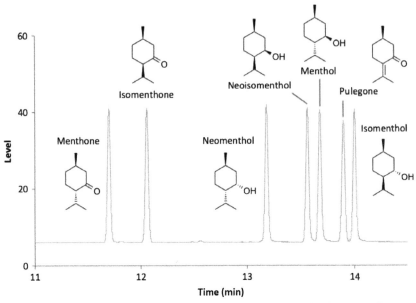

Fig. 4 Gas chromatography separation of seven monoterpenoids of the *Mentha* menthol biosynthetic pathway. Analysis was performed on a DB-WAX column (30 m; 0.32 mm; 0.25-μm film thickness; JW Scientific) on an Agilent Technologies 7890A GC system equipped with an FID detector. *Adapted from Toogood, H. S., Cheallaigh, A. N., Tait, S., Mansell, D. J., Jervis, A., Lygidakis, A., et al. (2015). "Enzymatic menthol production: One-pot approach using engineered* Escherichia coli.*" ACS Synthetic Biology 4(10), 1112–1123.*

Note: GC method: injector temperature was 220°C (split ratio of 20:1) and helium flow rate of 1 mL/min (5.1 psi). The program started at 40°C, with a 1 min hold followed by a 10°C/min increase of temperature, and a final hold at 210°C for 1 min. The FID detector temperature was 250°C with a hydrogen flow of 30 mL/min. GC–MS method: injector temperature was 220°C (split ratio of 10:1) and helium flow rate of 1 mL/min (5.1 psi). The program started at 40°C, with a 3 min hold followed by a 10°C/min increase of temperature, and a final hold at 210°C for 3 min. The mass spectra fragmentation patterns were analyzed by the NIST/EPA/NIH 11 mass spectral library database for compound identification (Toogood et al., 2015).

4.2 Alternative Protocols

The inherent volatility of the menthol pathway components allows alternative analytical methods to be employed. Early work to determine the native

Mentha metabolic pathways described the use of in vivo detection of volatile monoterpenoid metabolites via simultaneous steam distillation and solvent extraction techniques combined with GC–MS analysis (Ringer, Davis, & Croteau, 2005; Ringer, McConkey, Davis, Rushing, & Croteau, 2003). This in situ method could be applied to the detection of enzyme activity during expression trials of constructs under different optimization conditions. It has the advantage of eliminating nonvolatile compound background, allowing cleaner samples to be analyzed. However, the extraction process is lengthy compared to standard solvent extraction, and the apparatus limits the minimal size of sample volumes to be analyzed.

A simpler approach to detecting monoterpenoid components is headspace sampling of biotransformation reactions coupled to GC or GC–MS analysis. This enables the detection of very light volatile organics without the need to perform extractions on complex mixtures (eg, cell extracts). Purge and trap technology (Lee, Lee, Hsiang, & Chen, 1997) can effectively concentrate the volatile compounds from the headspace prior to GC injection, allowing the analysis of trace levels of compounds. A downside of this technique is the lack of detection of some semi- and higher boiling point volatiles. However, for simple compound identification, rather than in-depth analysis of all monoterpenoid analytes, this may be a suitable method for routine compound quantification and identification.

5. FURTHER STUDIES

After the successful demonstration of an active biosynthetic pathway in *E. coli*, the next stage is to undergo a variety of optimization trials to improve the efficiency and yields of the desired product(s). These studies are critical to move from a proof-of-concept state to ultimately a commercially viable process. The following Section 5.1 discusses some microbial and other factors that may play a role in limiting the industrially relevance of a newly established synthetic biology approach to valuable compound production.

5.1 Expression Constructs

Expression of multiple recombinant genes controlled by the same promoter does not guarantee equivalent production levels of each protein. Protein levels are dependent on factors such as genomic sequence, number of genes in the operon, promoter and SD sequence strengths, vector copy number, and even the order of genes in the operon. The screening of homologous genes for optimal functioning in *E. coli* may improve biocatalytic ability, as could a screen of variant libraries of existing biocatalysts. It is very important to look at the flux through the pathway to identify bottlenecks and specifically aims to modify the construct to overcome the problem. For example, The DM construct shows significantly higher NtDBR expression and activity than for MMR, as seen by incomplete menthol production, and an accumulation of the NtDBR reaction products menthone and isomenthone (Toogood et al., 2015).

An earlier study described the addition of four genes into *E. coli* encoding sequential steps of the biosynthesis of carvone from C5 isoprenoid precursors (Carter et al., 2003). Unfortunately, accumulation and cell export of the first enzymatic product limonene occurred due to a severe limitation in the availability of the precursors within central metabolism of the organism. An upregulation of the latter stage genes may improve this by reacting with the limonene before it is exported out of the cell. This may be accomplished by the addition of extra promoters within the construct to boost the levels of latter gene products. The synthesis of the Taxol precursor taxadiene by *E. coli* from glucose was achieved by partitioning the required genes into a native upstream MEP pathway and a downstream terpenoid-forming pathway (Ajikumar et al., 2010). The study described a systematic multivariate search was performed to identify conditions to balance the two pathways to maximize taxadiene production (Ajikumar et al., 2010).

5.2 Strain and Expression Optimization

The nature of the *E. coli* strain can have a dramatic impact on the expression levels of recombinant proteins. Different strains have been developed to overcome specific problems such as the expression of toxic proteins or the minimization of insoluble protein formation. The expression of the DMN construct was screened with 12 different strains of *E. coli* to determine the optimal host organism (Toogood et al., 2015). Dramatic, yet nonpredictive differences in menthol yields were obtained with the different strains (Fig. 5A). Additional improvements in protein expression can be achieved by varying the induction conditions (inducer concentration, induction time, and postinduction temperature) and growth medium (Fig. 5B).

Strain engineering is a crucial aspect in microbial monoterpenoid formation as these compounds are often cytotoxic, such as limonene (Chubukov et al., 2015; Chueca, Pagán, & García-Gonzalo, 2014). One approach to alleviate this toxicity is to incorporate terpenoid efflux pumps into *E. coli* to facilitate the excretion of the compounds into the culture medium.

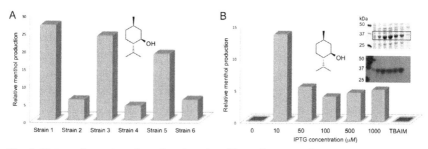

Fig. 5 Biotransformation data showing the effect of varying the (A) *E. coli* strain and (B) inducer concentration on menthol production. Inset in (B) shows the SDS-PAGE (*top*) and Western blot (*bottom*) analyses of the cultures. *TBAIM*, terrific broth autoinduction medium.

For example, a synthetic biology approach to microbial biofuel production included the screening of 43 known efflux pumps in bacteria for improved survivability of the cultures (Dunlop et al., 2011). This approach also has the added advantage of simplifying the product extraction and purification processes. Another bacterial efflux pump outer membrane protein AlkL from *Pseudomonas fluorescens* CHA0 in *E. coli* was shown to increase the dodecanoic acid methyl ester oxygenation activity 28-fold (Julsing et al., 2012). It was shown to be an efficient tool in boosting productivities of whole-cell biotransformations with hydrophobic and aliphatic substrates.

Microbial biocatalysis protocols can be complicated by the presence of competing pathways by constitutive *E. coli* enzymes, leading to unwanted by-products. For example, a synthetic biology approach to propane biosynthesis showed reduced titres due to a competition for the precursor molecule butyraldehyde by *E. coli* aldehyde dehydrogenases Ahr and YqhD (Menon et al., 2015). To overcome this, a double knockout strain Δahr and $\Delta yqhD$ was constructed, and propane production was increased significantly.

5.3 High-Throughput Optimization Studies

The successful redesign of in vivo biofactories to optimize monoterpenoid production is likely to require a combinatorial approach to overcome existing enzymatic and host bottlenecks and/or other constraints. In an ideal world, reliable prediction of the behavior of the organism during monoterpenoid pathway would enable a limited, rational approach to optimal biofactory redesign. In the real world, high-throughput optimization studies are often needed to overcome the often nonpredictive, competing factors involved in regulating secondary metabolite production. Advances in new computational techniques and algorithms have enabled the development of "-omics" technologies, significantly impacting on our understanding of the complex nature of gene regulation and metabolite production (Cascante & Marin, 2008). Metabolomic analyses focus on the identification and quantification of all the metabolites within an organism. This enables an understanding of the physiological state of an organism under defined growth conditions. Complementary flux analyses (fluxomics) looks at the dynamic nature of the metabolome, and how it changes under environmental and genetic pressures (Cascante & Marin, 2008). Additional regulatory control can be discerned by investigating the transcriptome; the complete set of RNA transcripts generated under certain growth conditions. Therefore, a true synthetic biology approach to novel pathway design must include a more holistic paradigm approach, where cell metabolism regulation in addition to enzymatic considerations informs on key metabolite production.

The current menthol-producing *E. coli* strains utilize pulegone as the initial substrate (Toogood et al., 2015). A more industrially useful process would be linking menthol generation to *E. coli* central metabolism through the isoprenoid biosynthesis pathway (MEP/mevalonate pathways). This would require an upregulation of biosynthesis of the isoprenoid precursor GPP and incorporation of plant genes to generate pulegone from (S)-limonene (Fig. 1). Therefore, the addition of more genes and its concomitant effect on central metabolism must be investigated to enable successful menthol production. Similar approaches have been taken to produce artemisinin in *E. coli*. Initial studies showed production was hampered by limited isoprenoid precursor supply (Chang & Keasling, 2006). To overcome this, the *E. coli* nonmevalonate (DXP) pathway for precursor production was bypassed, and the heterologous mevalonate pathway from *S. cerevisiae* was incorporated in two synthetic operons. Growth inhibition, due to an accumulation of toxic isoprenoid precursors, was overcome by the coexpression of the latter stage biocatalyst amorpha-4,11-diene (amorphadiene) synthase (Chang & Keasling, 2006).

Recently, the Centre for Synthetic Biology of Fine and Speciality Chemicals (SYNBIOCHEM; http://synbiochem.co.uk) was set up to develop new faster, more predictable, and reliable "greener" routes to sustainable fine and speciality chemicals production. It aims to accelerate the delivery of bespoke chemicals syntheses by adopting a modular "plug-and-play" platform approaches, using a "design–build–test–deploy" life cycle (Fig. 6). At the Design stage, in silico whole-cell metabolic modelling

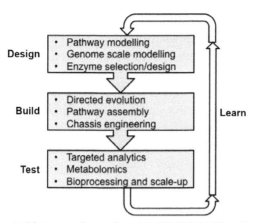

Fig. 6 The Design–Build–Test cycle employed in SYNBIOCHEM at Manchester University for the design of microbial biofactories for speciality biochemical production. Each part of the cycle is divided into three scales, parts, devices, and systems to represent the considerations in developing industrially relevant microbial production chassis.

and pathway flux modelling can produce strategies for engineering of the chassis to optimize flux through the pathway of interest, reduce competitive side reactions, and increase the supply of cofactors and precursors (Becker et al., 2007; Breitling, Achcar, & Takano, 2013; Sridhar, Suthers, & Maranas, 2010). To address poorly performing, or missing steps in a pathway, automated prediction and selection of enzyme alternatives using tools such as RetroPath (Carbonell, Parutto, Herisson, Pandit, & Jaulon, 2014) or PathPred (Moriya et al., 2010) can potentially identify enzymes displaying improved activity in vivo. Alternatively, directed evolution of biocatalysts can be employed to improve fitness toward the desired activity (Currin, Swainston, Day, & Kell, 2014; Currin, Swainston, Day, & Kell, 2015; Swainston, Currin, Day, & Kell, 2014).

These design tools can give rise to high numbers of test parameters including testing of regulatory features (eg, promoters, RBS, vector, gene order), alternative biocatalytic parts, addition of accessory pathways and genome modifications. Automated experimental procedures are therefore required to allow HTP acceleration of assembly and screening of production chassis. Robotic liquid handling platforms are ideally suited for DNA manipulations such as one-step pathway assembly but also for rapid screening of the resultant production chassis. Automated colony picking and culturing in microtiter plates or microfluidics (Pico droplet) followed by sample processing allow a flexible Build pipeline which can directly feed Test platforms. Typical Test systems include mass spectrometry for targeted analysis (GC–MS for monoterpenoids) (Alonso-Gutierrez et al., 2013; Toogood et al., 2015) or metabolomics (Chokkathukalam, Kim, Barrett, Breitling, & Creek, 2014; Muhamadali et al., 2015; Nguyen et al., 2012), proteomics, and transcriptomics. The output-of-Test data can then feedback into the Design platforms to improve accuracy and predictability of the in silico models and facilitates predictable downstream scale-up. This iterative cycle implements a multidisciplinary approach for the iterative development of industrially relevant production chassis.

6. CONCLUSIONS

The synthetic biology approach to secondary metabolite production within microorganisms heralds a new era in natural and "green," sustainable fine chemicals production. It goes beyond the classical biocatalysis technology of optimizing enzyme expression/functionality to include considerations for enabling compound production from cheap, renewable waste feed stocks via linking biosynthesis to central metabolism. Recent studies in the production of monoterpenoids in *E. coli* have demonstrated the commercial potential of such technologies, bringing fine chemical production in line with the increasing world-wide economic and political pressures for environmentally friendly and nonfossil fuel-derived sustainable production processes.

The method of construction of successful biosynthetic constructs is important to enable seamless and scar-less "stitching" of both genes and suitable regulatory elements. Adaptations of existing ligation existing cloning methodologies enable the successful incorporation of DNA parts containing repeating sequences with high annealing capability, compared to lengthy and possibly costly whole operon gene synthesis. Initial constructs can then undergo functional screening, followed by plug-and-play construct optimization to improve its biosynthetic ability. New and emerging technologies may further advance how we will regulate secondary metabolite production in vivo. For example, there is current interest in the use of both riboswitches (Morra et al., 2015) and light-activated promoter elements for novel regulatory control (Hardman, Hauck, Clark, Heyes, & Scrutton, 2014; Tabor, Levskaya, & Voigt, 2011).

Whole organism metabolic constraints must also be taken into consideration in construct design. The use of fluxomics/metabolomics and transcriptomics enables dynamic metabolic pathway modelling, and the impact of compound production on the overall performance of the microbial biofactory. These modelling technologies, combined with pathway design and in vitro biotransformation performance studies, enable the tuning of the host microorganism by upregulating/downregulating key genes to eliminate competing pathway, optimize flux balance and potentially overcome toxicity issues.

The future outlook for the synthetic biology approach could be to revolutionize the manufacture of fine chemicals and pharmaceuticals. We are in an exciting new era of not only reading and understanding DNA, but also now writing our own constructs and building nonnative metabolic pathways to existing or novel chemicals. This transition from understanding biology to informed editing and rewriting metabolic processes has exciting prospects for the future.

ACKNOWLEDGMENTS
This work was funded by the UK Biotechnology and Biological Sciences Research Council (BBSRC BB/J015512/1; BB/M017702/1) and GlaxoSmithKline. SYNBIOCHEM is funded by the BBSRC (BB/M017702/1). N.S.S. is an Engineering and Physical Sciences Research Council (EPSRC; EP/J020192/1) Established Career Fellow.

REFERENCES
Ajikumar, P. K., Tyo, K., Carlsen, S., Mucha, O., Phon, T. H., & Stephanopoulos, G. (2008). Terpenoids: Opportunities for biosynthesis of natural product drugs using engineered microorganisms. *Molecular Pharmaceutics, 5*(2), 167–190.

Ajikumar, P. K., Xiao, W. H., Tyo, K. E. J., Wang, Y., Simeon, F., Leonard, E., et al. (2010). Isoprenoid pathway optimization for taxol precursor overproduction in *Escherichia coli*. *Science, 330*(6000), 70–74.

Alonso-Gutierrez, J., Chan, R., Batth, T. S., Adams, P. D., Keasling, J. D., Petzold, C. J., et al. (2013). Metabolic engineering of *Escherichia coli* for limonene and perillyl alcohol production. *Metabolic Engineering, 19*, 33–41.

Atanasov, A. G., Waltenberger, B., Pferschy-Wenzig, E. M., Linder, T., Wawrosch, C., Uhrin, P., et al. (2015). Discovery and resupply of pharmacologically active plant-derived natural products: A review. *Biotechnology Advances, 33*(8), 1582–1614.

Becker, S. A., Feist, A. M., Mo, M. L., Hannum, G., Palsson, B. O., & Herrgard, M. J. (2007). Quantitative prediction of cellular metabolism with constraint-based models: The COBRA toolbox. *Nature Protocols, 2*(3), 727–738.

Blount, B. A., Weenink, T., & Ellis, T. (2012). Construction of synthetic regulatory networks in yeast. *FEBS Letters, 586*(15), 2112–2121.

Breitling, R., Achcar, F., & Takano, E. (2013). Modeling challenges in the synthetic biology of secondary metabolism. *ACS Synthetic Biology, 2*(7), 373–378.

Breitling, R., & Takano, E. (2015). Synthetic biology advances for pharmaceutical production. *Current Opinion in Biotechnology, 2*, 46–51.

Carbonell, P., Parutto, P., Herisson, J., Pandit, S. B., & Jaulon, J.-L. (2014). XTMS: Pathway design in an eXTended metabolic space. *Nucleic Acids Research, 42*, W389–W394.

Carter, O. A., Peters, R. J., & Croteau, R. (2003). Monoterpene biosynthesis pathway construction in *Escherichia coli*. *Phytochemistry*, *64*(2), 425–433.

Cascante, M., & Marin, S. (2008). Metabolomics and fluxomics approaches. *Essays in Biochemistry*, *45*, 67–81.

Chang, M. C. Y., & Keasling, J. D. (2006). Production of isoprenoid pharmaceuticals by engineered microbes. *Nature Chemical Biology*, *2*(12), 674–681.

Chemler, J. A., & Koffas, M. A. (2008). Metabolic engineering for plant natural product biosynthesis in microbes. *Current Opinion in Biotechnology*, *19*(6), 597–605.

Chokkathukalam, A., Kim, D. H., Barrett, M. P., Breitling, R., & Creek, D. J. (2014). Stable isotope-labeling studies in metabolomics: New insights into structure and dynamics of metabolic networks. *Bioanalysis*, *6*(4), 511–524.

Chubukov, V., Mingardon, F., Schackwitz, W., Baidoo, E. E. K., Alonso-Gutierrez, J., Hu, Q., et al. (2015). Acute limonene toxicity in Escherichia coli is caused by limonene-hydroperoxide and alleviated by a point mutation in alkyl hydroperoxidase (AhpC). *Applied and Environmental Microbiology*, *81*(14), 4690–4696.

Chueca, B., Pagán, R., & García-Gonzalo, D. (2014). Oxygenated monoterpenes citral and carvacrol cause oxidative damage in Escherichia coli without the involvement of tricarboxylic acid cycle and Fenton reaction. *International Journal of Food Microbiology*, *189*, 126–131.

Croteau, R. B., Davis, E. M., Ringer, K. L., & Wildung, M. R. (2005). (−)-Menthol biosynthesis and molecular genetics. *Naturwissenschaften*, *92*(12), 562–577.

Currin, A., Swainston, N., Day, P. J., & Kell, D. B. (2014). SpeedyGenes: An improved gene synthesis method for the efficient production of error-corrected, synthetic protein libraries for directed evolution. *Protein Engineering, Design & Selection*, *27*, 273–280.

Currin, A., Swainston, N., Day, P. J., & Kell, D. B. (2015). Synthetic biology for the directed evolution of protein biocatalysts: Navigating sequence space intelligently. *Chemical Society Reviews*, *44*(5), 1172–1239.

de Kok, S., Stanton, L. H., Slaby, T., Durot, M., Holmes, V. F., Patel, K. G., et al. (2014). Rapid and reliable DNA assembly via ligase cycling reaction. *ACS Synthetic Biology*, *3*(2), 97–106.

Duetz, W. A., Bouwmeester, H., van Beilen, J. B., & Witholt, B. (2003). Biotransformation of limonene by bacteria, fungi, yeasts, and plants. *Applied Microbiology and Biotechnology*, *61*(4), 269–277.

Dunlop, M. J., Dossani, Z. Y., Szmidt, H. L., Chu, H. C., Lee, T. S., Keasling, J. D., et al. (2011). Engineering microbial biofuel tolerance and export using efflux pumps. *Molecular Systems Biology*, *7*, 1–7.

Farmer, W. R., & Liao, J. C. (2001). Precursor balancing for metabolic engineering of lycopene production in *Escherichia coli*. *Biotechnology Progress*, *17*(1), 57–61.

Friedman, D. C., & Ellington, A. D. (2015). Industrialization of biology. *ACS Synthetic Biology*, *4*(10), 1053–1055.

Gibson, D. G., Young, L., Chuang, R. Y., Venter, J. C., Hutchison, C. A. r., & Smith, H. O. (2009). Enzymatic assembly of DNA molecules up to several hundred kilobases. *Nature Methods*, *6*(5), 343–345.

Hardman, S. J., Hauck, A. F., Clark, I. P., Heyes, D. J., & Scrutton, N. S. (2014). Comprehensive analysis of the green-to-blue photoconversion of full-length Cyanobacteriochrome Tlr0924. *Biophysical Journal*, *107*(9), 2195–2203.

Hunter, P. (2008). Harnessing Nature's wisdom. *EMBO Reports*, *9*(9), 838–840.

Hussain, M. S., Fareed, S., Ansari, S., Rahman, M. A., Ahmad, I. Z., & Saeed, M. (2012). Current approaches toward production of secondary plant metabolites. *Journal of Pharmacy & Bioallied Sciences*, *4*(1), 10–20.

Jackson, B. E., Hart-Wells, E. A., & Matsuda, S. P. T. (2003). Metabolic engineering to produce sesquiterpenes in yeast. *Organic Letters*, *5*(10), 1629–1632.

Julsing, M. K., Schrewe, M., Cornelissen, S., Hermann, I., Schmid, A., & Buhler, B. (2012). Outer membrane protein AlkL boosts biocatalytic oxyfunctionalization of hydrophobic substrates in Escherichia coli. *Applied and Environmental Microbiology, 78*(16), 5724–5733.

Kiyota, H., Okuda, Y., Ito, M., Hirai, M. Y., & Ikeuchi, M. (2014). Engineering of cyanobacteria for the photosynthetic production of limonene from CO_2. *Journal of Biotechnology, 185*, 1–7.

Lange, B. M., Mahmoud, S. S., Wildung, M. R., Turner, G. W., Davis, E. M., Lange, I., et al. (2011). Improving peppermint essential oil yield and composition by metabolic engineering. *Proceedings of the National Academy of Sciences of the United States of America, 108*(41), 16944–16949.

Lawrence, B. M. (Ed.), (2007). In *Medicinal and aromatic plants—Industrial profiles: Vol. 44. Mint: The genus Mentha* (pp. 156–158). Florida: CRC Press.

Lee, M.-R., Lee, J.-S., Hsiang, W.-S., & Chen, C.-M. (1997). Purge-and-trap gas chromatography–mass spectrometry in the analysis of volatile organochlorine compounds in water. *Journal of Chromatography A, 775*(1–2), 267–274.

Li, J. W. H., & Vederas, J. C. (2009). Drug discovery and natural products: End of an era or an endless frontier? *Science, 325*(5937), 161–165.

Mansell, D. J., Toogood, H. S., Waller, J., Hughes, J. M. X., Levy, C. W., Gardiner, J. M., et al. (2013). Biocatalytic asymmetric alkene reduction: Crystal structure and characterization of a double bond reductase from *Nicotiana tabacum*. *ACS Catalysis, 3*, 370–379.

Martin, V. J. J., Pitera, D. J., Withers, S. T., Newman, J. D., & Keasling, J. D. (2003). Engineering a mevalonate pathway in *Escherichia coli* for production of terpenoids. *Nature Biotechnology, 21*(7), 796–802.

Menon, N., Pásztor, A., Menon, B. R., Kallio, P., Fisher, K., Akhtar, M. K., et al. (2015). A microbial platform for renewable propane synthesis based on a fermentative butanol pathway. *Biotechnology for Biofuels, 8*(1), 61.

Mitchell, W. (2011). Natural products from synthetic biology. *Current Opinion in Chemical Biology, 15*(4), 505–515.

Moriya, Y., Shigemizu, D., Hattori, M., Tokimatsu, T., Kotera, M., Goto, S., et al. (2010). PathPred: An enzyme-catalyzed metabolic pathway prediction server. *Nucleic Acids Research, 38*, W138–W143.

Morra, R., Shankar, J., Robinson, C. J., Halliwell, S., Butler, L., Upton, M., et al. (2015). Dual transcriptional-translational cascade permits cellular level tuneable expression control. *Nucleic Acids Research, 4*(3), e21.

Muhamadali, H., Xu, Y., Ellis, D. I., Allwood, J. W., Rattray, N. J., Correa, E., et al. (2015). Metabolic profiling of *Geobacter sulfurreducens* during industrial bioprocess scale-up. *Applied and Environmental Microbiology, 81*(10), 3288–3298.

Nguyen, Q. T., Merlo, M. E., Medema, M. H. M., Jankevics, A., Breitling, R., & Takano, E. (2012). Metabolomics methods for the synthetic biology of secondary metabolism. *FEBS Letters, 586*, 2177–2183.

Nicolaou, K. C., Yang, Z., Liu, J. J., Ueno, H., & Nantermet, P. G. (1994). Total synthesis of taxol. *Nature, 367*(6464), 630–634.

Novy, R., & Morris, B. (2001). Use of glucose to control basal expression in the pET System. *Innovations Biotechniques, 12*, 1–3.

Ringer, K. L., Davis, E. M., & Croteau, R. (2005). Monoterpene metabolism. Cloning, expression, and characterization of (−)-isopiperitenol/(−)-carveol dehydrogenase of peppermint and spearmint. *Plant Physiology, 137*(3), 863–872.

Ringer, K. L., McConkey, M. E., Davis, E. M., Rushing, G. W., & Croteau, R. (2003). Monoterpene double-bond reductases of the (−)-menthol biosynthetic pathway: Isolation and characterization of cDNAs encoding (−)-isopiperitenone reductase and (+)-pulegone reductase of peppermint. *Archives of Biochemistry and Biophysics, 418*(1), 80–92.

Sell, C. (2003). *A fragrant introduction to terpenoid chemistry*. Cambridge, UK: Royal Society of Chemistry.
Sridhar, R., Suthers, P. F., & Maranas, C. D. (2010). OptForce: A optimization procedure for identifying all genetic manipulations leading to targeted overproductions. *PLoS Computational Biology*, *6*(4), e1000744.
Starmer, J., Stomp, A., Vouk, M., & Bitzer, D. (2006). Predicting Shine–Dalgarno sequence locations exposes genome annotation error. *PLoS Computational Biology*, *2*(5), e57.
Swainston, N., Currin, A., Day, P. J., & Kell, D. B. (2014). GeneGenie: Optimized oligomer design for directed evolution. *Nucleic Acids Research*, *12*, W395–W400.
Tabor, J. J., Levskaya, A., & Voigt, C. A. (2011). Multichromatic control of gene expression in *Escherichia coli*. *Journal of Molecular Biology*, *405*(2), 315–324.
Toogood, H. S., Cheallaigh, A. N., Tait, S., Mansell, D. J., Jervis, A., Lygidakis, A., et al. (2015). Enzymatic menthol production: One-pot approach using engineered *Escherichia coli*. *ACS Synthetic Biology*, *4*(10), 1112–1123.
Unger, T., Jacobovitch, Y., Dantes, A., Bernheim, R., & Peleg, Y. (2010). Applications of the restriction free (RF) cloning procedure for molecular manipulations and protein expression. *Journal of Structural Biology*, *172*(1), 34–44.
Weyler, C., & Heinzle, E. (2015). Multistep synthesis of UDP-glucose using tailored, permeabilized cells of E. coli. *Applied Biochemistry and Biotechnology*, *175*(8), 729–736.
Wilkinson, D. L., & Harrison, R. G. (1991). Predicting the solubility of recombinant proteins in *Escherichia coli*. *Bio/Technology*, *9*, 443–448.
Winter, J. M., & Tang, Y. (2012). Synthetic biological approaches to natural product biosynthesis. *Current Opinion in Biotechnology*, *23*(5), 736–743.
Woodman, M. E. (2008). Appendix 3D direct PCR of intact bacteria (colony PCR). *Protocols in Microbiology*, *9*(A.3D), 1–6.
Yadav, V. G., De Mey, M., Giaw Lim, C., Kumaran Ajikumar, P., & Stephanopoulos, G. (2012). The future of metabolic engineering and synthetic biology: Towards a systematic practice. *Metabolic Engineering*, *14*(3), 233–241.
Yonekura-Sakakibara, K., & Saito, K. (2009). Functional genomics for plant natural product biosynthesis. *Natural Product Reports*, *26*(11), 1466–1487.
Zhu, B., Cai, G., Hall, E., & Freeman, G. (2007). In-Fusion™ assembly: Seamless engineering of multidomain fusion proteins, modular vectors, and mutations. *Biotechniques*, *43*(3), 354–359.

CHAPTER TWELVE

High-Efficiency Genome Editing of *Streptomyces* Species by an Engineered CRISPR/Cas System

Y. Wang*, R.E. Cobb*,†, H. Zhao*,†,1

*University of Illinois at Urbana-Champaign, Urbana, IL, United States
†Carl R. Woese Institute for Genomic Biology, University of Illinois at Urbana-Champaign, Urbana, IL, United States
[1]Corresponding author: e-mail address: zhao5@illinois.edu

Contents

1. Introduction 271
2. Design of pCRISPomyces for Genome Editing in *Streptomyces* Species 273
3. pCRISPomyces for Single Gene Disruption in *S. lividans* 275
4. pCRISPomyces for Multiplex Gene Deletion in *S. lividans* 278
5. Conjugation, Genotype Screening, and pCRISPomyces Clearance 279
6. Evaluation of pCRISPomyces in Other *Streptomyces* Species 281
7. Conclusion 282
Acknowledgment 283
References 283

Abstract

Next-generation sequencing technologies have rapidly expanded the genomic information of numerous organisms and revealed a rich reservoir of natural product gene clusters from microbial genomes, especially from *Streptomyces*, the largest genus of known actinobacteria at present. However, genetic engineering of these bacteria is often time consuming and labor intensive, if even possible. In this chapter, we describe the design and construction of pCRISPomyces, an engineered Type II CRISPR/Cas system, for targeted multiplex gene deletions in *Streptomyces lividans*, *Streptomyces albus*, and *Streptomyces viridochromogenes* with editing efficiency ranging from 70% to 100%. We demonstrate pCRISPomyces as a powerful tool for genome editing in *Streptomyces*.

1. INTRODUCTION

Streptomyces strains are widely used hosts for heterologous expression of secondary metabolites produced by actinomycetes. These secondary metabolites are excellent starting scaffolds for the development of antibiotics, anticancer agents, immunomodulators, anthelmintic agents, and insect control agents (Baltz, 2010). In addition, as the largest genus of presently known actinobacteria, *Streptomyces* are important sources of natural products that contribute to around two-thirds of the useful antibiotics such as neomycin and chloramphenicol (Lucas et al., 2013). Due to recent advances in high-throughput DNA sequencing techniques, a tremendous number of cryptic gene clusters have been identified in *Streptomyces* species. These cryptic gene clusters encode unknown secondary metabolites that have never been detected under laboratory conditions, and they have revealed a much larger capacity for biosynthesis of specialized natural products than previously thought. This has led to the development of "genome mining" as a new strategy for natural product discovery by either: (i) activating transcriptionally silent cryptic gene clusters by genetic manipulation of the native organisms or (ii) heterologously expressing cryptic pathways in suitable hosts, such as *Streptomyces coelicolor*, *Streptomyces lividans*, and *Streptomyces albus* (Luo, Cobb, & Zhao, 2014; Luo, Enghiad, & Zhao, 2016; Luo, Li, et al., 2015).

Unfortunately, the current tools for genome editing of streptomycetes are time consuming and rather limited. For example, activation of cryptic gene clusters in native hosts is restricted to genetic modification of only single genes, such as deleting or overexpressing putative negative or positive transcriptional regulators (Gottelt, Kol, Gomez-Escribano, Bibb, & Takano, 2010; Laureti et al., 2011). The typical methods of gene disruption in *Streptomyces* are based on homologous recombination (HR) and can be classified into two categories: single-crossover integration and double-crossover integration. Single-crossover integration of a suicide plasmid can disrupt a gene of interest with a selectable marker (Kieser, Bibb, Buttner, Chater, & Hopwood, 2000). However, one of the biggest drawbacks is limited reusability of this approach for multiplex genome editing due to limited selectable markers. Additionally, the modified gene can be reverted back to the wild-type allele in the absence of selective pressure. Although the use of flanking recombinase target sites like *loxP* and expression of the corresponding recombinase enable the recycling of markers and improve mutation stability, this requires additional steps and leaves a scar sequence at the target site. Double-crossover integration can overcome some disadvantages of the single-crossover method, but it is time consuming and labor intensive (Kieser et al., 2000). First, single crossover of the disruption vector at a target site needs to be identified by a selective marker. Next, a second single crossover for clearance of the disruption vector needs to be identified by using counterselectable markers such as *rpsL* and *glkA* that are limited to use only in particular mutant hosts (Kieser et al., 2000).

Recently, double-strand break (DSB)-mediated genome editing has been achieved by the Type II clustered regularly interspaced short palindromic repeat (CRISPR)/CRISPR-associated proteins (Cas) system of *Streptococcus pyogenes* (Mali, Esvelt, & Church, 2013). The CRISPR/Cas system requires three functional units to detect and cleave foreign DNA: Cas9, the nuclease and scaffold for the recognition elements; CRISPR RNA (crRNA), short RNAs that are complementary to the intended cleavage site and encoded as spacers in CRISPR arrays; and trans-activating crRNA (tracrRNA), a short RNA that facilitates crRNA processing and recruitment to Cas9. The CRISPR/Cas system has been engineered for various genome manipulation purposes (Bikard et al., 2013; Cong et al., 2013; Maeder et al., 2013; Qi et al., 2013; Ran et al., 2013; Sander & Joung, 2014). Compared with other tools for site-specific genome engineering, such as zinc-finger nucleases and transcription activator-like effector nucleases, the engineered CRISPR/Cas system offers unprecedented modularity (Gaj, Gersbach, & Barbas, 2013). Its simplicity and customizability enable the development of host-independent gene-targeting platforms for both genome editing and transcriptional control in both prokaryotes and eukaryotes. It has been successfully applied to a wide variety of organisms such as *Saccharomyces cerevisiae*, *Escherichia coli*, tobacco plants, wheat plants, mouse embryonic cell lines, and mammalian cell lines (Genga, Kearns, & Maehr, 2015; Jakociunas, Jensen, & Keasling, 2016; Li, Gee, Ishida, & Hotta, 2015; Lin, Zhang, & Li, 2015; Mou, Kennedy, Anderson, Yin, & Xue, 2015; Schaart, Van De Wiel, Lotz, & Smulders, 2015).

In this chapter, we highlight our engineered CRISPR/Cas system, pCRISPomyces, for rapid multiplex genome editing of *Streptomyces* strains. We describe the detailed protocols for the design and construction of pCRISPomyces plasmids and their applications in single-gene deletion and multiplex gene deletion in *S. lividans*. In addition, we describe the procedures for conjugation, genotyping, and plasmid clearance. Finally, we briefly discuss the application of pCRISPomyces in other *Streptomyces* strains.

2. DESIGN OF pCRISPomyces FOR GENOME EDITING IN *STREPTOMYCES* SPECIES

The Type II CRISPR/Cas system for targeted genome editing contains three components: Cas9 nuclease, tracrRNA, and crRNA. The tracrRNA and crRNA can be either individually expressed or fused together as a chimeric guide RNA (gRNA). The former is more flexible and is most commonly used in multiple-gene targeting, while the latter is more simplified, has higher targeting efficiency, and is applied for single-gene targeting in most cases (Xu, 2014). To harness the CRISPR/Cas system for genome editing in *Streptomyces* species, we constructed two versions of the pCRISPomyces expression system: pCRISPomyces-1 and pCRISPomyces-2 (Fig. 1). pCRISPomyces-1 contains both tracrRNA and crRNA along with *cas9*, and pCRISPomyces-2 includes a gRNA expression cassette and *cas9*. In both cases, previously identified strong promoters—rpsLp(XC), the promoter of 30S ribosomal protein S12 in *Xylanimonas cellulosilytica*; rpsLp(CF), the promoter of 30S ribosomal protein S12 in *Cellulomonas flavigena*; and gapdhp(EL), the promoter of the GAPDH operon in *Eggerthella lenta*—were selected to drive expression of the CRISPR/Cas elements (Shao et al., 2013). The widely used terminators from phages fd, λ, and T7 were used for each expression cassette. Although *Streptococcus* and *Streptomyces* are both genera of Gram-positive eubacteria, they have quite different phylogeny (Muto & Osawa, 1987). *Streptococcus* has low average genomic G + C content of around 41.2%, while *Streptomyces* has high G + C content of up to 73% (Zhang et al., 2011). Analysis of the *S. pyogenes cas9* gene revealed the presence of several rare codons, including many *bldA* codons, translation-level regulators of secondary metabolism in *Streptomyces* species. To ensure the proper expression level of Cas9 in *Streptomyces* species, the *cas9* gene has been optimized to favor the *Streptomyces* codon bias.

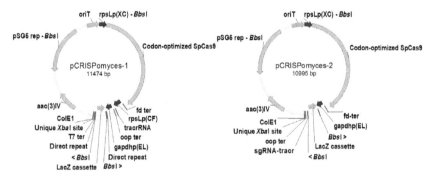

Fig. 1 pCRISPomyces plasmids for targeted genome editing in *Streptomyces* species. Reproduced with permission from Cobb, R. E., Wang, Y., & Zhao, H. (2014). High-efficiency multiplex genome editing of Streptomyces species using an engineered CRISPR/Cas system. ACS Synthetic Biology, 4, 723–728. (See the color plate.)

To facilitate scarless, one-step Golden Gate assembly of specific spacers into crRNA expression cassettes, a *lacZα* cassette flanked by *Bbs*I restriction sites, was incorporated between two direct repeat units in pCRISPomyces-1. A *Bbs*I-flanked *lacZα* cassette was also included in the sgRNA expression cassette for the same purpose. In both systems, a unique *Xba*I site was added for recombination-driven repair template insertion via Gibson assembly or simple restriction digestion–ligation. In addition, both pCRISPomyces systems contain the ColE1 origin for replication in *E. coli*, the RP4 origin of transfer *oriT* for conjugative transfer of pCRISPomyces plasmids from an *E. coli* donor strain to *Streptomyces* hosts (Mazodier, Petter, & Thompson, 1989), *aac(3)IV* for selection in both *E. coli* and *Streptomyces* strains, and the temperature-sensitive replication origin from pSG5 for fast clearance of pCRISPomyces plasmids from *Streptomyces* hosts (Muth, Nußbaumer, Wohlleben, & Pühler, 1989). Since pCRISPomyces-2 has much higher editing efficiency than pCRISPomyces-1, we will mainly discuss pCRISPomyces-2 in the following sections.

3. pCRISPomyces FOR SINGLE GENE DISRUPTION IN *S. lividans*

The engineered pCRISPomyces system has unprecedented modularity for *Streptomyces* genome editing. The first step to design a pCRISPomyces construct for deletion of a target gene is to select the proper protospacer sequence. A protospacer is a 20 nt sequence that the Cas9/tracrRNA/crRNA complex can target. It must contain the requisite protospacer adjacent motif (PAM) NGG at its 3′ end, where N is any nucleotide. Preference is given to:

1. sequence with purines occupying the last four (3′) bases of the protospacer (Wang, Wei, Sabatini, & Lander, 2014);
2. sequences on the noncoding strand; and
3. sequences in which the last 12 nt of the protospacer plus the 3 nt PAM are unique in the context of the genome (Cong et al., 2013).
   Note: To ensure uniqueness, BLAST analysis should be performed for all four possible 15 nt queries (12 nt + AGG, CGG, TGG, or GGG) against the complete genome sequence.

The spacer that will be assembled in pCRISPomyces via Golden Gate assembly method is the reverse complimentary sequence of identified 20 nucleotides led by CCN on coding sequence. To facilitate one-step Golden Gate assembly of a 20 nt spacer into pCRISPomyces, generate

two 24 nt oligonucleotides (4 nt 5′ sticky end + 20 nt spacer sequence) with the sticky ends ACGC (the last 4 nt of gapdp(EL) promoter) on the forward oligo and AAAC (complementary to the first 4 nt of direct repeat or sgRNA–tracr) on the reverse oligo. For example, if the spacer sequence is CTCACGGACGGAGACCAGGA, then the two primers are:
a. Spacer-for: 5′-**ACGC** CTCACGGACGGAGACCAGGA-3′
b. Spacer-rev: 5′-**AAAC** TCCTGGTCTCCGTCCGTGAG-3′

To anneal the forward and reverse spacer oligos:
1. Separately suspend oligos spacer-for and spacer-rev to 100 μM in nuclease-free water.
2. Mix 5 μL spacer-for and 5 μL spacer-rev with 90 μL 30 mM HEPES, pH 7.8.
3. Using thermal cycler, heat the mixture to 95°C and hold for 5 min, then ramp to 4°C at 0.1°C/s.

To insert annealed spacer by Golden Gate assembly:
1. Prepare the Golden Gate assembly reaction mixture in a 0.2-mL tube. For each 20 μL reaction, add 100 ng pCRISPomyces-1 or pCRISPomyces-2; 0.3 μL of the annealed spacer oligos, diluted 10-fold; 2 μL 10 × T4 DNA ligase buffer (New England Biolabs (NEB), Ipswich, MA); 1 μL T4 DNA ligase at 400 U/μL (NEB); and 1 μL *Bbs*I at 10 U/μL (NEB) to the necessary volume of nuclease-free water.
    Note: T4 DNA ligase should be added last.
2. Run the following Golden Gate assembly program on a thermal cycler: 10 min at 37°C; 10 min at 16°C; repeat first two steps nine additional times; 5 min at 50°C; 20 min at 65°C. Keep the finished reaction at 4°C until transformation.
3. Transform 3 μL of each reaction to NEB 5-alpha competent *E. coli* (high efficiency) or 1 μL from a threefold diluted reaction to *E. coli* NEB 5-alpha electrocompetent *E. coli* following the manufacturer's suggested protocol. Recover the cells in 1 mL SOC medium at 37°C for 1 h.
4. Plate 10% of the recovery culture on selective plates (50 μg/mL apramycin, 20 mg/mL LB broth, 20 mg/mL agar) with 10 μL of 0.5 M isopropyl-β-D-1-thiogalactopyranoside in sterile water and 40 μL of 20 mg/mL Bluo-gal in DMSO. Incubate the plate at 37°C overnight.
5. On the next day, pick white colonies to selective LB medium (50 μg/mL apramycin, 20 mg/mL LB broth) and recover the plasmid to check the sequence of the 20 nt insertion by sequencing. The recommend sequencing primers are:

a. pCm-seq-for: 5′-CATTCAGGCTGCGCAACTG-3′
b. pCm-seq-rev: 5′-CGTTTTACAACGTCGTGACTGG-3′

The DSB introduced at the target site can be repaired by either non-homologous end joining (NHEJ) or HR, the latter requiring an editing template with sequence homology to regions upstream and downstream of the DSB. Imperfect NHEJ usually introduces random sized deletions in actinomycetes, disrupting the protospacer to prevent further recognition and cleavage by Cas9 (Tong, Charusanti, Zhang, Weber, & Lee, 2015). For defined deletion with higher efficiency, a 2 kb editing template is used to enable HR between the chromosomal locus and the pCRISPomyces plasmid. The editing template consists of two 1 kb arms homologous to the corresponding sequences upstream and downstream of the protospacer, excluding a few nucleotides to partially or fully eliminate the protospacer sequence and introduce a frame shift.

To assemble two 1 kb homology arms into the pCRISPomyces plasmid:

1. Digest the spacer-containing pCRISPomyces plasmid with XbaI and dephosphorylate with FastAP Thermosensitive Alkaline Phosphatase (Thermo Fisher Scientific, Waltham, MA) to prevent self-ligation. The reaction mixture is prepared in a total volume of 50 μL containing 3 μg plasmid, 5 μL 10× CutSmart buffer (NEB), and 10 U of XbaI restriction enzyme. The reaction is incubated at 37°C for 1 h, after which 1 U of FastAP (Thermo) is added, and the reaction is incubated at 37°C for another 30 min. The lincarized plasmid is purified by DNA Clean & Concentrator kit (Zymo Research, Irvine, CA).

2. PCR amplify 1 kb homology arms from genomic DNA of the strain of interest. PCR can be conducted using 100 ng genomic DNA, 25 pmol of designed primers, 100 μM dNTPs, 5× Q5 reaction buffer (NEB), 5× Q5 High GC Enhancer (NEB), and 1 U Q5 High-Fidelity DNA polymerase (NEB). The PCR conditions are: 98°C for 5 min; 30 cycles of 98°C for 30 s; T_a (annealing temperature calculated for each pair of primers by T_m calculator from NEB) for 20 s; 72°C for 30 s (1 kb/30 s); and finally 72°C for 10 min.

3. The assembly can be performed by 3-piece Gibson assembly by designing overlaps (30–35 nt) at both ends of both homology arms. To perform the reaction, a 20 μL mixture consisting of 300 ng backbone, threefold molar excess of each of the two 1 kb homology arms, and 10 μL Gibson Assembly Master Mix (NEB) is incubated at 50°C for 60 min.

4. The assembly can also be performed by restriction digestion and ligation. In this case, design overlaps (20–30 nt) at the junction of the two

homology arms and *Xba*I cut sites at the opposite ends of the two arms. Then splice the two arms by overlap-extension PCR. The first step of overlap-extension PCR is performed in a total volume of 20 μL with 30 ng of each homology arms FailSafe PCR 2× PreMix G buffer (Epicentre) and 1 U Q5 High-Fidelity DNA polymerase (NEB). PCR conditions are: 98°C for 3 min; then 15 cycles of 95°C for 30s, 55°C for 20 s, and 72°C for 30 s; then extend at 72°C for 10 min. In the second step, 2 μL of the reaction mixture from step 1 is used as template. The PCR reaction is performed following the protocol explained in point 2. Digest the resulting 2 kb product with *Xba*I as explained in point 1, but without adding FastAP. To ligate the 2 kb editing template to the linearized plasmid, 100 ng of linearized plasmid and threefold molar excess of editing template are added to 10× T4 DNA Ligase Buffer (NEB) and 200 U T4 DNA ligase (NEB) in a total volume of 20 μL. Incubate the reaction at room temperature for 10 min prior to *E. coli* transformation.

5. Transform 3 μL of each reaction to NEB 5-alpha competent *E. coli* (high efficiency) or 1 μL from a threefold diluted reaction to *E. coli* NEB 5-alpha electrocompetent *E. coli* following the manufacturer's suggested protocol. Recover the cells in 1 mL SOC medium at 37°C for 1 h.
6. Plate 200 μL of recovered cells on selective plates (50 μg/mL apramycin, 20 mg/mL LB broth, 20 mg/mL agar) and incubate at 37°C overnight.
7. On the next day, pick colonies and grow overnight in selective LB medium (50 μg/mL apramycin, 20 mg/mL LB broth, 20 mg/mL agar).
8. On the next day, recover the plasmids and confirm the assembly by digestion or sequencing.

4. pCRISPomyces FOR MULTIPLEX GENE DELETION IN *S. lividans*

Additional constructs were also designed with two sgRNA cassettes for multiplex gene deletion in *Streptomyces* species. In these constructs, the pCRISPomyces-2 system was used, and tandem sgRNA cassettes were driven by two copies of the same strong promoter gapdhp(EL). Two sgRNA cassettes targeting two different positions can be included in a synthetic DNA fragment containing (from 5′ to 3′) the first gRNA sequence (spacer 1 and gRNAtracr), a terminator (fd), a promoter (gapdhp(EL)), and the second gRNA sequence (spacer 2 and gRNAtracr), all flanked by *Bbs*I sites. Note that the sticky ends for the *Bbs*I sites (5′ to 3′) are ACGC and GTTT.

To assemble the synthetic DNA fragment into pCRISPomyces-2, one-step Golden Gate assembly can be applied following the same program described in the previous section. To prepare the Golden Gate assembly reaction mixture in 0.2-mL tube, for each 20 μL reaction, add 100 ng pCRISPomyces 2; two- to threefold excess of insert and 2 μL 10 × T4 ligase buffer (NEB); 1 μL T4 ligase at 400 U/μL (NEB); and 1 μL *Bbs*I at 10 U/μL (NEB) to the necessary volume of nuclease-free water. Transformation and assembly confirmation can be performed as described previously.

If noncontiguous genes are the targets, an individual editing template for each gene should be added into the pCRISPomyces plasmid, with each editing template containing 1 kb arms homologous to the corresponding sequences upstream and downstream of each protospacer. If the whole gene cluster is the target, two gRNAs can be designed to target the first and last genes of the cluster, respectively, and a single 2 kb editing template containing 1 kb arms homologous to the corresponding sequences upstream and downstream of the cluster can be assembled into the pCRISPomyces plasmid as discussed before.

5. CONJUGATION, GENOTYPE SCREENING, AND pCRISPomyces CLEARANCE

The fully assembled pCRISPomyces plasmid is first transferred to the *E. coli* conjugation strain WM6026 by electroporation (Blodgett et al., 2007). Conjugation of plasmids into *Streptomyces* spores is performed by modified protocols described later:

1. Grow the donor strain WM6026 in 2 mL LB medium supplemented with 19 μg/mL 2,6-diaminopimelic acid at 37°C, 250 rpm until OD600 reaches 0.7.
2. Wash 100 μL WM6026 culture with 1 mL LB medium twice and suspend the cell with 1 mL LB medium supplemented with 19 μg/mL 2,6-diaminopimelic acid.
3. Mix 2 μL WM6026 culture from step 2 with 10 μL *Streptomyces* spores that are stored in sterilized nuclease-free water and spot onto R2 plates with each droplet containing 1.5 μL strain mixture.

   Note: To prepare R2 plate, in 600 mL H_2O add 0.15 g K_2SO_4, 6.07 g $MgCl_2 \cdot 6\ H_2O$, 6 g glucose, 0.06 g casamino acids, 3.44 g TES buffer, 12 g agar. After autoclave, add 1.6 μL (A) 50 mg/mL KH_2PO_4 and 3 mL (B) 2.94 g $CaCl_2 \cdot 2H_2O$, 3 g L-proline in 5 mL 1 M NaOH. Both (A) and (B) are filtered.

4. Incubate R2 plate in 30°C about 18 h and flood the conjugation plate with 2 mL of 1 mg/mL nalidixic acid and 10 mg/mL apramycin.
5. Incubate R2 plate in 30°C for another 3–5 days to check the exconjugants.

The exconjugants should be restreaked on apramycin-selective ISP2 plates (50 μg/mL apramycin, 10 g/L malt extract broth, 4 g/L yeast extract, 4 g/L glucose) to confirm reception of the plasmid. To screen for the desired editing event:

1. Inoculate *Streptomyces* strain growing on apramycin-selective ISP2 plates into 2 mL MYG medium and incubate the culture in 30°C, 250 rpm for 2–3 days.
2. Isolate genomic DNA of the *Streptomyces* strain from saturated MYG culture using the Wizard Genomic DNA Purification kit (Promega, Madison, WI) according to the manufacturer's suggested protocol.
3. PCR amplify the target locus on genomic DNA by the protocol described in Section 2. To avoid PCR amplification from the pCRISPomyces plasmid, primers should be designed to anneal slightly upstream and downstream of the editing template sequence.
4. To check for a short deletion at a single locus, sequence the PCR product from step 2 with internal primers.
5. To check for large gene cluster deletion, perform PCR with a pair of primers (1 and 3) that anneal outside of the 1 kb homology arms (Fig. 2). A 2.1 kb product should be observed if the deletion event has occurred. In addition, PCR amplification with a primer (2) annealing within the deleted region and a primer outside the deleted region should be performed. A 1.1 kb product should be amplified from the wild-type genomic DNA, but not from the strains with the target gene cluster deleted.

After confirmation of the desired genotype, the pCRISPomyces system can be cleared from the strain to allow reuse of the apramycin selection marker for future applications, such as further genome editing or introduction of heterologous genes. To achieve plasmid clearance:

1. Inoculate one colony with confirmed genotype in nonselective MYG medium (10 g/L malt extract both, 4 g/L yeast extract, 4 g/L glucose) and culture at elevated temperature (37–39°C) with 250 rpm agitation.
2. After growth to stationary phase, a fraction of the culture can be plated on solid MYG medium (10 g/L malt extract both, 4 g/L yeast extract, 4 g/L glucose, 10 g/L agar) to isolate single colonies.

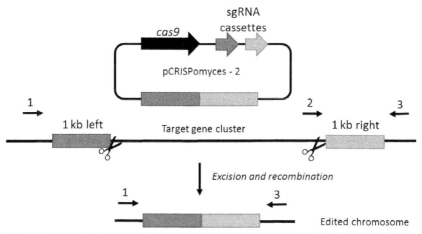

Fig. 2 Strategy for deletion of a large gene cluster and screening for the desired genotype by PCR. Two sgRNA transcripts guide Cas9 to introduce DSBs at both ends of the cluster, and a codelivered editing template bridges the gap via homologous recombination. (See the color plate.)

3. Randomly pick 8–12 colonies and replica plate on MYG with and without 50 μg/mL apramycin. The strains with restored apramycin sensitivity have successfully cleared of the temperature-sensitive pCRISPomyces plasmid.

The above method is only suitable for *Streptomyces* species that can grow under higher temperature, like *S. lividans*, *S. coelicolor*, and *S. viridochromogenes*. If the strain cannot grow at temperatures higher than 30°C, a serial culture method can be applied to clear the pCRISPomyces plasmid:

1. Inoculate one colony with confirmed genotype in 2 mL nonselective MYG medium (10 g/L malt extract both, 4 g/L yeast extract, 4 g/L glucose) and culture at 30°C with 250 rpm agitation.
2. After growth to stationary phase, aliquot 20 μL cell culture into 2 mL nonselective MYG medium and repeat culture at 30°C.
3. Repeat step 2 at least two times and then plate a fraction of the culture on solid MYG to isolate single colonies. Check for apramycin sensitivity as explained before. If no candidate shows apramycin sensitivity, additional rounds of nonselective culture may be applied.

6. EVALUATION OF pCRISPomyces IN OTHER *STREPTOMYCES* SPECIES

The pCRISPomyces-1 system with separate tracrRNA and CRISPR array elements only has moderate, around 24%, editing efficiency (Cobb,

Wang, & Zhao, 2014). However, pCRISPomyces-2 constructs with chimeric gRNA have high editing efficiency for single-gene disruption, two-gene disruption, and 30 kb large gene cluster deletion (Cobb et al., 2014). In addition, pCRISPomyces-2 and analogous systems have been applied in multiple *Streptomyces* species, including *S. coelicolor*, *S. lividans*, *S. viridochromogenes*, *S. albus* (Cobb et al., 2014; Tong et al., 2015), and *S. griseus* (unpublished data).

Compared with typical methods of gene disruption applied in *Streptomyces* species like single-crossover and double-crossover recombination, pCRISPomyces systems increase HR efficiency by creating a DSB. This method not only improves single-locus targeting efficiency but also enables multiplex editing. In addition, the modularity of the system enables high-throughput library construction via modern DNA assembly techniques. The rapid plasmid clearance method allows recycling of the selective marker and allows repeated usage of pCRISPomyces for serial genome editing events. Compared with the traditional single-crossover method, the pCRISPomyces system improves experimental efficiency by avoiding restoration of the wild-type allele. It also largely reduces the time and labor investment relative to the double-crossover method.

In addition, we have demonstrated use of pCRISPomyces-2 to insert strong promoters in front of silent gene clusters to activate the expression of cryptic natural product gene clusters in multiple *Streptomyces* species (unpublished data). Compared with gene delivery by integrases such as that of actinophage $\phi C31$, which only catalyze recombination at a certain recognition site, pCRISPomyces enables heterologous gene insertion site-specifically into the chromosome of the *Streptomyces* species at practically any site of interest. The only restriction is that the site must be near an NGG PAM sequence, which is particularly abundant in high $G+C$ *Streptomyces* species.

7. CONCLUSION

pCRISPomyces plasmids have been designed based on the type II CRISPR/Cas system of *S. pyogenes* to enable multiplex gene disruption with high efficiency in multiple *Streptomyces* species. To assemble pCRISPomyces plasmids that target a certain gene, the custom spacers and editing templates (homology donors) can be inserted into pCRISPomyces plasmids through well-developed DNA assembly techniques. These tools reduce the time and labor needed to perform precise genome manipulation compared with previous techniques. This system can facilitate a wide variety of studies in *Streptomyces* species, such as analysis of metabolic pathways, activation of silent natural product gene clusters for discovery of novel bioactive compounds, pathway/metabolic engineering, and host engineering.

ACKNOWLEDGMENT
This work was supported by the National Institutes of Health (GM077596).

REFERENCES
Baltz, R. H. (2010). Streptomyces and Saccharopolyspora hosts for heterologous expression of secondary metabolite gene clusters. *Journal of Industrial Microbiology & Biotechnology, 37*, 759–772.

Bikard, D., Jiang, W., Samai, P., Hochschild, A., Zhang, F., & Marraffini, L. A. (2013). Programmable repression and activation of bacterial gene expression using an engineered CRISPR-Cas system. *Nucleic Acids Research, 41*, 7429–7437.

Blodgett, J. A., Thomas, P. M., Li, G., Velasquez, J. E., Van Der Donk, W. A., Kelleher, N. L., et al. (2007). Unusual transformations in the biosynthesis of the antibiotic phosphinothricin tripeptide. *Nature Chemical Biology, 3*, 480–485.

Cobb, R. E., Wang, Y., & Zhao, H. (2014). High-efficiency multiplex genome editing of Streptomyces species using an engineered CRISPR/Cas system. *ACS Synthetic Biology, 4*, 723–728.

Cong, L., Ran, F. A., Cox, D., Lin, S., Barretto, R., Habib, N., et al. (2013). Multiplex genome engineering using CRISPR/Cas systems. *Science, 339*, 819–823.

Gaj, T., Gersbach, C. A., & Barbas, C. F. (2013). ZFN, TALEN and CRISPR/Cas-based methods for genome engineering. *Trends in Biotechnology, 31*, 397–405.

Genga, R. M., Kearns, N. A., & Maehr, R. (2015). Controlling transcription in human pluripotent stem cells using CRISPR-effectors. *Methods*. http://dx.doi.org/10.1016/j.ymeth.2015.10.014.

Gottelt, M., Kol, S., Gomez-Escribano, J. P., Bibb, M., & Takano, E. (2010). Deletion of a regulatory gene within the cpk gene cluster reveals novel antibacterial activity in *Streptomyces coelicolor* A3(2). *Microbiology, 156*, 2343–2353.

Jakociunas, T., Jensen, M. K., & Keasling, J. D. (2016). CRISPR/Cas9 advances engineering of microbial cell factories. *Metabolic Engineering, 34*, 44–59. http://dx.doi.org/10.1016/j.ymben.2015.12.003.

Kieser, T., Bibb, M. J., Buttner, M. J., Chater, K. F., & Hopwood, D. A. (2000). *Practical streptomyces genetics*. Norwich, UK: John Innes Foundation.

Laureti, L., Song, L., Huang, S., Corre, C., Leblond, P., Challis, G. L., et al. (2011). Identification of a bioactive 51-membered macrolide complex by activation of a silent polyketide synthase in *Streptomyces ambofaciens*. *Proceedings of the National Academy of Sciences of the United States of America, 108*, 6258–6263.

Li, H. L., Gee, P., Ishida, K., & Hotta, A. (2015). Efficient genomic correction methods in human iPS cells using CRISPR-Cas9 system. *Methods*. http://dx.doi.org/10.1016/j.ymeth.2015.10.015.

Lin, G., Zhang, K., & Li, J. (2015). Application of CRISPR/Cas9 technology to HBV. *International Journal of Molecular Sciences, 16*, 26077–26086.

Lucas, X., Senger, C., Erxleben, A., Grüning, B. A., Döring, K., Mosch, J., et al. (2013). StreptomeDB: A resource for natural compounds isolated from Streptomyces species. *Nucleic Acids Research, 41*, 1130–1136.

Luo, Y., Cobb, R. E., & Zhao, H. (2014). Recent advances in natural product discovery. *Current Opinion in Biotechnology, 30*, 230–237.

Luo, Y., Enghiad, B., & Zhao, H. (2016). New tools for reconstruction and heterologous expression of natural product biosynthetic gene clusters. *Natural Product Reports, 33*, 174–182. http://dx.doi.org/10.1039/C5NP00085H.

Luo, Y., Li, B. Z., Liu, D., Zhang, L., Chen, Y., Jia, B., et al. (2015). Engineered biosynthesis of natural products in heterologous hosts. *Chemical Society Reviews, 44*, 5265–5290.

Maeder, M. L., Linder, S. J., Cascio, V. M., Fu, Y., Ho, Q. H., & Joung, J. K. (2013). CRISPR RNA-guided activation of endogenous human genes. *Nature Methods, 10*, 977–979.

Mali, P., Esvelt, K. M., & Church, G. M. (2013). Cas9 as a versatile tool for engineering biology. *Nature Methods, 10*, 957–963.

Mazodier, P., Petter, R., & Thompson, C. (1989). Intergeneric conjugation between *Escherichia coli* and Streptomyces species. *Journal of Bacteriology, 171*, 3583–3585.

Mou, H., Kennedy, Z., Anderson, D. G., Yin, H., & Xue, W. (2015). Precision cancer mouse models through genome editing with CRISPR-Cas9. *Genome Medicine, 7*, 53.

Muth, G., Nußbaumer, B., Wohlleben, W., & Pühler, A. (1989). A vector system with temperature-sensitive replication for gene disruption and mutational cloning in streptomycetes. *Molecular Genetics and Genomics, 219*, 341–348.

Muto, A., & Osawa, S. (1987). The guanine and cytosine content of genomic DNA and bacterial evolution. *Proceedings of the National Academy of Sciences of the United States of America, 84*, 166–169.

Qi, L. S., Larson, M. H., Gilbert, L. A., Doudna, J. A., Weissman, J. S., Arkin, A. P., et al. (2013). Repurposing CRISPR as an RNA-guided platform for sequence-specific control of gene expression. *Cell, 152*, 1173–1183.

Ran, F. A., Hsu, P. D., Lin, C. Y., Gootenberg, J. S., Konermann, S., Trevino, A. E., et al. (2013). Double nicking by RNA-guided CRISPR Cas9 for enhanced genome editing specificity. *Cell, 154*, 1380–1389.

Sander, J. D., & Joung, J. K. (2014). CRISPR-Cas systems for editing, regulating and targeting genomes. *Nature Biotechnology, 32*, 347–355.

Schaart, J. G., Van De Wiel, C. C., Lotz, L. A., & Smulders, M. J. (2015). Opportunities for products of new plant breeding techniques. *Trends in Plant Science.* http://dx.doi.org/10.1016/j.tplants.2015.11.006.

Shao, Z., Rao, G., Li, C., Abil, Z., Luo, Y., & Zhao, H. (2013). Refactoring the silent spectinabilin gene cluster using a plug-and-play scaffold. *ACS Synthetic Biology, 2*, 662–669.

Tong, Y., Charusanti, P., Zhang, L., Weber, T., & Lee, S. Y. (2015). CRISPR-Cas9 based engineering of actinomycetal genomes. *ACS Synthetic Biology, 4*, 1020–1029.

Wang, T., Wei, J. J., Sabatini, D. M., & Lander, E. S. (2014). Genetic screens in human cells using the CRISPR-Cas9 system. *Science, 343*, 80–84.

Xu, T. (2014). Cas9-based tools for targeted genome editing and transcriptional control. *Applied and Environmental Microbiology, 80*, 1544–1552.

Zhang, A., Yang, M., Hu, P., Wu, J., Chen, B., Hua, Y., et al. (2011). Comparative genomic analysis of *Streptococcus suis* reveals significant genomic diversity among different serotypes. *BMC Genomics, 12*, 523.

CHAPTER THIRTEEN

Rapid Optimization of Engineered Metabolic Pathways with Serine Integrase Recombinational Assembly (SIRA)

C.A. Merrick*,[1], C. Wardrope*, J.E. Paget*, S.D. Colloms[†], S.J. Rosser*

*University of Edinburgh, Edinburgh, United Kingdom
[†]University of Glasgow, Glasgow, United Kingdom
[1]Corresponding author: e-mail address: christine.merrick@ed.ac.uk

Contents

1. Introduction	286
2. Mechanism of Serine Integrase Recombination	288
3. How SIRA Works	291
3.1 *attP* × *attB* Recombination	291
3.2 *attL* × *attR* Recombination	293
4. Applications of SIRA	294
4.1 Inserting a Single Piece of DNA into a Plasmid	294
4.2 Assembly of Multiple Genes in Predefined Orders for Compound Biosynthesis	294
4.3 Combinatorial Construct Libraries	295
4.4 Targeted Postassembly Modification for Enhanced Productivity	297
5. Design Principles for *att* Sites	297
5.1 Naming Convention for *att* Sites	298
5.2 Central Dinucleotides of *att* Sites Determine the Positions of DNA Parts in a Construct	298
5.3 Central Dinucleotides of *att* Sites Determine Polarity of Recombination	298
5.4 Elimination of Intramolecular Recombination	300
5.5 Assembled Products Contain Only *attL* Sites	300
6. Materials Preparation	300
6.1 DNA Components	300
6.2 Purification of ΦC31 Integrase and gp3	305
6.3 SIRA Reaction Buffer	307
6.4 Competent *E. coli* Strains	307
7. Example SIRA Protocols	308
7.1 SIRA Example 1: Inserting a Single Gene Using SIRA	308
7.2 SIRA Example 2: Assembly of Multiple DNA Parts in a Predefined Order	309
7.3 SIRA Example 3: Combinatorial Assembly of DNA Parts for a Library of Constructs with Different Gene Orders	310

7.4 SIRA Example 4: Targeted Postassembly Modification of a SIRA Construct   311
8. The Future of SIRA   312
Acknowledgments   314
References   314

Abstract

Metabolic pathway engineering in microbial hosts for heterologous biosynthesis of commodity compounds and fine chemicals offers a cheaper, greener, and more reliable method of production than does chemical synthesis. However, engineering metabolic pathways within a microbe is a complicated process: levels of gene expression, protein stability, enzyme activity, and metabolic flux must be balanced for high productivity without compromising host cell viability. A major rate-limiting step in engineering microbes for optimum biosynthesis of a target compound is DNA assembly, as current methods can be cumbersome and costly. Serine integrase recombinational assembly (SIRA) is a rapid DNA assembly method that utilizes serine integrases, and is particularly applicable to rapid optimization of engineered metabolic pathways. Using six pairs of orthogonal *attP* and *attB* sites with different central dinucleotide sequences that follow SIRA design principles, we have demonstrated that ΦC31 integrase can be used to (1) insert a single piece of DNA into a substrate plasmid; (2) assemble three, four, and five DNA parts encoding the enzymes for functional metabolic pathways in a one-pot reaction; (3) generate combinatorial libraries of metabolic pathway constructs with varied ribosome binding site strengths or gene orders in a one-pot reaction; and (4) replace and add DNA parts within a construct through targeted postassembly modification. We explain the mechanism of SIRA and the principles behind designing a SIRA reaction. We also provide protocols for making SIRA reaction components and practical methods for applying SIRA to rapid optimization of metabolic pathways.

1. INTRODUCTION

Rapid progress in synthetic biology and metabolic engineering has advanced the efficient heterologous biosynthesis of complex compounds such as pharmaceuticals, biomaterials, biofuels, and food additives in genetically modified microbial hosts in recent years (Hansen et al., 2009; Paddon & Keasling, 2014; Peralta-Yahya et al., 2011; Yu, Xia, Zhong, & Qian, 2014). This progress has been limited by uncharacterized variables within the engineered host system: eg, rates of gene expression, enzyme activity, protein degradation, and the build up of toxic intermediate metabolites. A number of genetic components can be incorporated into the design of a heterologous pathway to improve productivity including promoters, ribosome binding sites (RBSs), dynamic biosensors, scaffolds, and gene variants (Alper, Fischer, Nevoigt, & Stephanopoulos, 2005; Colloms et al., 2014; Dahl et al., 2013; Dueber et al., 2009; Ma et al., 2011; Nowroozi et al., 2014; Zhang, Carothers, & Keasling, 2012). However, due to a current lack of predictive modeling tools capable of simulating the effects of a designed metabolic pathway on product yield and cell growth, determining the optimum combination of genetic components often relies on iterative cycles of DNA assembly followed by performance testing. This cumbersome and costly trial-and-error approach is a major rate-limiting step in pathway optimization.

Current DNA assembly methods for building heterologous constructs can be categorized into three main groups: endonuclease-mediated assembly methods, homology-based methods, and site-specific recombination-based methods (Casini, Storch, Baldwin, & Ellis, 2015). Often, endonuclease-based approaches, such as BioBrick and Golden Gate methods, require DNA parts to be precloned for removal of forbidden restriction sites and then assembled in a hierarchical manner which limits generation of combinatorial libraries of constructs (Shetty, Endy, & Knight, 2008; Werner, Engler, Weber, Gruetzner, & Marillonnet, 2012). Homology-based approaches have limitations on the size of DNA parts that can be assembled and have difficulty with repetitive sequences. Methods such as Gibson Assembly that rely on exonuclease activity to generate complementary single-strand overhangs on DNA parts work best to assemble pieces of DNA that are larger than 250 bp in size (Gibson et al., 2009; Hillson, 2011). Polymerase chain reaction (PCR)-based methods such as circular polymerase extension cloning are restricted to DNA parts that are short enough to be amplified in vitro with a DNA polymerase (Quan & Tian, 2009). A major limitation of endonuclease- and homology-based DNA assembly methods is that once assembled, scar sequences at junctions between parts do not allow easy exchange of individual DNA parts within a construct.

Site-specific recombination-based DNA assembly methods recombine specific DNA sequences called "attachment" (*att*) sites added to DNA parts. These methods include popular Gateway™ Cloning that uses the tyrosine integrase from phage λ, and serine integrase-based methods: site-specific recombination-based tandem assembly (SSTRA) and serine integrase recombinational assembly (SIRA) (Cheo et al., 2004; Colloms et al., 2014; Hartley, Temple, & Brasch, 2000; Sasaki et al., 2004; Zhang, Zhao, & Ding, 2011). In comparison to Gateway™, *att* sites for serine integrases are small (both *attP* and *attB* are less than 50 bp) and no accessory proteins are required to bind and bend *att* sites. SSTRA and SIRA utilize orthogonal *att* sites to enable assembly of multiple DNA parts in a specific order in a one-pot reaction (Colloms et al., 2014; Zhang et al., 2011). SIRA *att* sites follow a set of design principles that enhance DNA assembly by prohibiting intramolecular recombination; enable targeted postassembly modification of a DNA construct; and can be applied to other serine integrases for additional orthogonal sites. The efficiency, flexibility, and simplicity of SIRA make it particularly well suited for rapid assembly of constructs for optimizing metabolic pathways.

In this chapter we briefly describe the current mechanistic model of serine integrase recombination and regulation, focussing on aspects with vital implications in SIRA. We explain the basis of the SIRA system and outline a number of applications for SIRA in metabolic pathway engineering including inserting a single piece of DNA, assembly of multiple DNA parts, generation of combinatorial libraries of constructs, and targeted postassembly modification of constructs. This chapter explains the principles for designing SIRA DNA parts and provides protocols for preparing SIRA reaction components including ΦC31 integrase. Finally, we provide detailed protocols for inserting a single piece of DNA, assembling multiple DNA parts in a predefined gene order, combinatorial assembly of multiple gene orders in one-pot reaction, and targeted postassembly modification of a construct in four examples of SIRA reactions.

2. MECHANISM OF SERINE INTEGRASE RECOMBINATION

Serine integrases catalyze integration of phage DNA into bacterial host chromosomes by recombining cognate *attP* (phage) and *attB* (bacteria) sites, generating *attL* (*left*) and *attR* (*right*) sites that flank the integrated prophage (Fig. 1A). This reaction is highly directional in that (the products of recombination between *attP* and *attB*) *attL* and *attR* sites do not recombine in the presence of integrase alone. For excision of the phage DNA from its host chromosome, recombination between *attL* and *attR* sites is stimulated by a small phage-encoded accessory protein called a recombination directionality factor (RDF). The RDF associates with the integrase protein and changes its conformation allowing it to synapse and recombine *attL* and *attR* (Bibb, Hancox, & Hatfull, 2005; Breüner, Brøndsted, & Hammer, 1999; Ghosh, Wasil, & Hatfull, 2006; Khaleel, Younger, McEwan, Varghese, & Smith, 2011). The *attP* and *attB* sites for serine integrases are imperfect inverted repeats composed of two half-sites, P and P' for *attP* and B and B' for *attB*, on either side of a 2-bp overlap sequence. During

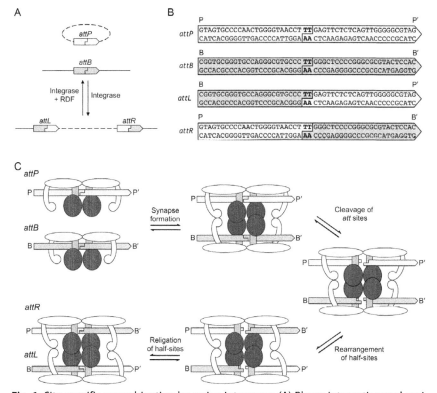

Fig. 1 Site-specific recombination by serine integrases. (A) Phage integration and excision. A serine integrase recombines an *attP* site (phage) and an *attB* site (bacteria) to integrate a phage genome into a bacterial host genome generating an *attL* (*left*) and an *attR* (*right*) which flank the integrated prophage. This reaction is directional because the *attL* and *attR* sites will not recombine in the presence of integrase alone. In the presence of an accessory protein called a recombination directionality factor (RDF), the serine integrase can recombine *attL* and *attR* sites to excise the phage DNA and restore the original *attP* and *attB* sites. (B) Sequences of ΦC31 integrase *att* sites. An *attP* site consists of two half-sites, P and P', on either side of a central dinucleotide core sequence, and an *attB* site consists of B and B' half-sites, on either side of a central dinucleotide core sequence. If the central dinucleotides in *attP* and *attB* sites match, they recombine to make *attL* which consists of B and P' half-sites on either side of a central dinucleotide, and *attR* which consists of P and B' half-sites on either side of the central dinucleotide. Central dinucleotides are in *bold*, and the cleavage sites are indicated by staggered *lines*. (C) Mechanism of serine integrase recombination. The serine integrase binds *attP* and *attB* sites as a dimer, and brings the sites together forming a tetrameric protein-DNA synapse. Next, nucleophilic serine residues in the integrase N-terminal domains catalyze cleavage of all DNA strands forming covalent phosphoserine linkages with the DNA backbones and generating 3' overhangs. Next, integrase subunits rotate in relation to each other rearranging *attP* and *attB* half-sites. If the 3' overhangs of the rearranged half-sites match, they are religated to form *attL* and *attR* sites. A current theory proposes that the C-terminal domains of integrase dimers bound the *attL* and *attR* sites form intramolecular interactions that prevent formation of an *attL* × *attR* synapse. The serine integrase N-terminal domain is in *dark gray*, the connecting α-helix is in *medium gray*, and the C-terminal domain is in *pale gray*. In the color version of this figure P and P' half-sites are *green*, and B and B' half-sites are *blue*. (See the color plate.)

recombination, *attP* and *attB* are cut on both strands either side of the overlap sequence; half-sites B and P' are joined to form *attL*, while P and B' are joined to form *attR* (Fig. 1B).

Serine integrases are comprised of a ~130 amino acid N-terminal catalytic domain which shares similarity with serine resolvases/invertases, and a large ~300 to >500 amino acid C-terminal domain connected to the catalytic domain by a long α-helix (Rutherford, Yuan, Perry, Sharp, & Van Duyne, 2013; Smith, Brown, McEwan, & Rowley, 2010). It has been recognized for some time that the C-terminal domain mediates *att* site recognition, DNA binding, synapsis, and controls the directionality of the recombination reactions (Ghosh, Pannunzio, & Hatfull, 2005; McEwan, Rowley, & Smith, 2009; Rowley, Smith, Younger, & Smith, 2008).

Serine integrases bind their cognate *attP* and *attB* sites and bring them together to form a synapse containing an integrase tetramer. Once in this synapse, a nucleophilic serine in each of the four subunits attacks the DNA backbone at staggered positions either side of the overlap sequence, forming a covalent 5' phosphoserine bond and freeing a 3' hydroxyl group. This cleavage results in double-strand breaks in both *att* sites, leaving half-sites with 3' dinucleotide overhangs. Two subunits of the tetramer then rotate in relation to the others and, if the overhangs of the *att* sites are complementary, the rearranged pieces of DNA religate to form *attL* and *attR* sites (Fig. 1C) (Smith et al., 2010). A current model of RDF-regulated L1 integrase recombination proposes that the C-terminal domains of integrase subunits bound to *attL* and *attR* sites form autoinhibited complexes (Rutherford et al., 2013). These authors also propose that the RDF binds to the C-terminal domains of integrase dimers on *attL* and *attR*, positioning them to make an *attL* × *attR* tetrameric synapse (Rutherford et al., 2013).

In the context of SIRA, it is important to note two aspects of the serine integrase recombination mechanism. The first is that recombination of *attP* and *attB* sites to make *attL* and *attR* sites is highly directional, meaning the reaction cannot be reversed in the absence of a RDF. The second is that the exact sequence of the overlap dinucleotide at the center of *attP* and *attB* is not important so long as they match; the 3' overhangs generated by the double-strand breaks in *att* sites must be complementary for recombination to occur. As these features of recombination are shared among serine integrases, any number of these enzymes could be applied in SIRA. For simplicity, this chapter focuses on the application of one serine integrase, ΦC31 integrase, to SIRA. ΦC31 integrase is a well-characterized serine integrase that has found several applications in genome engineering and biotechnology (Karow et al., 2011; Nimmo, Alphey, Meredith, & Eggleston, 2006; Venken, He, Hoskins, & Bellen, 2006; Ye et al., 2010). The cognate RDF for ΦC31 integrase, ΦC31 gp3, has been identified and characterized (Khaleel et al., 2011; Olorunniji et al., 2012).

3. HOW SIRA WORKS

The premise of SIRA lies in the ability to predetermine the position of DNA parts within a construct by engineering the cleaved *att* site overhang sequence. This enables multiple DNA parts to be assembled in a single reaction and specific DNA parts to be targeted for postassembly modification. SIRA reactions fall into two categories: *attP* × *attB* recombination for assembly of DNA parts to make a construct; and *attL* × *attR* recombination for targeted postassembly modification of a construct.

3.1 *attP* × *attB* Recombination

SIRA DNA parts are designed with *attB* or *attP* sites on each end. For SIRA with ΦC31 integrase, these *attP* and *attB* sites must have matching central dinucleotides for efficient recombination. This means that a DNA part flanked with an *attP* site with a TT central dinucleotide sequence will recombine with (and become joined to) a DNA part flanked with an *attB* site with a TT central dinucleotide; a DNA part flanked with an *attP* site with a CT central dinucleotide sequence will recombine with a DNA part flanked with an *attB* site with a CT central dinucleotide; and so on (Fig. 2A and B). These *att* sites with different central dinucleotide sequences will herein be referred to as orthogonal *att* sites. DNA parts are assembled by recombination of orthogonal *att* sites flanking each part. In a multipart assembly, DNA parts are incorporated at specific positions all separated by *attL* sites and with their order dependent on the 2-bp overlap sequences of the *attP* and *attB* sites used (Fig. 2B). Short linear pieces of DNA containing only *attR* sites are also produced as side products.

To assemble DNA parts in a SIRA substrate plasmid, the plasmid must contain at least a single target *att* site, but it is desirable that it should facilitate screening for inserts. Such plasmids carry a "SIRA insertion site," consisting of two *att* sites on either side of a marker gene that is replaced by the assembled DNA parts. Such a marker gene could simply encode a fluorescent protein or it could be a negative selection marker allowing selection for growth or survival. We use the negative selection gene *ccdB*, the product of which is

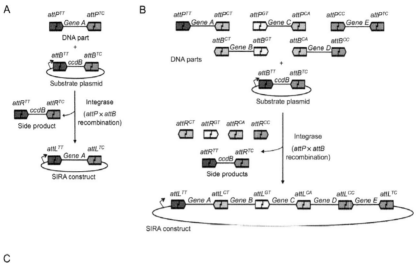

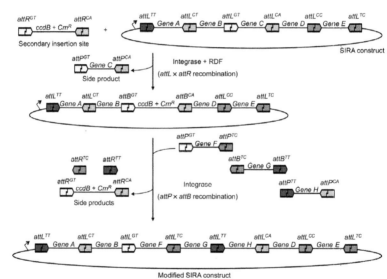

Fig. 2 The serine integrase recombinational assembly (SIRA) system. (A) Inserting a single piece of DNA into a substrate plasmid by $attP \times attB$ recombination. A DNA part containing *gene A* is flanked with $attP^{TT}$ and $attP^{TC}$ in inverted orientations. The substrate plasmid contains a "SIRA insertion site" in which $attB^{TT}$ and $attB^{TC}$ are in inverted orientations on either side of *ccdB*, a gene encoding a cytotoxic protein that causes cell death. Recombination of the $attP^{TT}$ with $attB^{TT}$, and $attP^{TC}$ and $attB^{TC}$, by a serine integrase inserts the DNA part into the substrate plasmid. The DNA part replaces the *ccdB* gene enabling selection of assembled constructs after transformation of *E. coli*. (B) Assembly of five DNA parts in a substrate plasmid by $attP \times attB$ recombination. DNA parts carrying arbitrary *genes A* to *E* are flanked with *attB* sites or *attP* sites in inverted orientations. In the presence of integrase, *attB* sites will recombine with *attP* sites with matching

a potent 101 amino acid cytotoxin from *Escherichia coli* that traps DNA gyrase in a cleaved complex and causes cell death (Bernard & Couturier, 1992). When *E. coli* cells are transformed with SIRA reactions, cells containing assembled plasmid products will grow whereas cells containing the original plasmid containing *ccdB* will not. A plasmid in which the insertion site has been replaced with assembled DNA parts is called a SIRA construct.

3.2 *attL* × *attR* Recombination

The *attL* sites separating the DNA parts in SIRA constructs allow targeted postassembly modification. A "secondary SIRA insertion site," containing genes for positive and negative selection and flanked by *attR* sites, can be recombined with matching *attL* sites using integrase and its cognate RDF (Fig. 2C). By engineering the central dinucleotides of *attR* sites to recombine with specific *attL* sites, the secondary insertion site can be targeted to replace a specific DNA part within the SIRA construct. Once inserted, the secondary SIRA insertion site is flanked with orthogonal *attB* sites. The positive selection gene, eg, an antibiotic resistance marker, is used to select for constructs containing the insertion site. The secondary insertion site can be replaced with additional DNA parts in a subsequent round of *attP* × *attB* recombination with integrase alone to make a modified SIRA construct (Fig. 2C). With this method, additional DNA parts can be inserted to tweak or increase the size of a SIRA construct. This *attP* × *attB* recombination removes the secondary insertion site and therefore the negative selection gene allowing selection for the desired construct; after transformation, cells

---

Fig. 2—Cont'd central dinucleotides assembling the DNA parts in the substrate plasmid, replacing the SIRA insertion site. This means that each DNA part will be incorporated at a specific position within an assembly (here giving the gene order A-B-C-D-E), and that these positions will be separated by *attL* sites with different central dinucleotides in alternating orientations. (C) Targeted postassembly modification of a SIRA construct carrying genes A to E by *attL* × *attR* recombination. A "secondary SIRA insertion site" carrying a chloramphenicol resistance gene for positive selection and the *ccdB* gene for negative selection, and flanked by $attR^{GT}$ and $attR^{CA}$, recombines with $attL^{GT}$ and $attL^{CA}$ sites in the SIRA construct in the presence of integrase and RDF. The result of this reaction is that gene C is replaced with the secondary SIRA insertion site that, once inserted, is flanked by $attB^{GT}$ and $attB^{CA}$. The plasmid carrying the secondary SIRA insertion site can be selected through resistance to chloramphenicol. Additional DNA parts such as those carrying genes F to H can be added to the construct by *attP* × *attB* recombination, removing the secondary SIRA insertion site. Loss of the lethal *ccdB* gene enables selection for the new, modified SIRA construct. Different colors represent the different central dinucleotide sequences of *att* sites. (See the color plate.)

containing the modified SIRA construct will grow, while those harboring the secondary insertion site will not.

4. APPLICATIONS OF SIRA

We have demonstrated that SIRA can be applied to the controlled assembly of single or multiple DNA parts, construction of combinatorial libraries and targeted postassembly modification, and that these features make SIRA an ideal technology with which to rapidly optimize engineered metabolic pathways.

4.1 Inserting a Single Piece of DNA into a Plasmid

The simplest application of SIRA is that of inserting a single DNA part into a SIRA substrate plasmid. A 5-min reaction containing (1) a linear DNA part flanked with orthogonal *attP* sites made by PCR, (2) a substrate plasmid with orthogonal *attB* sites, (3) reaction buffer, and (4) ΦC31 integrase, produced correctly assembled products in 20 out of 20 constructs analyzed (Colloms et al., 2014). It is easy to produce large numbers of colonies ($>10^6$) from a single reaction, making the method highly suitable for producing libraries of mutant variants. Because SIRA DNA parts are typically linear PCR products and do not require modification (eg, site-directed mutagenesis of incompatible restriction sites, or digestion for generation of "sticky ends"), they are well suited for efficient production of mutant gene libraries. We have demonstrated that SIRA can be used to insert a DNA part of less than 75 bp containing a multiple cloning site and a DNA part over 12 kb containing genes for conversion of acetyl-CoA to farnesyl pyrophosphate into a plasmid. In both cases 20 of 20 constructs analyzed were correctly assembled (data not shown). It is likely that SIRA with ΦC31 integrase could be used to insert even larger pieces of DNA because ΦC31 integrase integrates the ΦC31 bacteriophage genome, approximately 41.5 kb in length, into a genome of *Streptomyces* spp., and it has been used to integrate DNA fragments up to ~146 kb into the *Drosophila* genome (Smith, Burns, Wilson, & Gregory, 1999; Venken et al., 2006). A detailed protocol for using SIRA to insert a single piece of DNA is provided in SIRA Example 1.

4.2 Assembly of Multiple Genes in Predefined Orders for Compound Biosynthesis

We first demonstrated assembly of three, four, and five DNA parts using DNA parts with orthogonal *att* sites containing genes from the carotenoid biosynthesis pathway from *Pantoea ananatis*. Reaction products were used to transform *E. coli* and colonies were screened for the production of carotenoids; when three DNA parts containing genes for the biosynthesis of lycopene, a red compound, were assembled, 87% of transformant colonies were red; when four DNA parts containing genes for the biosynthesis of β-carotene, an orange compound, were assembled, 37% of transformant colonies were orange; and when five DNA parts containing genes for the biosynthesis of zeaxanthin, a yellow compound, were assembled, 18% of transformant colonies were yellow (Colloms et al., 2014). For a detailed protocol for using SIRA to assemble multiple DNA parts in a specific order, see SIRA Example 2.

4.3 Combinatorial Construct Libraries

It is recognized that expression of heterologous genes can impose significant burden on cells causing selection against certain genetic constructs (Ceroni, Algar, Stan, & Ellis, 2015). This emphasizes the importance of balancing variables such as plasmid copy number, gene expression levels, and enzyme activity to minimize diversion of cellular resources such as nucleotides, amino acids, and metabolites, and also to limit accumulation of toxic intermediates. Without detailed prior characterization of such variables it is nearly impossible to predict which genetic constructs will cause cellular burden. Increasingly, combinatorial approaches to metabolic engineering are being adopted for rapid optimization of cellular phenotypes (Ajikumar et al., 2010; Santos, Xiao, & Stephanopoulos, 2012; Wang et al., 2009). These approaches produce libraries of genotypic diversity and require little knowledge of the metabolic network's properties, but rely on a robust method to generate diversity.

4.3.1 Degenerate RBSs for Combinatorial Optimization of Expression Levels of Biosynthetic Pathway Genes

An effective approach to optimize heterologous metabolic pathways in microbial systems is to identify combinations of gene expression levels that balance pathway flux while increasing productivity and maintaining host viability. We used a single SIRA reaction to make a library of constructs for conversion of tryptophan to violacein, a purple compound. This library consisted of the five genes of the violacein biosynthesis pathway (*vioA*, *vioB*, *vioC*, *vioD*, and *vioE*) assembled with a wide range of RBS strengths for each gene in the pathway (Colloms et al., 2014). We used the RBS calculator to design degenerate RBS sequences with predicted expression levels ranging from ~100 to 10,000 a.u. for each gene (Salis, Mirsky, & Voigt, 2009). Because SIRA DNA parts can be linear PCR products, we were able to incorporate these degenerate RBS sequences and the required *att* sites into PCR primers, and to make libraries of DNA parts with different predicted expression levels in a single PCR for each gene of the pathway. We used these libraries of DNA parts to make one large library of constructs that generated different violacein production phenotypes in *E. coli* using the SIRA protocol for assembly of multiple parts in a predefined order (*vioA-vioB-vioC-vioD-vioE*; see SIRA Example 2) (Colloms et al., 2014). After transformation, we screened colonies for different levels of violacein production and sequenced the constructs they harbored. We used the sequenced RBSs to determine the predicted expression levels of the violacein biosynthesis genes in each of the constructs. The results suggested that low *vioA* expression levels contribute to increased violacein biosynthesis in *E. coli* (Colloms et al., 2014). This SIRA approach is valuable for quickly balancing expression levels of genes in metabolic pathways and identifying bottlenecks in pathway flux. Nowroozi et al. also demonstrated this in *E. coli* by cloning RBSs of different theoretical strengths 5′ of heterologous genes required for conversion of mevalonate to amorphadiene and assembling them into different constructs all with the same promoter (Nowroozi et al., 2014). Results from this study strongly indicate that changes in protein levels can alter flux through a pathway producing different product yields and growth rates.

4.3.2 Varying Biosynthetic Pathway Gene Order for Phenotypic Diversity

We also used a single SIRA reaction to generate a combinatorial library of constructs encoding the three-gene lycopene biosynthesis pathway with all possible gene orders. To do this, we made three DNA parts with different pairs of orthogonal *attP* or *attB* sites for each of the genes such that each gene could be incorporated at any position within the construct. When all of these parts were incubated together in a SIRA reaction with a high-copy SIRA plasmid with a weak constitutive promoter, a library of constructs was made with different gene orders and gene copy numbers. Based on the hypothesis that levels of gene expression are inversely related to the distance from the promoter (Nishizaki, Tsuge, Itaya, Doi, & Yanagawa, 2007), and that individual genes and DNA parts may contain internal sequences with promoter activity, we anticipated that different gene orders would produce different phenotypes. When the lycopene-producing colonies were analyzed, we observed that certain genotypes had a reduced growth phenotype at 37°C, possibly reflecting different levels of toxic intermediate build up for the different gene orders (Colloms et al., 2014). For a detailed protocol describing the use of SIRA to assemble DNA parts in different orders in a one-pot reaction, see SIRA Example 3.

4.4 Targeted Postassembly Modification for Enhanced Productivity

We demonstrated targeted modification of a construct for the purpose of enhancing zeaxanthin biosynthesis in *E. coli*. The construct contained an assembly of five DNA parts carrying genes for zeaxanthin biosynthesis (*crtB, crtE, crtI, crtY,* and *crtZ*). A secondary SIRA insertion site, carrying a chloramphenicol resistance gene and the *ccdB* gene, flanked with *attR* sites (*attR^{GT}* and *attR^{CA}*) was made to recombine with the *attL* sites (*attL^{GT}* and *attL^{CA}*) on either side of *crtI* in the zeaxanthin biosynthesis construct thereby replacing *crtI*. After *attL* × *attR* recombination in the presence of ΦC31 integrase and gp3, the secondary SIRA insertion site had been inserted correctly in 10 out of 10 transformants selected with chloramphenicol. In our first approach to enhancing zeaxanthin production, a library of *crtI* variants, flanked with suitable *attP* sites (*attP^{GT}* and *attP^{CA}*) were made by error-prone PCR and integrated into the secondary insertion site through *attP* × *attB* recombination. The effect of mutations in *crtI* on zeaxanthin production was examined and none of the *crtI* variants screened enhanced production. In a second approach, we replaced the secondary insertion site with a further three DNA parts carrying *crtI, dxs,* and *idi* genes because overexpression of *dxs* and *idi* had previously been shown to enhance carotenoid biosynthesis in *E. coli* (Jones, Kim, & Keasling, 2000; Kajiwara, Fraser, Kondo, & Misawa, 1997; Matthews & Wurtzel, 2000). *E. coli* cells harboring constructs of seven DNA parts including *dxs* and *idi* produced a stronger yellow color suggesting enhanced zeaxanthin production. For a detailed protocol using *attL* x *attR* recombination to modify assembled constructs, see SIRA Example 4.

5. DESIGN PRINCIPLES FOR *att* SITES

In order to maximize the efficiency of assembly, ensure stability of constructs, and enable postassembly modification, SIRA DNA parts adhere to some simple design principles. SIRA DNA parts have a distinctive structure in that they are flanked with *attP* or *attB* sites in inverted orientations with different central dinucleotides. These features ensure that each DNA part has a defined position within a construct, prevent intramolecular recombination, and produce *attL* sites in SIRA constructs that facilitate targeted exchange of DNA parts.

5.1 Naming Convention for *att* Sites

By convention, we refer to all sites according to the overlap central dinucleotide of the top strand of the sequences shown in Fig. 1B (with the B and P sequences on the *left* and B′ and P′ on the *right*). Thus for instance the wild-type ΦC31 *attP* site $attP^{TT}$ recombines with $attB^{TT}$ to form $attL^{TT}$ and $attR^{TT}$. Note that because *attP* sites generally face inward on our DNA parts, and *attB* sites are generally outward facing (Fig. 2B.), PCR primers used to add *attP* sites to DNA parts contain the named central dinucleotide, whereas those used to add *attB* sites contain the complement of the named central dinucleotide (see Table 1).

5.2 Central Dinucleotides of *att* Sites Determine the Positions of DNA Parts in a Construct

The central dinucleotide shared between an *attP* site and an *attB* site must match for efficient recombination. If the dinucleotide overhangs on half-sites do not match at the point of integrase subunit rotation, the integrase subunits will further rotate restoring the original *attP* and *attB* sites (Olorunniji et al., 2012). A DNA part therefore has predetermined *att* sites on each end designed to recombine with the specific DNA parts that precede and follow it. This enables the user to design constructs in which DNA parts have specific, predefined positions and to target these positions for postassembly modification.

5.3 Central Dinucleotides of *att* Sites Determine Polarity of Recombination

The description of serine integrase recombination in this chapter has been based on parallel synapsis of *attP* and *attB* sites meaning that the sites align in the same orientation in the tetrameric synapse. An important design consideration for SIRA is the occurrence of antiparallel synapsis in which *attP* and *attB* sites are brought together in opposing orientations in the tetrameric synapse. In the antiparallel conformation, *attP* and *attB* sites with identical central dinucleotides (eg, $attP^{TT}$ and $attB^{TT}$) will not recombine. The 3′

TT overhang of the P half-site will not complement the 3′ TT overhang of the B half-site, and the 3′ AA overhang of the P′ half-site will not complement the 3′ AA overhang of the B′ half-site. However, should *att* sites have central dinucleotides with inverted complementarity (eg, $attP^{TT}$ with $attB^{AA}$), only antiparallel recombination can occur, producing P-B and B′-P′ as opposed to the normal B-P′ (*attL*) and P-B′ (*attR*) products (Fig. 3A; Smith, Till, & Smith, 2004). Similarly, should the central dinucleotides have a palindromic dinucleotide sequence (eg, $attP^{TA}$ with $attB^{TA}$), both parallel and antiparallel recombination can occur. This antiparallel recombination will not join the DNA fragments together in the desired manner, but will instead just swap the terminal sequences between two fragments. Six orthogonal pairs of *attP* and *attB* sites exist with nonpalindromic central dinucleotide sequences that do not share inverted complementarity (TT, CT, GT, CA, CC, and TC) (Fig. 3B).

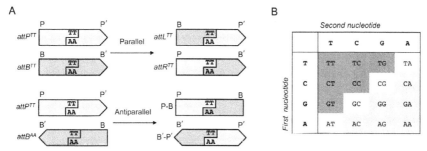

Fig. 3 Central dinucleotides in *att* sites determine the polarity of recombination. (A) Alignment of *att* sites in parallel and antiparallel conformations. In this diagram, *attP* and *attB* sites have central dinucleotides that are nonpalindromic. *attP* is shown with a TT central dinucleotide, *attB* is shown with either TT or AA, the reverse complement of TT. Cleaved half-sites carry either a 3′ TT or a 3′ AA overhang. When *attP* and *attB* sites align in a parallel conformation, half-sites with complementary overhangs ligate to make *attL* and *attR*. When sites align in an antiparallel conformation, half-sites with complementary overhangs ligate to make P-B and B′-P′. $attP^{TT}$ can only recombine with $attB^{TT}$ in the parallel conformation, whereas $attP^{TT}$ can only recombine with $attB^{AA}$ in the antiparallel conformation. In contrast, *att* sites with palindromic central dinucleotides, eg, $attP^{TA}$ and $attB^{TA}$, can recombine in both parallel and antiparallel conformations. In the color version of this figure P and P′ half-sites are *green*, and B and B′ half-sites are *blue*. (B) Nonpalindromic dinucleotides that can be used in SIRA *att* sites. The six nonpalindromic dinucleotides that are suitable for SIRA *att* sites are shaded in *dark gray*, and their reverse complement sequences are in *pale gray*. Palindromic dinucleotides that enable parallel and antiparallel recombination and are therefore not suitable for SIRA have a *white background*. (See the color plate.)

5.4 Elimination of Intramolecular Recombination

When located on the same piece of DNA, *attP* and *attB* sites will undergo efficient intramolecular synapsis in the presence of integrase. This would produce circular products if the *att* sites had matching central dinucleotides, but even with mismatched sites, this synapsis and subsequent cleavage will reduce the availability of DNA for the desired intermolecular recombination reactions. For this reason, as far as possible, DNA parts should be flanked either by two *attB* sites or two *attP* sites. It should be noted that one DNA part must be flanked by one *attP* site and one *attB* site for assembly of an even number of parts such as that of the four-gene β-carotene biosynthesis pathway (Colloms et al., 2014).

5.5 Assembled Products Contain Only *attL* Sites

For SIRA, sequential *att* sites on DNA parts are designed to have alternating orientations: *attP* sites point toward the inside of a DNA part; *attB* sites point toward the outside of a DNA part. It was found that PCR was more efficient with this arrangement of *att* sites within primers, probably because of reduced complementarity between *att* site sequences in forward and reverse primers. As a consequence of this, P′ and B half-sites are incorporated as *attL* sites between DNA parts in alternating orientations, and P and B′ half-sites form linear *attR* sites as reaction side products. This is an important design feature of SIRA for RDF-mediated postassembly modification because secondary SIRA insertion sites carry orthogonal *attR* sites designed to recombine with specific *attL* sites in an assembled construct.

6. MATERIALS PREPARATION

A minimal number of components are required for SIRA reactions: DNA parts and SIRA substrate plasmids, integrase and RDF proteins, reaction buffer, and cells for transformation. This section provides protocols for preparing these SIRA reaction components.

6.1 DNA Components

Focussing on the DNA components of SIRA reactions, this section contains information on how to make SIRA DNA parts by PCR, how SIRA substrate plasmids are made, and a protocol for purifying DNA components by ethanol precipitation.

6.1.1 DNA Parts

For the applications described in this chapter, DNA parts for SIRA are linear DNA fragments with *attB* or *attP* sites on each end. For the purpose of engineering metabolic pathways in microbial hosts, a DNA part contains one or more genes, each with a RBS. In this chapter DNA parts are made by PCR with a proofreading polymerase. As there are no topological constraints on the ΦC31 integrase *att* sites, DNA parts could instead be precloned into a circular entry vector.

6.1.1.1 Primers for DNA Parts to be Assembled in Predefined Gene Orders

To make DNA parts, primers are designed with specific *att* sites with central dinucleotides that determine the position of a DNA part within a construct (Table 1). The forward primer (5′–3′) should consist of an *att* site followed by

Table 1 Primers for Adding *att* Sites to Genes with One Round of PCR

Primer	att Site	5′–3′ Sequence[a]
1 (F)	*attP^{TT}*	AGCTCTAGAAGTGCCCCAACTGGGGTAACCTTTGAGTTCTCTCAGTTGGGGGCGTAGGAGGATTACAAAA(N)$_n$
2 (R)	*attP^{CT}*	AGACTAGTAGTGCCCCAACTGGGGTAACCTCTGAGTTCTCTCAGTTGGGGGCGTTTA(N)$_n$
3 (F)	*attB^{CT}*	AGCTCTAGAGGAGTACGCGCCCGGGGAGCCCAGGGGCACGCCCTGGCACCCGCACAGGAGGATTACAAA(N)$_n$
4 (R)	*attB^{GT}*	AGCACTAGTGGAGTACGCGCCCGGGGAGCCCACGGGCACGCCCTGGCACCCGCACTTA(N)$_n$
5 (F)	*attP^{GT}*	AGCTCTAGAAGTGCCCCAACTGGGGTAACCTGTGAGTTCTCTCAGTTGGGGGCGTAGGAGGATTACAAA(N)$_n$
6 (R)	*attP^{CA}*	GCACTAGTAGTGCCCCAACTGGGGTAACCTCAGAGTTCTCTCAGTTGGGGGCGTTTA(N)$_n$
7 (F)	*attB^{CA}*	AGCTCTAGAGGAGTACGCGCCCGGGGAGCCCTGGGGCACGCCCTGGCACCCGCACAGGAGGATTACAAA(N)$_n$
8 (R)	*attB^{CC}*	AGCACTAGTGGAGTACGCGCCCGGGGAGCCCGGGGGCACGCCCTGGCACCCGCACTTA(N)$_n$
9 (F)	*attP^{CC}*	AGCTCTAGAAGTGCCCCAACTGGGGTAACCTCCGAGTTCTCTCAGTTGGGGGCGTAGGAGGATTACAAA(N)$_n$
10 (R)	*attP^{TC}*	AGCACTAGTAGTGCCCCAACTGGGGTAACCTTCGAGTTCTCTCAGTTGGGGGCGTTTA(N)$_n$

[a]ΦC31 integrase *att* sites are underlined, central dinucleotides are shaded, restriction sites for *Xba*I (TCTAGA) and *Spe*I (ACTAGT) are in *italics*, RBS sequences and stop codons are in **bold**, and (N)$_n$ represents the sequence designed to anneal to a template gene.

a RBS sequence, a start codon and a "binding sequence" that anneals to the start of the template gene with a basic melting temperature (T_m) of approximately 60°C. The reverse primer (5′–3′) should consist of an *att* site followed by a stop codon and a binding sequence specific for the end of the template gene that has a T_m similar to that of the forward primer. To place the *att* sites in inverted repeat at both ends of the DNA part, and to reduce complementarity between forward and reverse primers, *att* sites have the same sequence (apart from the central dinucleotide) 5′–3′ in both forward and reverse primers.

6.1.1.2 Primers for Degenerate RBS Libraries
A library of DNA parts carrying the same gene but with different predicted expression levels can be made utilizing PCR with a forward primer containing a degenerate RBS. For this PCR, primers are structurally the same as those used for DNA parts assembled in predefined orders except that in place of the RBS, they carry a RBS with degenerate bases that will either enhance or reduce gene expression levels. As the number of degenerate bases in a RBS increases, so does the number of RBS variants in a PCR reaction. We use the "Predict: RBS Library" function of the RBS Calculator V2.0 to determine the distribution of predicted expression levels from a degenerate RBS sequence (Espah Borujeni, Channarasappa, & Salis, 2014; Salis et al., 2009). We then use this information to refine the level of degeneracy within the RBS, thereby reducing redundancy in the library while obtaining the same range in predicted expression levels.

6.1.1.3 Primers for Combinatorial Construct Libraries
When making combinatorial construct libraries of different gene orders using SIRA, multiple DNA parts are made for each gene so that the gene can occupy any position within a construct. To reduce the burden of primer design, a two-step PCR approach can be undertaken. In the first PCR step, truncated *attP* or *attB* sites are added to DNA parts. Primers used for this step are based on those in Table 1 but lack the P half-site of *attP*, the B′ half-site of *attB* and the central dinucleotide sequence (primers 11–14 in Table 2). An advantage of this approach is that optimization of PCR conditions for new template DNA is not complicated by secondary structure formation within *att* sites. In the second PCR step, the product from the first reaction is used as template for primers comprised of complete *att* sites with chosen central dinucleotide sequences (primers 15–24 in Table 2). These have been

Rapid Optimization of Engineered Metabolic Pathways

Table 2 Primers for Adding *att* Sites to Genes with Two Rounds of PCR

Primer	*att* Site	5′–3′ Sequence[a]
11 (F)	tr-*attP*	TTGGGGGCGT**AGGAGGATTACAAA**(N)$_n$
12 (R)	tr-*attP*	GAGTTCTCTCAGTTGGGGGCGT**TTA**(N)$_n$
13 (F)	tr-*attB*	CACCCGCAC**AGGAGGATTACAAA**(N)$_n$
14 (R)	tr-*attB*	GCCCTGGCACCCGCAC**TTA**(N)$_n$
15 (F)	*attP*^{TT}	AGC*TCTAGA*AGTGCCCCAACTGGGGTAACCTTTGAG TTCTCTCAGTTGGGGGCGT**AGGAGGATTAC**
16 (R)	*attP*^{CT}	AGA*CTAGT*AGTGCCCCAACTGGGGTAACCTCTGAGT TCTCTCAGTTGGGGGC
17 (F)	*attB*^{CT}	AGC*TCTAGA*GGAGTACGCGCCCGGGGAGCCCAGGG GCACGCCCTGGCACCCGCAC**AGGAGGATTAC**
18 (R)	*attB*^{GT}	AGC*ACTAGT*GGAGTACGCGCCCGGGGAGCCCACGG GCACGCCCTGGCACCCGCAC
19 (F)	*attP*^{GT}	AGC*TCTAGA*AGTGCCCCAACTGGGGTAACCTGTGAG TTCTCTCAGTTGGGGGCGT**AGGAGGATTAC**
20 (R)	*attP*^{CA}	GC*ACTAGT*AGTGCCCCAACTGGGGTAACCTCAGAGT TCTCTCAGTTGGGGGC
21 (F)	*attB*^{CA}	AGC*TCTAGA*GGAGTACGCGCCCGGGGAGCCCTGGG GCACGCCCTGGCACCCGCAC**AGGAGGATTAC**
22 (R)	*attB*^{CC}	AGC*ACTAGT*GGAGTACGCGCCCGGGGAGCCCGGGG GCACGCCCTGGCACCCGCAC
23 (F)	*attP*^{CC}	AGC*TCTAGA*AGTGCCCCAACTGGGGTAACCTCCGAG TTCTCTCAGTTGGGGGCGT**AGGAGGATTAC**
24 (R)	*attP*^{TC}	AGC*ACTAGT*AGTGCCCCAACTGGGGTAACCTTCGAG TTCTCTCAGTTGGGGGC

[a] Sequences from ΦC31 integrase *attP* and *attB* sites are underlined, "tr-" indicates a truncated *att* site, central dinucleotides are shaded, restriction sites for *Xba*I (TCTAGA) and *Spe*I (ACTAGT) are in *italics*, RBS sequences and stop codons are in **bold**, and (N)$_n$ represents the sequence designed to anneal to a template gene.

designed to anneal to templates with T_m of 60°C and PCR conditions have been established for efficient amplification.

6.1.1.4 Secondary SIRA Insertion Sites

To modify a part of an existing assembly, an insertion site is first used to replace the chosen part. This is carried out by *attL* × *attR* recombination

Table 3 Primers for Adding *att* Sites to Positive and Negative Selectable Markers to Create SIRA Insertion Sites

Primer	*att* Site	5′–3′ Sequence[a]
25 (F)	*attR^{TT}*	AAT*TCTAGA*AGTGCCCCAACTGGGGTAACCT**TT**GGG CTCCCCGGGCGCGTACTCC(N)$_N$
26 (R)	*attR^{CT}*	AT*CACTAGT*AGTGCCCCAACTGGGGTAACCT**CT**GGG CTCCCCGGGCGCGTACTCC(N)$_N$
27 (F)	*attB^{TT}*	AGC*TCTAGA*GTGCGGGTGCCAGGGCGTGCCC**TT**GG GCTCCCCGGGCGCGTACTCC(N)$_n$
28 (R)	*attB^{TC}*	AGC*ACTAGT*GTGCGGGTGCCAGGGCGTGCCC**TC**GG GCTCCCCGGGCGCGTACTCC(N)$_n$

[a]ΦC31 integrase *attB* and *attR* sites are underlined, central dinucleotides are shaded, restriction sites for XbaI (TCTAGA) and SpeI (ACTAGT) are in italics, RBS sequences and stop codons are in bold, and (N)$_n$ represents the sequence designed to anneal to a template gene.

using a positive–negative selectable cassette, called a "secondary SIRA insertion site," flanked by *attR* sites with central dinucleotides chosen to target the part to be replaced. To make this secondary insertion site, *attR* sites are added to genes that enable genetic selection by PCR (primers 25 and 26 in Table 3).

6.1.1.5 PCR Conditions for Making DNA Parts

We use Phusion high-fidelity DNA polymerase (New England Biolabs) for PCR to make DNA parts and insertion sites using primers in Tables 1–3. For a 50 μL reaction, combine 10 μL of 5 × HF Phusion buffer, 5 μL dimethyl sulfoxide (DMSO), 5 μL of 2 mM dNTPs, 5 μL of 5 μM forward primer, 5 μL of 5 μM reverse primer, between 0.5 and 10 ng of gel purified or plasmid DNA as template, and nuclease-free water to a volume of 49.5 μL in a sterile PCR tube. Before thermal cycling, 1 U (0.5 μL) Phusion polymerase is added, and reactions are thoroughly mixed and gently spun down. Note that 10% DMSO is included in reactions to minimize formation of *att* site secondary structures. DMSO lowers the primer T_m so accordingly an annealing temperature below 60°C is used. Our recommended PCR thermal cycle is as follows: initial denaturing at 98°C for 40 s; 30 cycles of denaturing at 98°C for 30 s, annealing at 55–58°C for 30 s, and extension at 72°C for 30 s/kb; followed by a final extension at 72°C for 10 min. PCR products can be stored at 4°C short term or at −20°C longer term. Before use in SIRA reactions, PCR products are separated from primer dimers by agarose gel electrophoresis, stained with ethidium bromide, excised from the gel under longwave UV illumination and purified using a gel extraction kit (QIAGEN).

6.1.2 SIRA Substrate Plasmids

SIRA substrate plasmids are made by ligating an insertion site into the multiple cloning site of any plasmid already containing a selectable marker such as an antibiotic resistance gene, an origin of replication and a promoter. The insertion site consists of a counterselectable gene positioned between inward facing $attB^{TT}$ and $attB^{TC}$ sites, and is made by PCR (primers 27 and 28 in Table 3). The gene we use for counterselection is *ccdB*, which encodes a cytotoxin from *E. coli* that causes cell death (Bernard & Couturier, 1992). This means that when *E. coli* cells are transformed with SIRA reactions, cells containing assembled products will grow and form colonies because the *ccdB* gene has been replaced with assembled DNA parts, whereas cells containing the original substrate plasmid will not. For propagation of plasmids containing the *ccdB* gene, we use *E. coli* DB3.1 which carries a *gyrA462* mutation which makes the strain resistant to the CcdB toxin (Bernard & Couturier, 1992).

6.1.3 DNA Purification by Ethanol Precipitation

When purifying SIRA reactions or DNA components by ethanol precipitation, we first add 3 M sodium acetate to a final concentration of 0.3 M and then add 20 μg of glycogen and 2.5 volumes of ice-cold 100% ethanol in a microfuge tube. For best results, we store this microfuge tube at $-20°C$ for at least an hour. The DNA is pelleted by centrifugation at $16,000 \times g$ at $4°C$ for 1 h and the supernatant is pipetted off carefully. This is followed by two washes with ice-cold 70% ethanol before the DNA pellet is resuspended in 15 μL nuclease-free water.

6.2 Purification of ΦC31 Integrase and gp3

We use the following protocols to obtain ΦC31 integrase and gp3.

6.2.1 Purification of ΦC31 Integrase

We have cloned the ΦC31 integrase gene, *int*, to make plasmid pEXP5-Int for high-level expression using a pEXP5-NT/TOPO® TA Expression Kit (ThermoFisher). When expressed from pEXP5-Int, ΦC31 integrase carries an N-terminal His-tag.

For ΦC31 integrase expression we inoculate a single BL21 (DE3) (ThermoFisher) *E. coli* colony containing pEXP5-Int into 50 mL terrific broth containing 100 μg/mL ampicillin and incubate with shaking at 37°C overnight. We use this overnight culture to inoculate 1 L terrific broth containing 100 μg/mL ampicillin which we then incubate in a 5-L conical flask at 37°C with shaking at 200 rpm until the OD_{600} reaches 0.8 (Biowave Cell Density Meter CO8000). At this point we chill the culture on ice for 5 min before inducing integrase expression with 0.2 mM isopropyl β-D-1-thiogalactopyranoside (IPTG). The culture is then incubated at 18°C with shaking at 200 rpm for 12–20 h. For harvesting the cells the culture is chilled on ice for 10 min before centrifugation at 15,000 × g for 20 min. The supernatant is removed and the cell pellets are stored at −80°C.

For ΦC31 integrase purification, cell pellets are resuspended in 100 mL buffer A (25 mM Tris–HCl, 400 mM NaCl, 50 mM imidazole, pH 7.4) containing 0.1 U/mL benzonase. Cells are lysed using a TS cell disrupter (Constant Systems) set at 25 kpsi. Cell debris and insoluble protein are removed from the extract by centrifuging at 25,000 × g (Beckman Coulter, JA-25.50 fixed angle rotor) for 1 h.

The supernatant is removed and filtered (Millipore 2.2 μm) prior to loading on to a 5 mL Ni column (HisTrap FF, GE Healthcare) which has been equilibrated with buffer A at a flow rate of 1 mL/min. Unbound proteins are removed by washing with 10 column volumes of buffer A at 2 mL/min prior to elution using a linear gradient between buffer A and buffer B (25 mM Tris–HCl, 400 mM NaCl, 500 mM imidazole, pH 7.4). Peak fractions containing integrase are pooled and concentrated to <5 mL.

Pooled fractions are loaded on to a 120 mL Sephacryl S200 size-exclusion column (HiPrep™ 16/60, GE Healthcare) that has been preequilibrated with buffer C (25 mM Tris–HCl, 400 mM NaCl, 1 mM DTT, pH 7.4) at a flow rate of 0.5 mL/min. Peak fractions containing pure integrase are concentrated (10,000 Da cutoff spin concentrator, Millipore) and stored at ~1.5 mg/mL in 50% glycerol at −80°C. Protein fractions are analyzed by SDS-PAGE, visualized by staining with InstantBlue™ (Expedeon) and quantified with a Bradford Assay (Bradford, 1976). Prior to assembly, ΦC31 integrase is diluted to ~135 μg/mL (2 μM) in integrase dilution buffer (25 mM Tris–HCl, 1 M NaCl, 1 mM DTT, 50% (v/v) glycerol, pH 7.5). The NaCl from the integration dilution buffer (100 mM final) is required for efficient recombination. Generally integrase is used in the reactions at a final concentration of 200 nM.

6.2.2 Purification of gp3

We cloned the *gp3* gene between the *Bam*HI and *Xho*I sites of pET28a to make plasmid pET28a-gp3. When expressed from this plasmid, gp3 carries an N-terminal His-tag. For gp3 expression we follow the same protocol as for ΦC31 integrase expression in BL21 (DE3) *E. coli* cells using 40 μg/mL kanamycin in place of ampicillin, and inducing with 0.5 mM IPTG. For gp3 purification, see the published protocol by Khaleel et al. (2011). Prior to use in a reaction, gp3 is diluted to ~110 μg/mL (4 μM) in integrase dilution buffer. Generally gp3 is used in reactions at a final concentration of 400 nM in combination with 200 nM integrase.

6.3 SIRA Reaction Buffer

DNA parts and plasmids for SIRA reactions are mixed together with one-tenth the final reaction volume of 10× SIRA buffer before adding one-tenth of the final volume of integrase for *attP* × *attB* recombination, or one-tenth of the reaction volume of an integrase-gp3 mixture for *attL* × *attR* reactions. 10× SIRA buffer contains 100 mM Tris–HCl, 1 mM EDTA, 50 mM spermidine 3HCl, and 1 mg/mL bovine serum albumin, pH 7.5. Ethylene glycol can be added to a final concentration of 5–10% to increase the efficiency of some assembly reactions.

6.4 Competent *E. coli* Strains

In this chapter we use different *E. coli* strains for different applications. For expression of ΦC31 integrase and gp3 we use BL21 (DE3) *E. coli* cells (ThermoFisher). For propagation of plasmids containing the *ccdB* gene we use DB3.1 *E. coli* cells (Bernard & Couturier, 1992). For screening SIRA constructs we use TOP10 (ThermoFisher) or DS941 *E. coli* cells (Summers & Sherratt, 1988). All of these strains can be made chemically competent or electrocompetent using the following protocols.

6.4.1 Chemically Competent E. coli

A single colony is used to inoculate 5 mL lysogeny broth (LB) with the appropriate antibiotic and grown at 37°C, shaking at 200 rpm for 5–16 h. 250 μL of this culture is used to inoculate 20 mL LB, again with the appropriate antibiotic, and grown at 37°C, shaking at 200 rpm for 90 min. This culture is held on ice for 10 min to halt growth before being transferred to chilled centrifuge tubes and pelleted by centrifugation at 7000 × *g*, at 4°C for 2 min (Beckman Coulter Avanti J-20XP centrifuge,

JA-20 rotor). The supernatant is removed immediately and the cells are gently resuspended in 10 mL ice-cold 50 mM CaCl$_2$ and then held on ice for 0.5–3 h (longer incubation improves competency). Again, cells are pelleted by centrifugation at 7000 x g, at 4°C for 1 min. The supernatant is immediately removed and the cells are gently resuspended in 1 mL ice-cold 50 mM CaCl$_2$ ready for transformation. For improved competency cells can be held at 0°C overnight before use.

6.4.2 Electrocompetent E. coli

A single colony is used to inoculate 5 mL LB with the appropriate antibiotic and grown at 37°C, shaking at 200 rpm for 5–16 h. 2 mL of this overnight culture is used to inoculate 400 mL LB and grown at 37°C, shaking at 200 rpm to an OD$_{600}$ of 0.3–0.5 (Biowave Cell Density Meter CO8000). Cells are held on ice for 20 min to halt growth before being distributed between chilled 50 mL centrifuge tubes and pelleted by centrifugation at 4000 × g at 4°C for 5 min (Sorvall Legend RT centrifuge). Cells are washed three times by resuspending in 400, 200, and 20 mL of ice-cold, filter-sterilized (0.2 μm) 10% glycerol and centrifugation as above. Cells are resuspended in a final volume of 1 mL ice-cold sterile 10% glycerol ready for immediate transformation by electroporation, or frozen with dry ice in ethanol and stored at −80°C for later use.

7. EXAMPLE SIRA PROTOCOLS

7.1 SIRA Example 1: Inserting a Single Gene Using SIRA

The aim of this reaction is to insert one gene flanked by $attP^{TT}$ and $attP^{TC}$ into a SIRA substrate plasmid containing $attB^{TT}$ and $attB^{TC}$ sites using $attP \times attB$ recombination catalyzed by ΦC31 integrase (Fig. 2A).

1. Add inward orientated $attP^{TT}$ and $attP^{TC}$ sites to the gene by PCR using primers 1 and 10 in Table 1 and the PCR protocol in the previous section.
2. Purify PCR products by ethanol precipitation or using commercially available gel extraction kits. Quantify the purified PCR product and SIRA substrate plasmid by measuring UV absorbance at 260 nm, or by agarose gel electrophoresis against DNA standards of known size and concentration, and calculate their concentrations in molar units.
3. For a 20 μL $attP \times attB$ SIRA reaction, add 2 μL of 10 × SIRA buffer, and 1 μL ethylene glycol to a sterile microfuge tube. Add substrate plasmid to a final concentration of 3 nM, and purified PCR product to a final

concentration of 3 nM. Make to 18 μL with nuclease-free water. To start the reaction, add 2 μL of 2 μM ΦC31 integrase and gently mix. Incubate the reaction at 30°C for 30 min. Stop the reaction by heating to 85°C for 10 min.

4. Once cooled on ice, mix 1 μL of the SIRA reaction with 50 μL chemically competent *E. coli* cells (cells should be *ccdB*-sensitive to allow selection against the SIRA substrate plasmid) in a sterile, ice-cold microfuge tube, and leave on ice for 30 min. Heat shock the cells for 1 min at 42°C and return to ice for 5 min. Add 200 μL LB or SOC media, and incubate in a shaking 37°C incubator at 200 rpm. After 1 h, plate 200 μL on a LB agar plate containing an appropriate concentration of antibiotic for selection of the SIRA construct. Incubate overnight (~16 h) at 37°C.

5. Isolate SIRA constructs from *E. coli* colonies using a commercially available plasmid DNA preparation kit (eg, QIAGEN QIAprep® Spin Miniprep Kit) following the manufacturer's instructions. Use diagnostic restriction digests of these isolated SIRA constructs to verify the presence of the gene, and confirm the accuracy of PCR by DNA sequencing.

7.2 SIRA Example 2: Assembly of Multiple DNA Parts in a Predefined Order

The aim of this reaction is to assemble five DNA parts containing *genes A to E* in the predefined gene order *A-B-C-D-E* in a SIRA plasmid using ΦC31 integrase in an *attP* × *attB* recombination reaction (Fig. 2B). The substrate plasmid in this example carries an insertion site with $attB^{TT}$ and $attB^{TC}$ sites downstream of an inducible promoter.

1. Add inward orientated $attP^{TT}$ and $attP^{CT}$ sites to *gene A* by PCR using primers 1 and 2 in Table 1, outward orientated $attB^{CT}$ and $attB^{GT}$ sites to *gene B* using primers 3 and 4, inward orientated $attP^{GT}$ and $attP^{CA}$ sites to *gene C* using primers 5 and 6, outward orientated $attB^{CA}$ and $attB^{CC}$ sites to *gene D* using primers 7 and 8, and inward orientated $attP^{CC}$ and $attP^{TC}$ sites to *gene E* using primers 9 and 10. These DNA parts are designed so that *gene A* assembles in position 1 in the final construct, *gene B* in position 2, *gene C* in position 3, *gene D* in position 4, and *gene E* in position 5.

2. Purify PCR products and quantify DNA as in SIRA Example 1.

3. For a 20 μL *attP* × *attB* multipart SIRA reaction, add 2 μL of 10× SIRA buffer and 1 μL ethylene glycol to a sterile microfuge tube. Add substrate plasmid to a final concentration of 3 nM, and each of the purified PCR products to a final concentration of 3 nM. Make to 18 μL with

nuclease-free water. To start the reaction, add 2 µL of 2 µM ΦC31 integrase and mix gently. Incubate the reaction at 30°C for at least 2 h or overnight (~16 h) for a greater number of constructs. Stop the reaction by heating to 85°C for 10 min.

4. Once cooled on ice, use 1 µL of the SIRA reaction to transform chemically competent *E. coli* cells as in SIRA Example 1, or to obtain a greater number of colonies transform electrocompetent *E. coli* cells. To transform electrocompetent cells with SIRA reactions, first purify the DNA by ethanol precipitation and resuspend in 15 µL water. Add 1 µL of this DNA to 25 µL electrocompetent cells and electroporate. We conduct electroporations in 1 mm cuvettes with 2.5 kV using a Biorad Gene Pulser Xcell™ Electroporation System. Immediately add 1 mL SOC media to the cells, transfer to a 15-mL centrifuge tube and incubate in a shaking 37°C incubator at 200 rpm. After 1 h, plate 200 µL of this culture on LB agar containing an appropriate concentration of antibiotic for selection of SIRA constructs. Incubate the plates overnight at an appropriate temperature.

5. If possible, screen colonies for a phenotype characteristic of accurate assembly of all DNA parts. For example, when screening for accurate assembly of genes for zeaxanthin biosynthesis, colonies can be transferred to solid or liquid media containing an inducer that switches on gene expression. Cells containing constructs of correctly assembled parts will express all of the zeaxanthin biosynthesis genes and turn a distinctive yellow color.

6. Whether or not phenotypic screening of colonies is possible, plasmid DNA should be isolated, digested and sequenced to confirm that constructs are correct.

7.3 SIRA Example 3: Combinatorial Assembly of DNA Parts for a Library of Constructs with Different Gene Orders

The aim of this reaction is to assemble three genes, *A*, *B*, and *C*, into a SIRA substrate plasmid in every possible gene order in a one-pot assembly reaction. Again, the SIRA substrate plasmid contains $attB^{TT}$ and $attP^{TC}$.

1. In the first of two rounds of PCR reactions, add truncated *attP* sites (using primers 11 and 12 in Table 2) and truncated *attB* sites (using primers 13 and 14) to each of the three genes in six separate PCR reactions. Purify these PCR products as in SIRA Example 1, and use 0.5–10 ng as template for the second round of PCR in step 2.

2. Add three different pairs of orthogonal *attP* or *attB* sites to *gene A* in the second round of PCR reactions. Primers in these PCR reactions contain complete *att* sites including orthogonal central dinucleotides and bind to the truncated *att* sequences added to genes in step 1. The first pair of *att* sites, $attP^{TT}$ and $attP^{CT}$, place *gene A* at position 1 within the construct (primers 15 and 16 in Table 2); the second pair of *att* sites, $attB^{CT}$ and $attB^{GT}$, place *gene A* at position 2 within the construct (primers 17 and 18); and the third pair, $attP^{GT}$ and $attP^{TC}$, place *gene A* at position 3 within a construct (primers 19 and 24). Repeat these PCR reactions for *genes B* and *C* to generate a total of nine PCR products.
3. Purify PCR products and quantify DNA as in SIRA Example 1.
4. For a 20 μL *attP* × *attB* SIRA reaction for generating different gene orders, add 2 μL of 10× SIRA buffer, and 1 μL ethylene glycol to a sterile microfuge tube. Add substrate plasmid to a final concentration of 3 n*M*, and each of the purified PCR products to a final concentration of 1 n*M*. Make this up to 18 μL with nuclease-free water. To start the reaction, add 2 μL of 2 μ*M* ΦC31 integrase and gently mix. Incubate the reaction at 30°C for 60 min. Stop the reaction by heating to 85°C for 10 min.
5. For a combinatorial library of this size (with 27 possible three-gene constructs), transformation of chemically competent cells, as in SIRA Example 1, is sufficient to generate enough colonies to represent the diversity of the library. For larger combinatorial libraries (eg, five genes assembled in all 3125 possible gene orders) electrocompetent cells are better suited for transformation. In our experience, electrocompetent *E. coli* cells produce approximately 50 times more colonies per μg DNA than do chemically competent cells. For transformation of electrocompetent cells with SIRA reactions, see SIRA Example 2. For large libraries plate the full 1 mL transformation culture distributing it evenly over five 90-mm LB agar plates containing an appropriate antibiotic for selection of SIRA constructs and incubate the plates overnight.
6. As with SIRA Example 2, if possible, screen for a phenotype distinctive of the genotypes assembled.
7. Isolate, digest, and sequence plasmid DNA to confirm the presence of all genes and the order in which they have been assembled.

7.4 SIRA Example 4: Targeted Postassembly Modification of a SIRA Construct

The aim of this reaction is to replace a specific DNA part in a SIRA construct with a secondary SIRA insertion site through *attL* × *attR* recombination. This protocol replaces a DNA part flanked by $attL^{GT}$ and $attL^{CA}$ sites with a secondary insertion site flanked by $attR^{GT}$ and $attR^{CA}$ sites. Once this secondary insertion site is integrated into a construct, it is flanked with $attB^{GT}$ and $attB^{CA}$ sites. It can therefore be replaced with more DNA parts through $attP \times attB$ recombination using integrase on its own so that the remaining $attL$ sites in the assembly do not participate in the reaction (Fig. 2C). Using alternative $attR$ sites with appropriate central dinucleotides allows replacement of any chosen segment in an existing assembly. Note that careful attention has to be paid to the orientation of the $attR$ sites to match the $attL$ sites in the assembly.

1. To make the secondary insertion site, add inward orientated $attR^{GT}$ and $attR^{CA}$ sites to a piece of DNA carrying a gene for positive selection (such as chloramphenicol resistance) and a gene for negative selection (such as $ccdB$ encoding the lethal cytotoxin CcdB) by PCR using primers 25 and 26 in Table 3 and the PCR protocol in the previous section.
2. Purify the PCR product and quantify DNA as in SIRA Example 1.
3. Dilute ΦC31 integrase and gp3 in integrase dilution buffer (25 mM Tris–HCl, 1 M NaCl, 1 mM DTT, 50% (v/v) glycerol, pH 7.5) to make a mixture containing 2 µM ΦC31 integrase and 4 µM gp3.
4. For a 20 µL $attL \times attR$ SIRA reaction, add 2 µL of 10× SIRA buffer to a sterile microfuge tube. Add the assembled SIRA construct plasmid to a final concentration of 3 nM, and the purified secondary insertion site to a final concentration of 3 nM. Make up to 18 µL with nuclease-free water. To start the reaction, add 2 µL of the integrase-gp3 mixture. Incubate the reaction at 30°C for 1 h. Stop the reaction by heating to 85°C for 10 min.
5. Once cooled on ice, mix 1 µL of the SIRA reaction with 50 µL chemically competent *E. coli* cells in a sterile, ice-cold microfuge tube, and follow the transformation procedure in SIRA Example 1. If the insertion site carries the negative selection gene $ccdB$, use chemically competent DB3.1 *E. coli* cells. When plating transformations, use the positive selection gene to select for constructs that contain the secondary insertion site, eg, if the insertion site contains a chloramphenicol resistance gene, plate transformants on LB agar containing chloramphenicol.
6. Isolate, digest, and sequence plasmid DNA to confirm integration of the new insertion site at the desired position within the construct.

8. THE FUTURE OF SIRA

In this chapter we have described how SIRA with ΦC31 integrase can be applied to rapidly optimize metabolic pathways through multipart assembly, generating combinatorial libraries of constructs and targeted post-assembly modification. A compelling direction for SIRA is to incorporate more serine integrases into assembly of DNA parts that follow the design principles outlined in this chapter. This would enable assembly of more DNA parts and rapid optimization of larger metabolic pathways for production of more complex compounds. The application of serine integrases in synthetic biology, in particular genetic memory storage, has led to a renaissance in putative serine integrase discovery (Yang et al., 2014). A surge in the discovery of cognate RDFs will likely follow as more is realized about the mechanisms of known RDFs in controlling the direction of recombination (Bibb et al., 2005; Breüner et al., 1999; Ghosh et al., 2006; Khaleel et al., 2011; Olorunniji et al., 2012). Additional RDFs will enable targeted modification of constructs encoding larger metabolic pathways.

It is becoming increasingly beneficial to express different genetic components from different plasmids and inducible expression systems as DNA constructs in metabolic pathway engineering become larger and more complex (Ajikumar et al., 2010; Jones et al., 2000; Martin, Pitera, Withers, Newman, & Keasling, 2003; Redding-Johanson et al., 2011). There is therefore a requirement for a library of SIRA substrate plasmids with different copy numbers, host ranges, antibiotic resistances, and inducible promoters. This is particularly useful for combinatorial approaches to assembling new metabolic pathways for which plasmid features are unknown variables. It should be noted that as complexity and combinatorial diversity of these systems increases, so too do requirements for robust, high-throughput genotyping and phenotypic screens.

SIRA could easily be adapted to function in vivo as a system for insertion and modification of constructs in a host chromosome. Serine integrases have in vivo activity in a variety of hosts including prokaryotes, invertebrates, and vertebrates (Fogg, Colloms, Rosser, Stark, & Smith, 2014). ΦC31 integrase is capable of engineering genomes both through pseudo *attP* sites and "landing sites" designed for DNA integration (Karow et al., 2011; Nimmo et al., 2006; Venken et al., 2006; Ye et al., 2010). With no need for plasmid maintenance there would be less metabolic burden on the host and, in eukaryotic systems, this would offer a controlled system for targeting constructs to areas of the chromosome that are not subject to gene silencing or high mutation rates. An in vivo SIRA system could be applied to making designer strains for specific purposes such as drug trials and crop enhancement.

SIRA is an adaptable system that could be applied to emerging technologies in synthetic biology. In genetic circuitry serine integrases have been used to build permanent amplifying logic gates (Bonnet, Yin, Ortiz, Subsoontorn, & Endy, 2013) and, using serine integrases, the Voigt group demonstrated that a 2 kb construct containing 11 pairs of orthogonal *att* sites has the capacity to store 1.375 bytes of data (Yang et al., 2014). SIRA could easily be applied to assembling such genetic systems. As part of the yeast synthetic genome project, chromosomes are being assembled containing attachment sites for Cre recombinase to enable synthetic chromosome rearrangement and modification by loxP-mediated evolution (SCRaMbLE), a tool for genome shuffling (Annaluru et al., 2014; Dymond & Boeke, 2012). As serine integrases are capable of moving BAC-sized pieces of DNA, SIRA offers an alternative system for building synthetic chromosomes enabling both rapid DNA assembly and postassembly rearrangements. SIRA genetic circuitry and genome shuffling could ultimately enable the rapid generation of self-adapting biological systems capable of detecting biological inputs and producing bespoke biological outputs. Such a system was recently developed in implantable synthetic cytokine converter cells to detect and treat psoriasis in mice (Schukur, Geering, Charpin-El Hamri, & Fussenegger, 2015).

To conclude, SIRA is an adaptable DNA assembly method with robust capabilities that make it particularly well suited to rapid optimization of metabolic pathways: one-pot assembly of multiple DNA parts, generation of combinatorial assembly libraries; and targeted postassembly modification of constructs. In this chapter we have explained the enzyme mechanism and design principles that underpin the SIRA system. We have also provided detailed protocols to enable widespread use of this technology. Ultimately, SIRA has massive potential to drive progress in metabolic pathway engineering and synthetic biology.

ACKNOWLEDGMENTS

Funding was provided by the UK Engineering and Physical Sciences Research Council [EP/H019154/1 and EP/K034359/1].

REFERENCES

Ajikumar, P. K., Xiao, W. H., Tyo, K. E., Wang, Y., Simeon, F., Leonard, E., et al. (2010). Isoprenoid pathway optimization for Taxol precursor overproduction in *Escherichia coli*. *Science*, *330*(6000), 70–74.

Alper, H., Fischer, C., Nevoigt, E., & Stephanopoulos, G. (2005). Tuning genetic control through promoter engineering. *Proceedings of the National Academy of Sciences of the United States of America*, *102*(36), 12678–12683.

Annaluru, N., Muller, H., Mitchell, L. A., Ramalingam, S., Stracquadanio, G., Richardson, S. M., et al. (2014). Total synthesis of a functional designer eukaryotic chromosome. *Science*, *344*(6179), 55–58.

Bernard, P., & Couturier, M. (1992). Cell killing by the F plasmid CcdB protein involves poisoning of DNA-topoisomerase II complexes. *Journal of Molecular Biology*, *226*(3), 735–745.

Bibb, L. A., Hancox, M. I., & Hatfull, G. F. (2005). Integration and excision by the large serine recombinase ΦRv1 integrase. *Molecular Microbiology*, *55*(6), 1896–1910.

Bonnet, J., Yin, P., Ortiz, M. E., Subsoontorn, P., & Endy, D. (2013). Amplifying genetic logic gates. *Science*, *340*(6132), 599–603.

Bradford, M. M. (1976). A rapid and sensitive method for the quantitation of microgram quantities of protein utilizing the principle of protein-dye binding. *Analytical Biochemistry*, *72*, 248–254.

Breüner, A., Brøndsted, L., & Hammer, K. (1999). Novel organization of genes involved in prophage excision identified in the temperate lactococcal bacteriophage TP901-1. *Journal of Bacteriology*, *181*(23), 7291–7297.

Casini, A., Storch, M., Baldwin, G. S., & Ellis, T. (2015). Bricks and blueprints: Methods and standards for DNA assembly. *Nature Reviews Molecular Cell Biology*, *16*(9), 568–576.

Ceroni, F., Algar, R., Stan, G. B., & Ellis, T. (2015). Quantifying cellular capacity identifies gene expression designs with reduced burden. *Nature Methods*, *12*(5), 415–418.

Cheo, D. L., Titus, S. A., Byrd, D. R., Hartley, J. L., Temple, G. F., & Brasch, M. A. (2004). Concerted assembly and cloning of multiple DNA segments using *in vitro* site-specific recombination: Functional analysis of multi-segment expression clones. *Genome Research*, *14*(10B), 2111–2120.

Colloms, S. D., Merrick, C. A., Olorunniji, F. J., Stark, W. M., Smith, M. C., Osbourn, A., et al. (2014). Rapid metabolic pathway assembly and modification using serine integrase site-specific recombination. *Nucleic Acids Research*, *42*(4), e23.

Dahl, R. H., Zhang, F., Alonso-Gutierrez, J., Baidoo, E., Batth, T. S., Redding-Johanson, A. M., et al. (2013). Engineering dynamic pathway regulation using stress-response promoters. *Nature Biotechnology*, *31*(11), 1039–1046.

Dueber, J. E., Wu, G. C., Malmirchegini, G. R., Moon, T. S., Petzold, C. J., Ullal, A. V., et al. (2009). Synthetic protein scaffolds provide modular control over metabolic flux. *Nature Biotechnology*, *27*(8), 753–759.

Dymond, J., & Boeke, J. (2012). The *Saccharomyces cerevisiae* SCRaMbLE system and genome minimization. *Bioengineered Bugs*, *3*(3), 168–171.

Espah Borujeni, A., Channarasappa, A. S., & Salis, H. M. (2014). Translation rate is controlled by coupled trade-offs between site accessibility, selective RNA unfolding and sliding at upstream standby sites. *Nucleic Acids Research*, *42*(4), 2646–2659.

Fogg, P. C., Colloms, S., Rosser, S., Stark, M., & Smith, M. C. (2014). New applications for phage integrases. *Journal of Molecular Biology*, *426*(15), 2703–2716.

Ghosh, P., Pannunzio, N. R., & Hatfull, G. F. (2005). Synapsis in phage Bxb1 integration: Selection mechanism for the correct pair of recombination sites. *Journal of Molecular Biology*, *349*(2), 331–348.

Ghosh, P., Wasil, L. R., & Hatfull, G. F. (2006). Control of phage Bxb1 excision by a novel recombination directionality factor. *PLoS Biology*, *4*(6), e186.

Gibson, D. G., Young, L., Chuang, R. Y., Venter, J. C., Hutchison, C. A., 3rd, & Smith, H. O. (2009). Enzymatic assembly of DNA molecules up to several hundred kilobases. *Nature Methods*, *6*(5), 343–345.

Hansen, E. H., Møller, B. L., Kock, G. R., Bünner, C. M., Kristensen, C., Jensen, O. R., et al. (2009). De novo biosynthesis of vanillin in fission yeast (*Schizosaccharomyces pombe*) and baker's yeast (*Saccharomyces cerevisiae*). *Applied and Environmental Microbiology*, *75*(9), 2765–2774.

Hartley, J. L., Temple, G. F., & Brasch, M. A. (2000). DNA cloning using in vitro site-specific recombination. *Genome Research, 10*(11), 1788–1795.

Hillson, N. (2011). DNA assembly method standardization for synthetic biomolecular circuits and systems. In H. Koeppl, G. Setti, M. di Bernardo, & D. Densmore (Eds.), *Design and analysis of biomolecular circuits* (pp. 295–314). New York: Springer.

Jones, K. L., Kim, S. W., & Keasling, J. D. (2000). Low-copy plasmids can perform as well as or better than high-copy plasmids for metabolic engineering of bacteria. *Metabolic Engineering, 2*(4), 328–338.

Kajiwara, S., Fraser, P. D., Kondo, K., & Misawa, N. (1997). Expression of an exogenous isopentenyl diphosphate isomerase gene enhances isoprenoid biosynthesis in *Escherichia coli*. *Biochemical Journal, 324*(Pt. 2), 421–426.

Karow, M., Chavez, C. L., Farruggio, A. P., Geisinger, J. M., Keravala, A., Jung, W. E., et al. (2011). Site-specific recombinase strategy to create induced pluripotent stem cells efficiently with plasmid DNA. *Stem Cells, 29*(11), 1696–1704.

Khaleel, T., Younger, E., McEwan, A. R., Varghese, A. S., & Smith, M. C. (2011). A phage protein that binds ΦC31 integrase to switch its directionality. *Molecular Microbiology, 80*(6), 1450–1463.

Ma, S. M., Garcia, D. E., Redding-Johanson, A. M., Friedland, G. D., Chan, R., Batth, T. S., et al. (2011). Optimization of a heterologous mevalonate pathway through the use of variant HMG-CoA reductases. *Metabolic Engineering, 13*(5), 588–597.

Martin, V. J., Pitera, D. J., Withers, S. T., Newman, J. D., & Keasling, J. D. (2003). Engineering a mevalonate pathway in *Escherichia coli* for production of terpenoids. *Nature Biotechnology, 21*(7), 796–802.

Matthews, P. D., & Wurtzel, E. T. (2000). Metabolic engineering of carotenoid accumulation in *Escherichia coli* by modulation of the isoprenoid precursor pool with expression of deoxyxylulose phosphate synthase. *Applied Microbiology and Biotechnology, 53*(4), 396–400.

McEwan, A. R., Rowley, P. A., & Smith, M. C. (2009). DNA binding and synapsis by the large C-terminal domain of ΦC31 integrase. *Nucleic Acids Research, 37*(14), 4764–4773.

Nimmo, D. D., Alphey, L., Meredith, J. M., & Eggleston, P. (2006). High efficiency site-specific genetic engineering of the mosquito genome. *Insect Molecular Biology, 15*(2), 129–136.

Nishizaki, T., Tsuge, K., Itaya, M., Doi, N., & Yanagawa, H. (2007). Metabolic engineering of carotenoid biosynthesis in *Escherichia coli* by ordered gene assembly in *Bacillus subtilis*. *Applied and Environmental Microbiology, 73*(4), 1355–1361.

Nowroozi, F. F., Baidoo, E. E., Ermakov, S., Redding-Johanson, A. M., Batth, T. S., Petzold, C. J., et al. (2014). Metabolic pathway optimization using ribosome binding site variants and combinatorial gene assembly. *Applied Microbiology and Biotechnology, 98*(4), 1567–1581.

Olorunniji, F. J., Buck, D. E., Colloms, S. D., McEwan, A. R., Smith, M. C., Stark, W. M., et al. (2012). Gated rotation mechanism of site-specific recombination by ΦC31 integrase. *Proceedings of the National Academy of Sciences of the United States of America, 109*(48), 19661–19666.

Paddon, C. J., & Keasling, J. D. (2014). Semi-synthetic artemisinin: A model for the use of synthetic biology in pharmaceutical development. *Nature Reviews Microbiology, 12*(5), 355–367.

Peralta-Yahya, P. P., Ouellet, M., Chan, R., Mukhopadhyay, A., Keasling, J. D., & Lee, T. S. (2011). Identification and microbial production of a terpene-based advanced biofuel. *Nature Communications, 2*, 483.

Quan, J., & Tian, J. (2009). Circular polymerase extension cloning of complex gene libraries and pathways. *PLoS One, 4*(7), e6441.

Redding-Johanson, A. M., Batth, T. S., Chan, R., Krupa, R., Szmidt, H. L., Adams, P. D., et al. (2011). Targeted proteomics for metabolic pathway optimization: Application to terpene production. *Metabolic Engineering, 13*(2), 194–203.

Rowley, P. A., Smith, M. C., Younger, E., & Smith, M. C. (2008). A motif in the C-terminal domain of ΦC31 integrase controls the directionality of recombination. *Nucleic Acids Research, 36*(12), 3879–3891.

Rutherford, K., Yuan, P., Perry, K., Sharp, R., & Van Duyne, G. D. (2013). Attachment site recognition and regulation of directionality by the serine integrases. *Nucleic Acids Research, 41*(17), 8341–8356.

Salis, H. M., Mirsky, E. A., & Voigt, C. A. (2009). Automated design of synthetic ribosome binding sites to control protein expression. *Nature Biotechnology, 27*(10), 946–950.

Santos, C. N., Xiao, W., & Stephanopoulos, G. (2012). Rational, combinatorial, and genomic approaches for engineering L-tyrosine production in *Escherichia coli*. *Proceedings of the National Academy of Sciences of the United States of America, 109*(34), 13538–13543.

Sasaki, Y., Sone, T., Yoshida, S., Yahata, K., Hotta, J., Chesnut, J. D., et al. (2004). Evidence for high specificity and efficiency of multiple recombination signals in mixed DNA cloning by the Multisite Gateway system. *Journal of Biotechnology, 107*(3), 233–243.

Schukur, L., Geering, B., Charpin-El Hamri, G., & Fussenegger, M. (2015). Implantable synthetic cytokine converter cells with AND-gate logic treat experimental psoriasis. *Science Translational Medicine, 7*(318), 318ra201.

Shetty, R. P., Endy, D., & Knight, T. F., Jr. (2008). Engineering BioBrick vectors from BioBrick parts. *Journal of Biological Engineering, 2*, 5.

Smith, M. C., Brown, W. R., McEwan, A. R., & Rowley, P. A. (2010). Site-specific recombination by ΦC31 integrase and other large serine recombinases. *Biochemical Society Transactions, 38*(2), 388–394. D.

Smith, M. C., Burns, R. N., Wilson, S. E., & Gregory, M. A. (1999). The complete genome sequence of the Streptomyces temperate phage straight ΦC31: Evolutionary relationships to other viruses. *Nucleic Acids Research, 27*(10), 2145–2155.

Smith, M. C., Till, R., & Smith, M. C. (2004). Switching the polarity of a bacteriophage integration system. *Molecular Microbiology, 51*(6), 1719–1728.

Summers, D. K., & Sherratt, D. J. (1988). Resolution of ColE1 dimers requires a DNA sequence implicated in the three-dimensional organization of the cer site. *EMBO Journal, 7*(3), 851–858.

Venken, K. J., He, Y., Hoskins, R. A., & Bellen, H. J. (2006). P[acman]: A BAC transgenic platform for targeted insertion of large DNA fragments in *D. melanogaster*. *Science, 314*(5806), 1747–1751.

Wang, H. H., Isaacs, F. J., Carr, P. A., Sun, Z. Z., Xu, G., Forest, C. R., et al. (2009). Programming cells by multiplex genome engineering and accelerated evolution. *Nature, 460*(7257), 894–898.

Werner, S., Engler, C., Weber, E., Gruetzner, R., & Marillonnet, S. (2012). Fast track assembly of multigene constructs using Golden Gate cloning and the MoClo system. *Bioengineered Bugs, 3*(1), 38–43.

Yang, L., Nielsen, A. A., Fernandez-Rodriguez, J., McClune, C. J., Laub, M. T., Lu, T. K., et al. (2014). Permanent genetic memory with >1-byte capacity. *Nature Methods, 11*(12), 1261–1266.

Ye, L., Chang, J. C., Lin, C., Qi, Z., Yu, J., & Kan, Y. W. (2010). Generation of induced pluripotent stem cells using site-specific integration with phage integrase. *Proceedings of the National Academy of Sciences of the United States of America, 107*(45), 19467–19472.

Yu, J. L., Xia, X. X., Zhong, J. J., & Qian, Z. G. (2014). Direct biosynthesis of adipic acid from a synthetic pathway in recombinant *Escherichia coli*. *Biotechnology and Bioengineering, 111*(12), 2580–2586.

Zhang, F., Carothers, J. M., & Keasling, J. D. (2012). Design of a dynamic sensor-regulator system for production of chemicals and fuels derived from fatty acids. *Nature Biotechnology, 30*(4), 354–359.

Zhang, L., Zhao, G., & Ding, X. (2011). Tandem assembly of the epothilone biosynthetic gene cluster by *in vitro* site-specific recombination. *Science Reports, 1*, 141.

CHAPTER FOURTEEN

Rewiring Riboswitches to Create New Genetic Circuits in Bacteria

C.J. Robinson[*,†], D. Medina-Stacey[*], M.-C. Wu[*], H.A. Vincent[*], J. Micklefield[*,†,1]

[*]School of Chemistry, Manchester Institute of Biotechnology, The University of Manchester, Manchester, United Kingdom
[†]Centre for Synthetic Biology of Fine and Speciality Chemicals (SYNBIOCHEM), Manchester Institute of Biotechnology, The University of Manchester, Manchester, United Kingdom
[1]Corresponding author: e-mail address: jason.micklefield@manchester.ac.uk

Contents

1. Introduction   320
2. Overview   324
3. Methods   324
   3.1 Construction of Inducible-Riboswitch Controlled Gene Expression Vectors for *E. coli*   326
   3.2 CheZ Motility Assays in *E. coli*   332
   3.3 Construction of Repressible-Riboswitch Controlled Constructs for Targeted Chromosomal Integration in *B. subtilis*   333
   3.4 β-Galactosidase Assays to Validate Synthetic Riboswitch Function   337
   3.5 Western Blot Analysis to Quantify Target Gene Expression Levels   340
   3.6 MreB Morphology Assays in *B. subtilis*   342
   3.7 Construction of Chimeric Riboswitches to Couple Orthogonal Aptamers to Host-Specific Expression Platforms   343
4. Concluding Remarks   346
Acknowledgments   346
References   347

Abstract

Riboswitches are RNA elements that control the expression of genes through a variety of mechanisms in response to the specific binding of small-molecule ligands. Since their discovery, riboswitches have shown promise for the artificial control of transcription or translation of target genes, be it for industrial biotechnology, protein expression, metabolic engineering, antimicrobial target validation, or gene function discovery. However, natural riboswitches are often unsuitable for these purposes due to their regulation by small molecules which are already present within the cell. For this reason, research has focused on creating riboswitches that respond to alternative biologically inert ligands or to molecules which are of interest for biosensing. Here we present methods for the development of artificial riboswitches in Gram-negative and

Gram-positive bacteria. These methods are based on reengineering natural aptamers to change their ligand specificity toward molecules which do not bind the original aptamer (ie, that are orthogonal to the original). The first approach involves targeted mutagenesis of native riboswitches to change their specificity toward rationally designed synthetic ligand analogs. The second approach involves the fusion of previously validated orthogonal aptamers with native expression platforms to create novel chimeric riboswitches for the microbial target. We establish the applicability of these methods both for the control of exogenous genes as well as for the control of native genes.

1. INTRODUCTION

Riboswitches are *cis*-acting RNA elements that regulate the expression of genes in response to the binding of specific small molecules. While the majority of riboswitches discovered to date have evolved to sense coenzymes and other primary metabolites (Breaker, 2011; Serganov & Nudler, 2013), some have been shown to respond to changes in the concentration of inorganic ions (Baker et al., 2012; Barrick & Breaker, 2007; Cromie, Shi, Latifi, & Groisman, 2006; Du Toit, 2015). Gene expression is regulated through the ligand-conditional adoption of alternative secondary structures which either facilitate or inhibit expression. Riboswitches are found across all domains of life, although most examples identified to date are from bacteria (Cheah, Wachter, Sudarsan, & Breaker, 2007). The sequence of riboswitches is typically divided into two regions: the aptamer domain and the expression platform. The association of the aptamer domain with its cognate ligand is accompanied by structural rearrangements that alter the conformation of the expression platform, which results in gene regulatory control by transcriptional, translational, RNA cleavage, or alternative splicing mechanisms. Transcriptional control is suggested to involve the riboswitch-mediated modulation of either Rho-dependent or intrinsic bacterial transcriptional termination mechanisms. Translation in bacteria can also be controlled by riboswitches that influence the binding of the ribosome to the ribosome-binding site (RBS), usually by sequestration of the Shine–Dalgarno sequence in a base-paired stem that is either formed or disrupted when the ligand binds to the aptamer domain. Alternative splicing mechanisms are found in eukaryotes (Cheah et al., 2007) and fall out of the scope of the methods we present here.

Due to their protein-independent mechanism, which simplifies cross-species application, riboswitches are of particular interest in biotechnology and basic research as tools for artificial gene regulation. For example, it has been demonstrated that natural or artificial riboswitches can be used as biosensors, to detect key intermediates or products in cells and thereby optimize engineered biosynthetic pathways (Michener, Thodey, Liang, & Smolke, 2012; Zhou & Zeng, 2015) or whole-cell biocatalysts (Meyer et al., 2015). The tunability and orthogonality of engineered riboswitches can also be exploited for protein production (Dixon et al., 2010; Morra et al., 2016), or in gene functional analysis and target validation (Robinson et al., 2014; Wu et al., 2015). Notwithstanding their promising potential, the use of natural riboswitches as artificial gene regulatory tools presents a number of shortcomings that can limit their utility. Natural riboswitches generally regulate genes involved in the metabolism or transport of their particular ligand. The fact that the metabolic pathways involving said ligands require such tight regulation may indicate that excessive concentrations are disadvantageous to the organism, making exogenous addition of the ligand problematic. Also, these natural ligands are frequently ubiquitous biological compounds which are expected to be present in the target organism at any given time, interfering with artificial attempts at regulation. In light of these issues, research has focused on the development of engineered riboswitches using aptamers that respond to artificial or nonnative ligands that do not interfere with the normal metabolism or function to the host organism.

One approach to create new riboswitches has involved the iterative selection of artificial RNA aptamers that bind a specific ligand using SELEX (Stoltenburg, Reinemann, & Strehlitz, 2007). These artificial aptamers may be introduced into the 5′-UTR of an mRNA to impede translation when bound to the ligand (Harvey, Garneau, & Pelletier, 2002), or alternatively can be fused with natural ribozymes (Tang & Breaker, 1997) or riboswitch expression platforms (Ceres, Garst, Marcano-Velázquez, & Batey, 2013) to create gene regulatory devices. Insertion of aptamers generated by SELEX into the 5′-UTR of eukaryotic mRNAs has been shown to provide ligand-dependent inhibition of translation, based on the eukaryotic ribosome mechanism of scanning for the start codon along the mRNA. However, given the targeted binding of the ribosome to the RBS in prokaryotes, the insertion of SELEX-derived aptamers into the 5′-UTR of prokaryotic mRNAs is less likely to provide translational control (Suess, Fink, Berens, Stentz, & Hillen, 2004). Additionally, while SELEX can greatly optimize the binding affinity of the aptamer to the desired ligand, this does not necessarily mean that the aptamer can be used to generate a functional riboswitch. In addition to high affinity the aptamer must also undergo significant change in secondary structure to induce a regulatory effect.

Our laboratory introduced an alternative approach to create artificial riboswitches by repurposing natural riboswitches to respond to artificial ligands in an orthogonal fashion, that is, that the artificial riboswitch would be insensitive to the natural ligand and vice versa (Dixon et al., 2010). This work was carried out originally on the *add* A-riboswitch from the Gram-negative bacterium *Vibrio vulnificus*, a translational activator switch which responds to adenine (Fig. 1A). Residues in the *add* riboswitch ligand-binding site were subjected to saturation mutagenesis, with the resulting sequences being placed upstream of a chloramphenicol resistance gene. Strains of *Escherichia coli* carrying these sequences were screened for resistance to chloramphenicol against a library of artificial purine and pyrimidine analogs. Promising riboswitch–ligand pairs were further characterized, and their orthogonality confirmed, by placing the mutant and natural riboswitches upstream of a *gfp* gene and carrying out fluorescence assays in the presence of adenine or the artificial ligand. From this procedure we derived two sets of mutants orthogonal to the parental *add* riboswitch, one pair (M6/M6″) which responded to ammeline, and another (M6C/M6C″) which responded to both ammeline and azacytosine. Structural analysis of the ligand–aptamer pairs revealed that the change in ligand specificity was mostly due to the reversal of a few hydrogen bonds through $U \rightarrow C$ transitions and in retrospect could have been rationally predicted.

Based on these insights, further work led to the development of more effective ligands, the pyrimidopyrimidine compounds PPDA and PPAO, which were specific to M6/M6″ (Fig. 1B) and M6C/M6C″, respectively. Using the inducible M6″ translational riboswitch, we successfully controlled the expression of the native motility gene *cheZ* in *E. coli* (Robinson et al., 2014). In addition to this, exploiting the inherent modularity of riboswitches (Ceres, Garst, et al., 2013; Ceres, Trausch, & Batey, 2013), we set out to combine the orthogonal M6″ aptamer with a natural transcription-repressing expression platform from *Bacillus subtilis* to generate a chimeric sequence. By combining the M6″ aptamer and the expression platform of the *xpt* G-riboswitch, we were able to artificially repress the expression of the morphology gene *mreB* in *B. subtilis*. This gene is required for the characteristic rod shape of *Bacillus*, and its repression could be observed by the

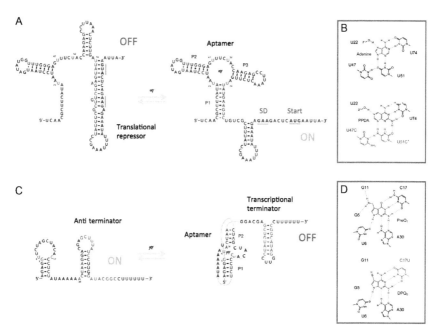

Fig. 1 Riboswitches used in this study to create orthogonal gene expression tools. (A) The adenine responsive translational ON riboswitch from the *add* gene of *V. vulnificus*. Adenine binding causes a conformational switch from an OFF state in which a translational repressor stem sequesters the Shine–Dalgarno sequence and start codon, to an ON state where these sequences are free to initiate translation. (B) Key mutations introduced into the *add* riboswitch change the H-bonding pattern (*dashed lines*) in the ligand-binding site to create an orthogonal riboswitch, termed *add*(M6″), which no longer recognizes adenine but instead can be induced by the synthetic ligand PPDA. (C) The PreQ$_1$ responsive transcriptional OFF riboswitch from the *queC* gene of *B. subtilis*. Binding of PreQ$_1$ causes a conformational switch from an ON state through which the RNA polymerase can pass unimpeded, to an OFF state in which a transcriptional terminator stem aborts transcription. (D) Key mutations introduced into the *queC* riboswitch change the H-bonding pattern (*dashed lines*) in the ligand-binding site to create an orthogonal riboswitch, termed *queC*(M1), which no longer recognizes PreQ$_1$ but instead can be repressed by the synthetic ligand DPQ$_0$.

irregular shape adopted by the transformed strains in the presence of PPDA. This work demonstrated cross-Phylum applicability of orthogonal aptamers, as well as the possibility of combining them with expression platforms to generate chimeric riboswitches with alternative mechanisms and opposite regulatory behaviors.

Also based on this experience, our laboratory set out to rationally reengineer a second riboswitch (Wu et al., 2015), the transcription repressing

PreQ$_1$ class I riboswitch from *B. subtilis* (Fig. 1C), which responds to natural queuosine precursors. By screening a set of just seven mutants against a library of six synthetic analogs, we found that the M1 (C17U) mutant riboswitch could regulate *lacZ* expression in response to the PreQ$_0$ analog DPQ$_0$ (Fig. 1D), while being insensitive to PreQ$_0$. The ability to control the native gene *mreB* with the orthogonal M1 riboswitch was then demonstrated.

2. OVERVIEW

Based on our previous work, we present here two alternative approaches for the development of artificial riboswitches to control gene expression in bacteria, complementing those previously published (Vincent, Robinson, Wu, Dixon, & Micklefield, 2014). The first approach is aimed at organisms for which there exist known riboswitches with aptamers that appear susceptible to rational reengineering, which we have successfully applied in both Gram-negative and Gram-positive bacteria. The second approach is intended for organisms in which there are riboswitches with expression platforms that operate through predictable mechanisms, and so may be combined with previously validated orthogonal aptamers to create chimeric riboswitches, which we used with success in Gram-positive bacteria. If need be, riboswitches identified in closely related species may also meet with success with either approach. These methods potentially allow for both the activation and repression of relevant genes and may be quickly validated through a reporter gene strategy (eg, β-galactosidase assay) before being deployed to control native genes (such as *cheZ* and *mreB* described) (Fig. 2A).

3. METHODS

The following protocols assume that the user has a good working knowledge of standard molecular cloning techniques (Sambrook & Russell, 2001) and access to standard molecular biology and microbiology laboratory equipment. Cloning protocols are presented as general workflows, illustrated with specific examples, but capable of being adapted to study any gene(s) of interest in a range of host organisms. More detail is provided for riboswitch validation and gene functional assays, where the reader is assumed to be less familiar with the techniques.

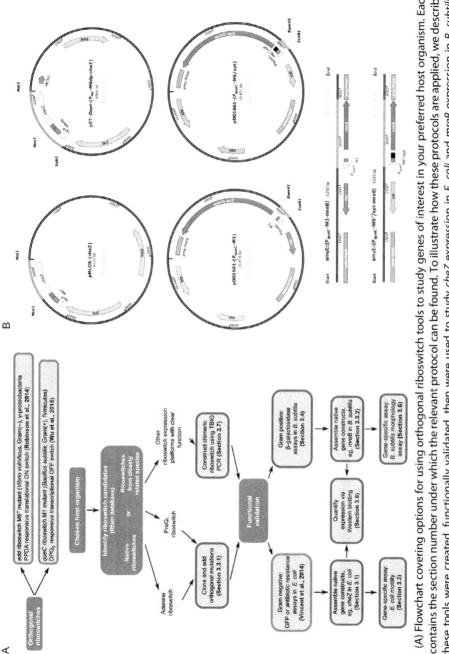

Fig. 2 (A) Flowchart covering options for using orthogonal riboswitch tools to study genes of interest in your preferred host organism. Each node contains the section number under which the relevant protocol can be found. To illustrate how these protocols are applied, we describe how these tools were created, functionally validated, then were used to study *cheZ* expression in *E. coli* and *mreB* expression in *B. subtilis*. This approach is described in detail in the text. (B) Plasmid vectors used in this study. Promoters, riboswitches, genes, and homologous recombination sites are labeled. Cloned sequences are shown as red arcs, with relevant restriction sites indicated. Underneath each plasmid name is the section number where it is described in more detail. (See the color plate.)

3.1 Construction of Inducible-Riboswitch Controlled Gene Expression Vectors for *E. coli*

The orthogonal M6″ mutant of the *add* riboswitch from *V. vulnificus* was reengineered through a targeted mutagenesis and genetic selection strategy, using a library of nonnatural synthetic ligand candidates (Dixon et al., 2010). The creation and functional validation of this riboswitch (using a chloramphenicol resistance selection and GFP reporter gene assays) has been covered in detail previously (Vincent et al., 2014). Subsequently a second-generation ligand, PPDA (pyrimido[4,5-*d*]pyrimidine-2,4-diamine), with superior gene regulatory properties was discovered (Robinson et al., 2014). The M6″-PPDA pairing allows for sensitive and precise titration of gene expression, making it suitable for synthetic biology applications and gene function studies. The section later describes our general strategy for creating M6″-controlled expression constructs, using as an example the native *E. coli* motility gene *cheZ* (Fig. 3). The promoter and M6″ riboswitch were subcloned from an in-house plasmid (pMLOS, available on request) but could easily be synthesized de novo by "Thermodynamically balanced inside-out" (TBIO) PCR-based gene synthesis (see Section 3.7).

1. *Primer design*: To PCR amplify the *cheZ* gene from the motile *E. coli* strain RP437 (Coli Genetic Stock Center, Yale CT #12122), primers were designed on the basis of the *cheZ* sequence annotated in the *E. coli* MG1655 genome (GenBank accession: U00096.3). The forward primer (cheZ-f) had a 5′-extension to add an *Nco*I cleavage site and two additional base pairs in front of the start codon (so that the gene was cloned in the correct reading frame). The reverse primer (cheZ-r) had a 5′-extension to add a *Not*I site after the stop codon (Table 1). A third primer was designed (cheZ-f2) with an *Sph*I site, to allow subcloning of the P_{lac}-M6″-*cheZ* construct from the pMLOS vector to pET-Duet-1 (see later). All primers were used to BLAST search the MG1655 genome to ensure unique sequences were selected.

2. *gDNA preparation*: Inoculate 2 mL Lysogeny broth (LB) (Sambrook & Russell, 2001) in a 15-mL Falcon tube with the glycerol stock of *E. coli* RP437, using sterile technique. Grow overnight at 37°C with shaking at 180 rpm in a shaker-incubator. Purify genomic DNA using a commercial DNA isolation kit (eg, Qiagen DNeasy Kit, #69581), following the manufacturer's protocol. Determine the concentration and purity of the genomic DNA sample using a NanoDrop (ThermoFisher) or equivalent spectrophotometer. [Note: Pure DNA should have an $A_{260}/_{280}$ ratio of ~1.8, and an $A_{260}/_{230}$ ratio of 2.0–2.2.]

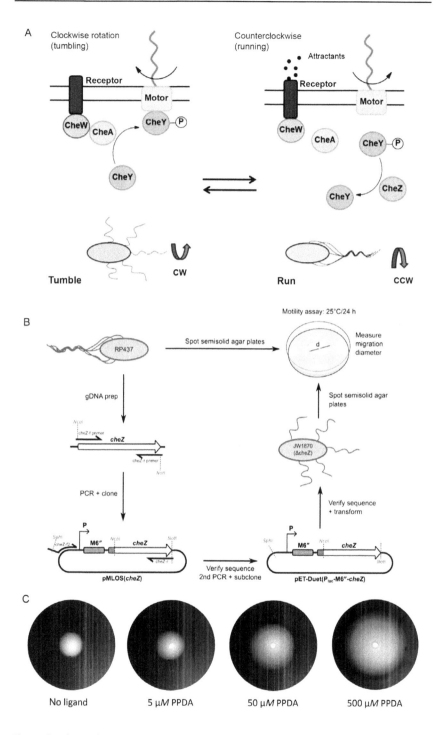

Fig. 3 See legend on opposite page.

3. *PCR amplification*: Set up a PCR using 50–500 ng of gDNA, the cheZ-f and cheZ-r primers, and a high-fidelity DNA polymerase (eg, Phusion, NEB #M0530). Purify the PCR product using a commercial kit (eg, Qiagen QIAquick PCR Purification Kit, #28104).
4. *Restriction digests*: Set up two double digests using *Nco*I and *Not*I (NEB) to cleave the *cheZ* PCR product and the pMLOS vector (*in-house* plasmid carrying the M6″ riboswitch under transcriptional control of the *lac* promoter). Resolve the two digests on a 1% agarose gel with a suitable ladder (eg, NEB 2-Log DNA Ladder, #N3200), visualize the gel on a UV transilluminator, and carefully excise the correct bands using a scalpel blade. Recover the DNA using a commercial kit (eg, Qiagen QIAquick Gel Extraction Kit, #28704) and NanoDrop for concentration and purity.
5. *Ligation*: Set up a 10-μL ligation reaction using T4 DNA ligase (NEB, #M0202) at 16°C overnight, with a 3:1 molar ratio of insert (*cheZ*) to vector (pMLOS), and keeping the total DNA concentration under 100 μg.
6. *Transform cells*: Transform 100 μL of your preferred cloning strain (we use DH5α, ThermoFisher #18258-012) with 0.2 μL of ligation reaction by heat shock. Spread 20 and 200 μL of cells onto fresh LB-Amp agar plates and grow overnight at 37°C.
7. *Sequencing and glycerol stocks*: Select five colonies from the transformation plates to inoculate five 2 mL aliquots of LB media supplemented with 100 μg/mL ampicillin (LB-Amp). Grow overnight at 37°C with shaking at 180 rpm in a shaker-incubator. Aliquot 0.5 mL of each culture into a 2-mL CryoTube, mix with 0.5 mL glycerol by pipetting gently, then freeze immediately in liquid nitrogen, and store at −80°C. Perform

Fig. 3 Controlling expression of the native *E. coli* motility gene *cheZ*. (A) CheZ is a phosphatase which acts upon CheY. In the absence of CheZ, phosphorylated CheY associates with the flagellar motor to induce clockwise rotation, flagella separate from one another and the cell tumbles in place (nonmotile phenotype). When CheZ is expressed, dephosphorylated CheY dissociates from the flagellar motor leading to counterclockwise rotation, flagella bundle together pushing the cell forward through the media (motile phenotype). (B) Overview of the workflow to create riboswitch controlled CheZ expression constructs and conduct motility assays. (C) Cells are spotted onto the center of semisolid agar plates containing the riboswitch ligand. After incubation at 25°C for 24 h, the diameter of the outermost ring of migration is measured. Representative plate images of the *E. coli* strain JW1870 (Δ*cheZ*), transformed with the pET-Duet-(P$_{lac}$-M6″-*cheZ*) plasmid, demonstrate dose-dependent motility in response to PPDA. (See the color plate.)

Table 1 Primers Described in This Chapter, Consult the Listed Sections for Specific Details

Primer	Description	Sequence	RE Sites
Cloning *cheZ* into a riboswitch expression construct (Section 3.1)			
cheZ-f	gDNA PCR	5′ AAA AAA CCA TGG AAA TGA TGC AAC CAT CAA TCA AA 3′	*Nco*I
cheZ-r	gDNA PCR and subcloning	5′ AAA AAA GCG GCC GCT CAA AAT CCA AGA CTA TCC AAC A 3′	*Not*I
cheZ-f2	Subcloning	5′ AAA AAA GCA TGC CTC ACT CAT TAG GCA CCG AAT 3′	*Sph*I
Cloning and mutagenesis of the PreQ₁ riboswitch (Section 3.3.1)			
qc5u-f	Flanking forward	5′ AAA AAA GAA TTC ATA ATG AAA CGA ACC GTC ACT ATA G 3′	*Eco*RI
qc5u-r	Flanking reverse	5′ AAA AAA GGA TCC GAC AAT TGC TTT TTC TTT TTT CAT G 3′	*Bam*HI
pDG1661-f	Sequencing	5′ AGT AAC TTC CAC AGT AGT TCA 3′	
pDG166x-r	Sequencing	5′ TAG GAG GCT TAC TTG TCT GCT T 3′	
qmut-f	Mutagenesis	5′ TTT ATA GAG GAT GTA GCT AGA ACC 3′	
qmut-r	Mutagenesis	5′ AAA CTA AGG ACG AGC TGT ATC 3′	
Cloning *mreB* and assembly of B. *subtilis* integrative constructs (Section 3.3.2)			
mreB-f	gDNA PCR	5′ AAA AAA GAA TTC GTG TTT GGA ATT GGT GCT AGA GA 3′	*Eco*RI
mreB-r	gDNA PCR	5′ AAA AAA GGA TCC TTA TCT AGT TTT CCC TTT GAA AAG AT 3′	*Bam*HI
mreB-f2	Gene-specific PCR assembly	5′ GTG GTG AAC TAC TGT GTT TGG AAT TGG TGC TAG AGA 3′	
amyE-f	PCR assembly and sequencing	5′ ATG TTT GCA AAA CGA TTC AA 3′	
amyE-r	PCR assembly and sequencing	5′ TCA ATG GGG AAG AGA ACC G 3′	
Spacer1	PCR assembly	5′ AGT AGT TCA CCA CCT TTT CCC TA 3′	
Spacer2	PCR assembly	5′ CCA GCT TGT TGA TAC ACT AAT GCT TTT ATA TAG GAA AAA GGT GGT GAA CTA CT 3′	
Spacer2-2	PCR assembly	5′ CCA GCT TGT TGA TAC ACT AAT GCT 3′	

Continued

Table 1 Primers Described in This Chapter, Consult the Listed Sections for Specific Details—cont'd

Primer	Description	Sequence	RE Sites
pDG1662-f	Sequencing	5′ ACT TAG GTA CTG AAC AAG AA 3′	
pDG166x-r	Sequencing	5′ TAG GAG GCT TAC TTG TCT GCT T 3′	
Synthesis of a constitutive promoter–chimeric riboswitch construct (Section 3.7)			
S1	TBIO synthesis	5′ CTG TCA AAC ATG AGA ATT CAT AAT GAA 3′	EcoRI
S2	TBIO synthesis	5′ CTG TCA AAC ATG AGA ATT CAT AAT GAA ACG AAC CGT CAC TAT AGA GCG ATA ATT GAT CTA 3′	EcoRI
S3	TBIO synthesis	5′ GTC ACT ATA GAG CGA TAA TTG ATC TAG GAA CCG GGG ACT GAC TTT TTT ATT GTC AAG AAA 3′	
S4	TBIO synthesis	5′ GGA CTG ACT TTT TTA TTG TCA AGA AAA ACA TCA TTA TGG TAA GAT AGC AGA AAT ATA ATA 3′	
S5	TBIO synthesis	5′ ATC ATT AG GTA AGA TAG CAG AAA TAT AAT AGG AAC ACT CAT ATA ATC CGA ATG ATA TGG T 3′	
AS5	TBIO synthesis	5′ CAA GAG TTT AAG GCT CTT GGT GGA AGC TCC GAA ACC ATA TCA TTC GGA TTA TAT GAG TG 3′	
AS4	TBIO synthesis	5′ AAA ACC GTA CAC GTG CGG TTC CAT TGC TCA CCC ATA ATC AAG AGT TTA AGG CTC TTG GT 3′	
AS3	TBIO synthesis	5′ TTG CCC GCT CAA ATA AAG AGC AAG CAA TGC TGA TAT CAC AAA AAA CCG TAC ACG TGC GG 3′	
AS2	TBIO synthesis	5′ TCC ATC CTG TCT ACC TCC GTT ATG AGA ATA AAA AAA GCA TTG CCC GCT CAA ATA AAG AG 3′	
AS1	TBIO synthesis	5′ CAG TGG ATC CAT CCT GTC TAC CTC CG 3′	BamHI
amyE-f	PCR check	5′ ATG TTT GCA AAA CGA TTC AA 3′	
amyE-r	PCR check	5′ TCA ATG GGG AAG AGA ACC G 3′	
pDG1661-f	Sequencing	5′ AGT AAC TTC CAC AGT AGT TCA 3′	
pDG166x-r	Sequencing	5′ TAG GAG GCT TAC TTG TCT GCT T 3′	

Restriction endonuclease (RE) cleavage sequences are underlined. For the TBIO synthesis primers, the P$_{queC}$ promoter sequence is highlighted in blue (light gray in the print version), and the M6″ aptamer sequence is in red (dark gray in the print version). All primers were purchased from Sigma-Aldrich.

plasmid mini-preps with the remaining 1.5 mL of each culture using a commercial kit (eg, Qiagen QIAprep Spin Miniprep Kit, #27104). Send samples for DNA sequencing using the *cheZ*-f2 primer.

The purpose of cloning *cheZ* into pMLOS is to place the gene under the control of the M6″ riboswitch and *lac* promoter (which are carried by this plasmid). The restriction cloning strategy places *cheZ* (or any gene cloned by this method) in frame with a 5′-fragment of GFP, to create a gene fusion (109 bp of GFP with 642 bp of *cheZ*). Cloning genes in this manner leads to predictable riboswitch behavior, as the 5′-region of the GFP gene does not interfere with M6″ riboswitch folding (Dixon et al., 2012). The resulting plasmid can be used directly to study riboswitch control over most genes, however; pMLOS is a high-copy number plasmid (~500 per cell). Overexpressing CheZ can cause cells to become embedded in the semisolid agar media, thereby reducing migration distance. Therefore, it is necessary to subclone the P_{lac}-M6″-*cheZ* construct into a lower copy plasmid, such as pET-Duet-1 (Novagen, #71146-3) which has a copy number of ~50 per cell.

8. *Subcloning*: Subcloning P_{lac}-M6″-*cheZ* from pMLOS to pET-Duet-1 requires steps 3–7 above to be repeated, but with the following amendments:
   a. *PCR amplification*: Use 1 ng of the pMLOS-*cheZ* plasmid as a template and the primers cheZ-f2 and cheZ-r (Table 1). As a precaution, perform a *Dpn*I (NEB) digest after the PCR to cleave the plasmid template.
   b. *Restriction digests*: Set up two double restriction digests using *Sph*I and *Not*I (NEB) to cleave the P_{lac}-M6″-*cheZ* PCR product and the pET-Duet-1 vector.
   c. *Ligation*: Use the digested P_{lac}-M6″-*cheZ* and pET-Duet-1 as the insert and vector, respectively.
   d. *Transform cells*: Prepare chemically competent cells (Sambrook & Russell, 2001) using the *cheZ* knockout *E. coli* strain JW1870 (Baba et al., 2006) (CGSC) and transform with the above ligation.
   e. *Sequencing and glycerol stocks*: As earlier, sequence mini-preps prepared from transformants using the *cheZ*-f2 primer.

Sequence-verified transformants are ready to be used in motility assays as described. A workflow for the whole cloning procedure is presented in Fig. 3B, and maps of the pMLOS-*cheZ* and pET-Duet-(P_{lac}-M6″-*cheZ*) plasmids can be found in Fig. 2B.

3.2 CheZ Motility Assays in *E. coli*

To demonstrate how an orthogonal riboswitch can be used to study the function of native genes in *E. coli*, we chose the *cheZ* gene, which is associated with a clear visible motile phenotype. CheZ is a protein phosphatase that acts upon CheY, a motility regulator which transmits chemotactic signals to the flagellar motor component (Fig. 3A). Phosphorylation of CheY causes flagella to rotate clockwise in a uncoordinated fashion, leaving the cell to "tumble" in place. In contrast, dephosphorylation of CheY, through the activity of CheZ, leads to counterclockwise rotation that causes adjacent flagella to bundle and rotate together, pushing the cell in a straight path through the media (run). Cells with insufficient levels of CheZ become locked in the tumbling mode and are essentially nonmotile. Orthogonal riboswitch constructs can be used to exert titratable control over CheZ expression, and therefore *E. coli* cell motility. The method below describes how we conduct motility assays on semisolid agar plates (Robinson et al., 2014).

1. Inoculate 2 mL aliquots of LB-Amp media in 15-mL falcon tubes using glycerol stocks of *E. coli* RP437 (wild type), JW1870 ($\Delta cheZ$), and JW1870 carrying the *cheZ* expression construct.
2. Grow cultures overnight at 37°C with shaking at 180 rpm in a shaker-incubator.
3. Inoculate fresh 2 mL aliquots of LB-Amp with 20 µL of overnight culture and incubate for 8 h at 37°C with shaking at 180 rpm in a shaker incubator.
4. Prepare fresh motility plates immediately before use:
    a. Microwave 400 mL of semisolid agar media (4 g tryptone, 2 g NaCl, and 1 g agar in 400 mL MilliQ H_2O, autoclaved at 121°C for 20 min) on "defrost" until liquefied. [Note: Semisolid agar is necessary to allow *E. coli* cells to migrate efficiently.]
    b. Cool media in a 50°C water bath for at least 2 h, swirling occasionally. Keep media in the water bath to prevent agar from resetting.
    c. Add 400 µL of ampicillin stock solution (50 µg/mL final concentration) and mix thoroughly by swirling.
    d. Split media into two 200 mL portions and add 2 µL of sterile 1 *M* IPTG (isopropyl β-D-1-thiogalactopyranoside) stock solution (10 µ*M* final concentration) to one.
    e. For each plate (35 × 10 mm), aliquot 3.04 mL of plate media and add 160 µL DMSO (5%, v/v) containing varying concentrations of

riboswitch ligand (0, 0.1, 0.5, 1, 5, 10, 25, 50, 75, 100, 150, 200, 300, 400, and 500 µM final PPDA concentration in media). Mix the media and DMSO thoroughly by swirling. Prepare equivalent plates using the media with and without IPTG.

   f. Allow the semisolid agar plates to set for at least 30 min under a laminar flow hood.
5. Spot 2 µL (approximately 2×10^9 cells) of cell culture onto the center of each motility plate and allow the spot to air-dry for 15 min at room temperature.
6. Incubate plates right-side up in an incubator at 25°C for 24 h.
7. Carefully measure the diameter of the outermost ring of migration for each ligand concentration.
8. Capture images of the migration plates using suitable photographic equipment (Fig. 3C).
9. *Analysis*: Experiments should ideally be performed in triplicate over consecutive days, using fresh overnight cultures, with migration diameters presented as mean values with standard deviations. For our analyses of translational ON riboswitches, scatter plots of *migration diameter* vs the logarithm of *inducer concentration* can be fit with a four-parameter logistic function (eg, using SigmaPlot 13, Systat Software Inc.) to determine a half-maximal effective concentration (EC_{50}) for each ligand-construct pairing.

3.3 Construction of Repressible-Riboswitch Controlled Constructs for Targeted Chromosomal Integration in B. subtilis

Reengineered riboswitches are ideally suited for studying native gene expression in their microbial source, since the folding kinetics and expression platform have evolved to function specifically with the host cell transcriptional/translational machinery. To demonstrate this principle, we reengineered an orthogonal M1 mutant of type-I $PreQ_1$ riboswitch from the *queC* gene of *B. subtilis*, which we then used in *B. subtilis* to establish conditional mutants of the native morphology gene *mreB* (Wu et al., 2015). In general, when assembling new orthogonal riboswitch constructs a suitable vector and reporter gene strategy should be chosen to validate riboswitch function before deployment with your gene of interest. For *B. subtilis*, we take advantage of the pDG1661 plasmid (Fig. 2B), available from the Bacillus Genetic Stock Center (BGSC, Columbus, OH). This plasmid contains a *spoVG–lacZ* gene fusion (for β-galactosidase assays), along with front and back fragments of the *amyE* gene to enable stable integration into the *B. subtilis* chromosome through homologous recombination (Guérout-Fleury, Frandsen, & Stragier, 1996). Once titratable control over *lacZ* reporter gene expression has been demonstrated, the promoter–riboswitch construct can be subcloned along with your gene of interest into the related vector pDG1662 (BGSC) which contains the *amyE* fragments without the *lacZ* reporter. The sections later describe this general strategy using the *queC* promoter–riboswitch and *mreB* gene from *B. subtilis*. Suggested kits, retailers, and catalogue numbers introduced in Section 3.1 have been omitted.

3.3.1 lacZ *Reporter Gene Constructs*

1. *Primer design*:
    a. *Genomic DNA PCR*: To PCR amplify the promoter and PreQ$_1$ riboswitch from the *queC* gene of *B. subtilis* strain 168 (Bacillus Genetic Stock Center (BGSC), Columbus, OH #1A168), primers were designed on the basis of the published 168 genome (GenBank accession: AL009126.3). The forward (qc5u-f) and reverse (qc5u-r) primers had 5′-extensions to add *Eco*RI and *Bam*HI cleavage sites, respectively (Table 1). All primers were used to BLAST search the 168 genome to ensure unique sequences were selected.
    b. *Riboswitch mutagenesis*: To incorporate the C17U (M1) mutation into the cloned PreQ$_1$ riboswitch, primers were designed using the NEBaseChanger online tool (http://nebaschanger.neb.com/).
2. *gDNA preparation*: Inoculate 2 mL LB media in a 15-mL Falcon tube with the glycerol stock of *B. subtilis* 168 using sterile technique. Grow overnight at 37°C with shaking at 180 rpm in a shaker-incubator. Purify genomic DNA using a commercial kit and then determine the concentration and purity using a spectrophotometer. This gDNA should be retained for PCR amplification of *B. subtilis* genes, such as *mreB* discussed later (see Section 3.3.2).
3. *PCR amplification*: Set up a PCR using 50–500 ng of gDNA, the qc5u-f and qc5u-r primers, and a high-fidelity DNA polymerase. Purify the PCR product.
4. *Restriction digests*: Set up two double restriction digests using *Eco*RI and *Bam*HI (NEB) to digest 2 μg of the PCR product and the pDG1661 vector (BGSC, #ECE112). Resolve the two digests on a 1% agarose

gel, visualize the gel on a UV transilluminator, and carefully excise the correct bands using a scalpel blade. Recover the DNA and then determine concentration and purity.

5. *Ligation*: Set up a 10-μL ligation reaction using T4 DNA ligase at 16°C overnight, with a 3:1 molar ratio of insert (PCR product) to vector (pDG1661), and keeping the total DNA concentration under 100 μg.

6. *E. coli transformation*: Transform 100 μL of your preferred cloning strain (we use DH5α) with 0.2 μL of reaction by heat shock. Spread 20 and 200 μL of cells onto fresh LB agar plates supplemented with 100 μg/mL ampicillin and grow overnight at 37°C.

7. *Sequencing and glycerol stocks*: Select five colonies from the transformation plates to inoculate five 2 mL aliquots of LB-Amp media. Grow overnight at 37°C with shaking at 180 rpm in a shaker-incubator. Aliquot 0.5 mL of each culture into a 2-mL CryoTube, mix with 0.5 mL glycerol by pipetting gently, then freeze immediately in liquid nitrogen, and store at −80°C. Perform plasmid mini-preps with the remaining 1.5 mL of each culture. Send samples for DNA sequencing using the primers pDG1661-f and pDG166X-r.

8. *Mutagenesis*: Use the remaining mini-prep with a commercial site-directed mutagenesis kit (eg, Q5 Site-Directed Mutagenesis Kit NEB, #E0554S) to introduce the M1 (C17U) mutation into the cloned PreQ$_1$ riboswitch using the primers qmut-f and qmut-r. Repeat steps 6–7 above to confirm successful mutagenesis by DNA sequencing.

9. *B. subtilis transformation*: Add 500 ng of the sequence-verified plasmid to 500 μL of competent *B. subtilis* 168 cells (BGSC). [Note: We prepare competent *B. subtilis* cells using the standard "Paris method" (Harwood & Cutting, 1990).] Incubate cells with DNA for 30 min at 37°C with shaking at 180 rpm. Add 300 μL of warm LB media and continue incubation for another 30 min and then spread the transformation media onto agar plates of fresh LB media supplemented with 5 μg/mL chloramphenicol (LB-Cam).

10. *Verification of chromosomal integration*:
    a. Aliquot 2 mL of LB-Cam media into ten 15-mL Falcon tubes and then add 50 μg/mL of spectinomycin to five of the tubes (LB-Cam/Spc). Select five colonies from the transformation plate to inoculate the LB-Cam and LB-Cam/Spc media. Grow overnight at 37°C with shaking at 180 rpm in a shaker-incubator. Cells with chromosomally integrated vector will grow in LB-Cam but not in LB-Cam/Spc.

b. Purify genomic DNA from the above cultures using a commercial kit. Set up a PCR using 50–500 ng of gDNA, the amyE-f and amyE-r primers, and a high-fidelity DNA polymerase. Resolve the PCR on a 1% agarose gel. Successfully integrated products will give a DNA band 6393 bp in length, whereas cells without insert will give a 1977 bp band corresponding to the undisrupted *amyE* gene.
   c. Send gDNA samples for sequencing using the primers pDG1661-f and pDG166X-r.

B. subtilis cells which pass the verification steps above are ready for functional validation of riboswitch activity using the β-galactosidase assay. Prepare glycerol stocks by aliquoting 0.5 mL of overnight LB-Cam culture into 2-mL CryoTubes, mix with 0.5 mL glycerol by pipetting gently, then freeze immediately in liquid nitrogen, and store at −80°C.

3.3.2 Native Gene Expression Constructs

1. *Primer design*: To PCR amplify your gene of interest from *B. subtilis* 168 (BGSC, #1A168), primers should be designed on the basis of the annotated sequence in the 168 genome (GenBank accession: AL009126.3). The forward primer should have the sequence 5′-AAAAAAGAATTC immediately before the start codon (to introduce an *Eco*RI site), while the reverse primer should have 5′-AAAAAAGGATCC immediately before the triplet complementary to the stop codon (to introduce a *Bam*HI site). In our example later we have selected the *mreB* gene, and therefore designed the primers mreB-f and mreB-r (Table 1). All primers were used to BLAST search the 168 genome to ensure unique sequences were selected.
2. *PCR amplification*: Set up a PCR using 50–500 ng of gDNA, the mreB-f and mreB-r primers, and a high-fidelity DNA polymerase. Purify the PCR product.
3. *Cloning and sequence verification*: Following steps 3–7 of the protocol above (see Section 3.3.1), clone the PCR product into the integrative plasmid pDG1662 (BGSC, #ECE112), which is identical to pDG1661 but without the *spoVG-lacZ* fusion.
4. *Overlap extension PCR*:
   a. Set up a PCR reaction using 1 ng of the pDG1661-P$_{queC}$-M1 plasmid as template, with the primers amyE-r and spacer1, to amplify a sequence encompassing the amyE-back fragment, the *cat* gene, and

the P_{queC}-M1 region. Digest the template with *Dpn*I (NEB) and PCR purify.

b. Set up a second PCR reaction using 1 ng of the pDG1662-*mreB* plasmid as template, with the primers amyE-f and mreB-f2, to amplify a sequence encompassing the amyE-front fragment and *mreB* gene (while adding a 5′-GTGGTGAACTACT extension). Digest the template with *Dpn*I and PCR purify.

c. Set up a third PCR using 50 ng of the product from step 4b, with the primers amyE-f, spacer2, and spacer2-2, to add a 53-bp region of overlap with the PCR product of step 4a. PCR purify.

d. Set up a fourth PCR using 50 ng each of the product from step 4a and c, with the primers amyE-f and amyE-r. PCR purify and run an aliquot on a 1% agarose gel. The desired product should be a linear fragment 4.2–4.3 kb in size (if using *mreB* with the riboswitches described in this chapter).

5. *B. subtilis transformation and verification*: Transform the *B. subtilis* 168 strain following steps 9–10 of the protocol above (see Section 3.3.1). Successfully integrated products will give a DNA band 4.2–4.3 kb in size, whereas cells without insert will give a 1977-bp band corresponding to the undisrupted *amyE* gene. For DNA sequencing the primers pDG1662-f and pDG166X-r should be used.

Prepare glycerol stocks of sequence-verified cell lines. These cells are ready for use in morphology assays as described.

3.4 β-Galactosidase Assays to Validate Synthetic Riboswitch Function

β-Galactosidase, encoded by the *lacZ* gene, hydrolyzes lactose to glucose and galactose. Miller described a standardized protocol for measuring β-galactosidase activity using the synthetic substrate *o*-nitrophenyl-β-D-galactoside (Miller, 1972). Hydrolysis of ONPG yields galactose and *o*-nitrophenol (ONP), which has a yellow color that can be measured spectrophotometrically. When ONPG is present in excess, ONP yield per unit time is proportional to β-galactosidase concentration, allowing *lacZ* gene expression levels to be quantified. The β-galactosidase assay is a popular strategy for studying the function of gene regulatory elements and quantifying their relative effects on gene expression. As a reporter gene assay, *lacZ* offers greater sensitivity than *GFP*-based techniques, which is important for quantifying expression from low- or single-copy gene constructs. The β-galactosidase assay described (Fig. 4A) was modified from the high-throughput microplate method described by Lynch, Desai, Sajja, and Gallivan (2007) and is used in our lab

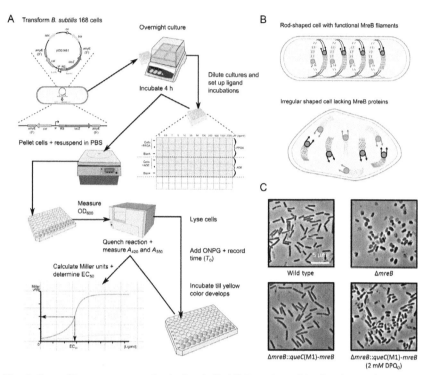

Fig. 4 Controlling gene expression in *B. subtilis*. (A) Overview of the β-galactosidase assay to validate novel riboswitch function using chromosomally integrated *lacZ* expression constructs. The orthogonal *queC*(M1) and chimeric *add*(M6″)/*xpt* riboswitches were both validated by this method prior to deployment with native *B. subtilis* genes. (B) Function of the native *B. subtilis* morphology gene *mreB*. Wild-type cells have a rod-shaped morphology maintained in part by dynamic filaments of MreB (*red lines*), which are deposited in a helical orientation around the circumference of the cell, recruiting the peptidoglycan synthetic complexes (*blue circles*) to deposit the cell wall (*green crosshatch*). In the absence of MreB, the cell wall is deposited in a uncoordinated manner leading to swollen amorphous cells. (C) Phase-contrast microscopy shows the typical rod-shaped morphology of wild-type cells (*top left*) and the aberrant amorphous phenotype of Δ*mreB* cells (*top right*). Chromosomal integration of the repressible *queC*(M1)-*mreB* construct restores the rod-shape morphology to Δ*mreB* cells (*bottom left*), but the mutant phenotype reappears when the DPQ$_0$ ligand is added to repress *mreB* expression. (See the color plate.)

to validate the gene regulatory function of newly constructed orthogonal riboswitches (Robinson et al., 2014; Wu et al., 2015).

1. Aliquot 200 μL of LB–Cam media into wells on a sterile 96-well microplate (one well for each strain to be tested). [Note: We use Greiner U-bottom microplates, #650161.]

2. Inoculate LB-Cam media using glycerol stocks of *B. subtilis* strains carrying the *lacZ* expression constructs.
3. Grow cultures overnight at 37°C with shaking at 500 rpm in a microplate shaker.
4. Dilute 150 μL of each overnight culture with 2.7 mL of fresh LB-Cam, mix by pipetting up and down gently, and then aliquot 190 μL into each of the 12 wells across a row of a new sterile microplate (Fig. 4A).
5. Aliquot 190 μL of sterile LB-Cam into each of the 12 wells across a row to serve as media controls.
6. Add 10 μL of the appropriate ligand solution ($20 \times$ concentration in MilliQ H_2O) to each well. Each ligand concentration is pipetted into all eight wells down a microplate column, allowing 12 different concentrations of inducer to be assayed (0, 0.5, 1, 5, 10, 25, 50, 100, 250, 500, 1000, and 2000 μM final ligand concentration in media (Fig. 4A)).
7. Incubate in a microplate shaker for 4 h at 37°C with shaking at 500 rpm.
8. Pellet cells by centrifugation at $1750 \times g$ for 20 min at 4°C and then resuspend in 200 μL of phosphate-buffered saline (PBS). [Note: To pellet cells in microplates, we use an Eppendorf 5804R centrifuge fitted with a A-2-DWP microplate rotor.]
9. Pipette 100 μL of each sample into a clear-bottomed 96-well microplate and measure the absorbance at 600 nm wavelength using a microplate reader. [Note: For UV–vis measurements we use Greiner F-bottom μCLEAR® plates, #780270.]
10. Lyse these cells by adding 100 μL of freshly prepared PBS containing 2 μg lysozyme (40 U/mL) and $2 \times$ PopCulture reagent (Merck-Millipore, #71092), mix by pipetting then incubate at room temperature for 15 min.
11. Pipette 10 μL of each lysate into fresh wells on a clear-bottomed 96-well microplate and then add 132.5 μL Z buffer [60 mM Na_2HPO_4, 40 mM NaH_2PO_4, 10 mM KCl, 1 mM $MgSO_4$, 50 mM β-mercaptoethanol, pH 7.0].
12. Start the enzyme reaction by adding 29 μL of room temperature ONPG (Sigma-Aldrich, #N1127) made up fresh at 4 mg/mL in Z buffer, record the precise time of addition.

13. Incubate at room temperature for 45 min. [Note: Reactions should be left until the yellow color of ONP is visible in all wells; for these particular constructs this takes 30–45 min.]
14. Quench the reaction by adding 75 μL of 1 M Na_2CO_3, again recording the precise time of addition.
15. Measure the absorbance at 420 and 550 nm wavelengths using a microplate reader.
16. Calculate the β-galactosidase activity (in Miller units) for each well using the following equation:

$$\frac{1000 * (A_{420} - 1.75 * A_{550})}{OD_{600} * t * v}$$

where:

t = total reaction time in minutes between ONPG addition and quenching (steps 12–14).

v = the volume of lysate in mL (0.01 mL in this instance; step 11).

OD_{600} = the optical density at 600 nm (step 9) corrected by subtracting the equivalent measurement from the media controls.

A_{420} and A_{550} = the absorbance at 420 and 550 nm (step 15) corrected by subtracting the equivalent measurements from the media controls.

17. *Analysis*: Experiments should ideally be performed in triplicate over consecutive days, using fresh overnight cultures, with Miller units presented as mean values with standard deviations. For our analyses of transcriptional OFF riboswitches, Miller units for cultures grown in the presence of ligand are divided by those grown in the absence of ligand, then multiplied by 100 to determine the *percentage repression*. A scatter plot of *percentage repression* vs the logarithm of *inducer concentration* can be fit with a four-parameter logistic function (eg, using SigmaPlot 13, Systat Software Inc.) to determine half-maximal inhibitory concentration (IC_{50}) for the ligand-construct pairing.

3.5 Western Blot Analysis to Quantify Target Gene Expression Levels

Once your gene of interest is placed under orthogonal riboswitch control in your preferred bacterial host, it is desirable to characterize the dose–response relationship in terms of the dynamic range of gene expression (maximal/minimal output), the dynamic range of ligand sensing (the range over which gene expression is sensitive to changes in riboswitch ligand concentration), and the EC_{50} (half-maximal effective concentration). However, most genes do not produce a simply quantifiable phenotype, therefore it is necessary to assay the levels of expression via other means. For these cases we typically use Western blotting, together with a chemiluminescent detection kit, to quantify target protein yields over a range of riboswitch ligand concentrations. A range of commercial vendors can supply antibodies/antisera when provided with the amino acid sequence of your target protein. The protocol below describes how we prepare samples for Western blot analysis to characterize dose–response relationships, using the M1-*mreB* construct as an example.

1. Aliquot 1900 μL of NB media into twelve 15-mL Falcon tubes and then add 50 μL of autoclaved 1 M $MgCl_2$ to each (25 mM final concentration), using sterile technique.
2. Use glycerol stocks of the *B. subtilis* strains 168 (wild type), 3725 (Δ*mreB*), and 3725 cells transformed with the riboswitch-*mreB* construct, to inoculate four tubes each of NB media.
3. Add 50 μL of sterile DMSO containing varying concentrations of riboswitch ligand to the NB media for each strain (we would typically test ligand at 0, 0.5, 1, and 2 mM final concentration in media).
4. Grow cultures overnight at 37°C with shaking at 180 rpm in a shaker-incubator.
5. Pellet 400 μL of each culture in 1.5-mL tubes at $12,000 \times g$ for 5 min.
6. Resuspend cell pellets in lysis buffer comprised of 40 μL of Y-per plus reagent (ThermoFisher, #78999), 10 μL of 10% SDS solution, and 1 μL of 50× protease inhibitor cocktail (Promega, #G6521).
7. Incubate lysis reactions at 45°C for 40 min in a water bath.
8. Microcentrifuge at $12,000 \times g$ for 5 min.
9. Recover the supernatant and determine total protein concentration by your preferred method (eg, Bicinchoninic acid assay kit, Sigma-Aldrich, #BCA-1).
10. Mix 15 μg of protein with loading buffer (eg, NuPAGE LDS Sample Buffer, ThermoFisher, #NP0008) and dilute to 10 μL final volume with distilled H_2O. Heat denature for 10 min at 95°C.
11. Perform Western blot using the standard protocol for your preferred method/equipment. For reference, we use the Amersham ECL system of Western blotting products according to the manufacturer's instructions. This system includes: precast 12% gels (GE Healthcare, #28-

9901-56), Hybond-C Extra nitrocellulose blotting membranes (GE Healthcare, #RPN303E), the ECL Gel Box (GE Healthcare, #28-9906-08), and ECL Semidry Transfer Unit (GE Healthcare, #TE 77). For MreB detection, blotted membranes are incubated in blocking buffer containing 1:5000 (v/v) anti-MreB antiserum (kindly provided by Prof. Jeff Errington) and 1:1000 (v/v) antielongation factor Tu (EF-Tu) monoclonal antibody (Hycult Biotech). We develop our blots using the chemiluminescent Amersham ECL detection kit (GE Healthcare, #RPN2108) and image them on a Typhoon 9210 variable mode imager (GE Healthcare). Band intensity is quantified using the associated ImageQuant TL analysis software (GE Healthcare), with target protein levels normalized using the EF-Tu loading control bands.

3.6 MreB Morphology Assays in *B. subtilis*

MreB is an actin homolog with a critical role in maintaining rod-shaped cell morphology in a wide range of bacterial species. MreB forms dynamic helical filaments around the periphery of the cell, perpendicular to the long axis (Fig. 4B). These filaments act to recruit and guide peptidoglycan synthetic complexes as they deposit the cell wall. Wild-type cells have rod-shaped morphology when visualized by microscope (Fig. 4C), whereas *mreB* knockout cells are swollen and amorphous as a result of uncoordinated cell wall synthesis (Errington, 2015). These clear visual phenotypes make MreB a suitable target for demonstrating orthogonal riboswitch tools. The protocol below describes how we grow and image *B. subtilis* cells to assess the effects of *mreB* expression (Wu et al., 2015). The *B. subtilis mreB* knockout strain 3725 was a gift from Prof. Jeff Errington (Newcastle University, UK).

1. Aliquot 1850 mL of NB media into six 15-mL Falcon tubes and then add 50 µL of autoclaved 1 M MgCl$_2$ to each (25 mM final concentration), using sterile technique. [Note: Media is supplemented with magnesium (or 0.5% xylose) to allow normal growth of Δ*mreB* cells.]
2. Add 100 µL of sterile 40 mM riboswitch ligand in H$_2$O to three tubes (2 mM final media concentration) and 100 µL of sterile H$_2$O to the rest.
3. Inoculate NB media using glycerol stocks of the *B. subtilis* strains 168 (wild type), 3725 (Δ*mreB*), and 3725 transformed with the riboswitch-*mreB* construct.
4. Grow cultures for 5 h at 37°C with shaking at 180 rpm in a shaker-incubator.
5. Pellet cells by centrifugation at $2400 \times g$ for 5 min.

6. Gently resuspend cells in 2 mL of fresh NB media (without $MgCl_2$ or ligands).
7. Repeat steps 5–6 twice to remove residual magnesium.
8. Incubate resuspended cultures for 30 min at 37°C with shaking at 180 rpm in a shaker-incubator.
9. Remove cultures from shaker-incubator and incubate at 37°C for 6 h without shaking, to allow morphological phenotypes to develop.
10. Pellet cells by centrifugation at $2400 \times g$ for 5 min.
11. Gently resuspend in 100 μL of media.
12. Transfer 5 μL to a poly-L-lysine coated glass slide (eg, Sigma-Aldrich, #P0425).
13. Visualize cells at $60 \times$ magnification using a phase-contrast microscope (eg, Olympus BX51) to take photographic images (Fig. 4C).
14. *Analysis*: Cell dimensions can be measured in captured images using suitable software (eg, ImageJ, NIH USA). By measuring a sufficient number of cells ($n = > 100$) for each incubation, mean cellular dimensions (length/width ratios) and standard deviations can be presented, and significance tests conducted.

3.7 Construction of Chimeric Riboswitches to Couple Orthogonal Aptamers to Host-Specific Expression Platforms

The *add*(M6″) and *queC*(M1) orthogonal riboswitches detailed earlier have been demonstrated to work in *E. coli* (Gram negative) and *B. subtilis* (Gram positive), respectively. They should also function in closely related bacteria, which will have similar gene expression machinery operating in similar cellular contexts (indeed the *add* switch was originally taken from *V. vulnificus*). The purine and type-I $PreQ_1$ riboswitches have wide species distributions (Barrick & Breaker, 2007; McCown, Liang, Weinberg, & Breaker, 2014) and could be targeted with the same orthogonal mutations if found in your species of interest. However, an alternative solution would be to exploit the inherent modular architecture of riboswitches, to fuse existing orthogonal aptamers with expression platforms which are native to your target bacterial host, discoverable through the Rfam database (Nawrocki et al., 2015). Not only does this expand the transferability of our existing functionally validated orthogonal aptamer parts, but it also offers the freedom to combine them with expression platforms with diverse regulatory outputs (ON or OFF switches, transcriptional or translational mechanisms, RNA self-cleavage, etc.). To demonstrate this novel approach, we created a chimeric riboswitch comprised of the orthogonal M6″ aptamer fused to the

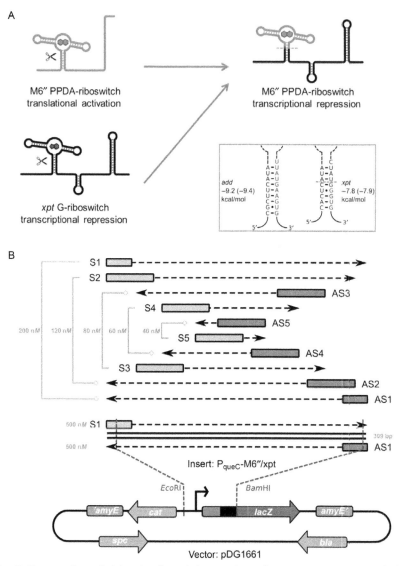

Fig. 5 Construction of chimeric riboswitches as bespoke gene expression tools for diverse bacterial species. (A) A repressible chimeric riboswitch was created by fusing the orthogonal add(M6″) aptamer with the transcriptional OFF expression platform from the native xpt riboswitch of B. subtilis. The parent riboswitches have different P1 stem strengths (*inset*), but by preserving the switching sequence (*blue*) and P1 stem strength of the xpt riboswitch, a functional chimeric riboswitch was constructed (termed add(M6″)/xpt). (B) Synthesis of a constitutive promoter-driven chimeric riboswitch construct by the TBIO PCR method (Gao, Yo, Keith, Ragan, & Harris, 2003). The final PCR product was restriction cloned directly into the B. subtilis integrative vector pDG1661 for functional validation by the β-galactosidase assay. Following validation, the promoter–riboswitch construct was fused upstream of the native B. subtilis morphology gene mreB by overlap extension PCR. (See the color plate.)

expression platform from the guanine-sensing *xpt* riboswitch from *B. subtilis* (Mandal, Boese, Barrick, Winkler, & Breaker, 2003). Effectively transforming a "translational ON" switch from a Gram-negative bacterium into a "transcriptional OFF" switch which functions in Gram positives (Robinson et al., 2014; Fig. 5A). Provided that the switching sequence shared by the expression platform is maintained, and the helical strength of the P1 stem is close to that of the aptamer being replaced, this strategy can be applied to the creation of chimeric riboswitches from the wide pool of aptamer and expression platform parts so far characterized, as has been elegantly demonstrated by the Batey group (Ceres, Garst, et al., 2013; Ceres, Trausch, et al., 2013). A construct comprised of the constitutive *queC* promoter and chimeric M6″/*xpt* riboswitch was created through the TBIO PCR synthesis method (Gao et al., 2003; Fig. 5B) described later. Following synthesis, the construct was functionally validated using the *lacZ* assay and then used to control the native *B. subtilis* morphology gene *mreB*. Suggested kits, retailers, and catalogue numbers introduced in Section 3.1 have been omitted.

Design: A 309-bp construct comprised of the constitutive promoter from the *B. subtilis queCDEF* operon (Reader, Metzgar, Schimmel, & de Crécy-Lagard, 2004), the orthogonal M6″ aptamer domain, and the *B. subtilis xpt* riboswitch expression platform was designed with flanking *Eco*RI and *Bam*HI cleavage sites. The DNA Works server (Hoover & Lubkowski, 2002) was used to design a series of overlapping primers with matching T_m values (60 nt in length with 25–30 nt overlapping regions; see Table 1), for PCR synthesis of this construct (Fig. 5).

1. *First-round PCR*: Set up a 50-µL PCR reaction using a high-fidelity DNA polymerase and the following primer pairs S1-AS1 (200 n*M* each), S2-AS2 (120 n*M*), S3-AS3 (80 n*M*), S4-AS4 (60 n*M*), and S5-AS5 (40 n*M*).
2. *Second-round PCR*: Set up a second 50-µL PCR reaction using primers S1 and AS1 alone (500 n*M* each) and 1 µL of the first-round PCR as template. Purify the PCR product and then determine the PCR product yield using a spectrophotometer.
3. *Cloning, sequence verification, B. subtilis transformation, and confirmation of chromosomal integration*: The PCR-synthesized product above is cloned into the pDG1661 vector upstream of the *spoVG-lacZ* gene exactly as described in steps 4–7 then steps 9–10 of Section 3.3.1 (ie, without the mutagenesis step). The successfully integrated construct will give a

DNA band 6350 bp in length, when PCR amplified from a gDNA prep using primers amyE-f and amyE-r.

B. *subtilis* cells which pass the verification steps are ready for functional validation of riboswitch activity using the β-galactosidase assay. Prepare glycerol stocks by aliquoting 0.5 mL of overnight LB-Cam culture into 2-mL CryoTubes, mix with 0.5 mL glycerol by pipetting gently, then freeze immediately in liquid nitrogen, and store at −80°C. The cloned P_{queC^-}M6″/*xpt* region can be subcloned in front of any gene of interest (such as *mreB*), to derive B. *subtilis* integrative constructs, by following the protocol outlined earlier (see Section 3.3.2). This simple process requires redesign of just one gene-specific primer.

4. CONCLUDING REMARKS

Riboswitches are coming of age as tools for the regulation of gene expression. However, for many applications it is desirable to create orthogonal riboswitches that function independently of the cellular environment of the target organism and that are not affected by the native ligands which are present in cells. We have already demonstrated that natural aptamers with known binding sites may be reengineered to respond to synthetic ligands (Dixon et al., 2010, 2012; Vincent et al., 2014; Wu et al., 2015), and in this article we delve further into the methods that may be used for reengineering aptamers following a rational and less resource-intensive procedure. Furthermore, we have shown that orthogonal reengineered aptamers may be combined with natural expression platforms from distantly related species and used to control the expression of both heterologous and native genes, making them useful tools for gene functional analysis and antimicrobial target validation. Our methods may be used when: (i) there is a known natural riboswitch in the target organism, or a closely related species, whose aptamer is amenable to rational mutation or (ii) to combine a previously validated orthogonal aptamer from a different organism with a native expression platform to create a novel chimeric riboswitch tailored for the microbial target.

ACKNOWLEDGMENTS

This work was supported by BBSRC Grants BB/I012648/1 and BB/M017702/1 (SYNBIOCHEM).

REFERENCES

Baba, T., Ara, T., Hasegawa, M., Takai, Y., Okumura, Y., Baba, M., ... Mori, H. (2006). Construction of *Escherichia coli* K-12 in-frame, single-gene knockout mutants: The Keio collection. *Molecular Systems Biology, 2*(1), 0008.

Baker, J. L., Sudarsan, N., Weinberg, Z., Roth, A., Stockbridge, R. B., & Breaker, R. R. (2012). Widespread genetic switches and toxicity resistance proteins for fluoride. *Science, 335*(6065), 233–235.

Barrick, J. E., & Breaker, R. R. (2007). The distributions, mechanisms, and structures of metabolite-binding riboswitches. *Genome Biology, 8*(11), R239.

Breaker, R. R. (2011). Prospects for riboswitch discovery and analysis. *Molecular Cell, 43*(6), 867–879.

Ceres, P., Garst, A. D., Marcano-Velázquez, J. G., & Batey, R. T. (2013). Modularity of select riboswitch expression platforms enables facile engineering of novel genetic regulatory devices. *ACS Synthetic Biology, 2*(8), 463–472.

Ceres, P., Trausch, J. J., & Batey, R. T. (2013). Engineering modular 'ON' RNA switches using biological components. *Nucleic Acids Research, 41*(22), 10449–10461.

Cheah, M. T., Wachter, A., Sudarsan, N., & Breaker, R. R. (2007). Control of alternative RNA splicing and gene expression by eukaryotic riboswitches. *Nature, 447*(7143), 497–500.

Cromie, M. J., Shi, Y., Latifi, T., & Groisman, E. A. (2006). An RNA sensor for intracellular Mg^{2+}. *Cell, 125*(1), 71–84.

Dixon, N., Duncan, J. N., Geerlings, T., Dunstan, M. S., McCarthy, J. E. G., Leys, D., & Micklefield, J. (2010). Reengineering orthogonally selective riboswitches. *Proceedings of the National Academy of Sciences of the United States of America, 107*(7), 2830–2835.

Dixon, N., Robinson, C. J., Geerlings, T., Duncan, J. N., Drummond, S. P., & Micklefield, J. (2012). Orthogonal riboswitches for tuneable coexpression in bacteria. *Angewandte Chemie, International Edition, 51*(15), 3620–3624.

Du Toit, A. (2015). Bacterial genetics: Metalloregulatory riboswitches. *Nature Reviews. Microbiology, 13*(5), 249.

Errington, J. (2015). Bacterial morphogenesis and the enigmatic MreB helix. *Nature Reviews. Microbiology, 13*(4), 241–248.

Gao, X., Yo, P., Keith, A., Ragan, T. J., & Harris, T. K. (2003). Thermodynamically balanced inside-out (TBIO) PCR-based gene synthesis: A novel method of primer design for high-fidelity assembly of longer gene sequences. *Nucleic Acids Research, 31*(22), e143.

Guérout-Fleury, A.-M., Frandsen, N., & Stragier, P. (1996). Plasmids for ectopic integration in *Bacillus subtilis*. *Gene, 180*(1–2), 57–61.

Harvey, I., Garneau, P., & Pelletier, J. (2002). Inhibition of translation by RNA-small molecule interactions. *RNA, 8*(4), 452–463.

Harwood, C. R., & Cutting, S. M. (Eds.), (1990). *Molecular biological methods for* Bacillus. Chichester, N.Y.: Wiley.

Hoover, D. M., & Lubkowski, J. (2002). DNAWorks: An automated method for designing oligonucleotides for PCR-based gene synthesis. *Nucleic Acids Research, 30*(10), e43.

Lynch, S. A., Desai, S. K., Sajja, H. K., & Gallivan, J. P. (2007). A high-throughput screen for synthetic riboswitches reveals mechanistic insights into their function. *Chemistry & Biology, 14*(2), 173–184.

Mandal, M., Boese, B., Barrick, J. E., Winkler, W. C., & Breaker, R. R. (2003). Riboswitches control fundamental biochemical pathways in *Bacillus subtilis* and other bacteria. *Cell, 113*(5), 577–586.

McCown, P. J., Liang, J. J., Weinberg, Z., & Breaker, R. R. (2014). Structural, functional, and taxonomic diversity of three PreQ1 riboswitch classes. *Chemistry & Biology, 21*(7), 880–889.

Meyer, A., Pellaux, R., Potot, S., Becker, K., Hohmann, H.-P., Panke, S., & Held, M. (2015). Optimization of a whole-cell biocatalyst by employing genetically encoded product sensors inside nanolitre reactors. *Nature Chemistry*, *7*(8), 673–678.

Michener, J. K., Thodey, K., Liang, J. C., & Smolke, C. D. (2012). Applications of genetically-encoded biosensors for the construction and control of biosynthetic pathways. *Metabolic Engineering*, *14*(3), 212–222.

Miller, J. H. (1972). *Experiments in molecular genetics*. Cold Spring Harbor, N.Y.: Cold Spring Harbor Laboratory.

Morra, R., Shankar, J., Robinson, C. J., Halliwell, S., Butler, L., Upton, M., ... Dixon, N. (2016). Dual transcriptional-translational cascade permits cellular level tuneable expression control. *Nucleic Acids Research*, *44*(3), e21.

Nawrocki, E. P., Burge, S. W., Bateman, A., Daub, J., Eberhardt, R. Y., Eddy, S. R., ... Finn, R. D. (2015). Rfam 12.0: Updates to the RNA families database. *Nucleic Acids Research*, *43*(D1), D130–D137.

Reader, J. S., Metzgar, D., Schimmel, P., & de Crécy-Lagard, V. (2004). Identification of four genes necessary for biosynthesis of the modified nucleoside queuosine. *Journal of Biological Chemistry*, *279*(8), 6280–6285.

Robinson, C. J., Vincent, H. A., Wu, M.-C., Lowe, P. T., Dunstan, M. S., Leys, D., & Micklefield, J. (2014). Modular riboswitch toolsets for synthetic genetic control in diverse bacterial species. *Journal of the American Chemical Society*, *136*(30), 10615–10624.

Sambrook, J., & Russell, D. W. (2001). *Molecular cloning: A laboratory manual* (3rd ed.). Cold Spring Harbor, N.Y.: Cold Spring Harbor Laboratory Press.

Serganov, A., & Nudler, E. (2013). A decade of riboswitches. *Cell*, *152*(1–2), 17–24.

Stoltenburg, R., Reinemann, C., & Strehlitz, B. (2007). SELEX—A (r)evolutionary method to generate high-affinity nucleic acid ligands. *Biomolecular Engineering*, *24*(4), 381–403.

Suess, B., Fink, B., Berens, C., Stentz, R., & Hillen, W. (2004). A theophylline responsive riboswitch based on helix slipping controls gene expression in vivo. *Nucleic Acids Research*, *32*(4), 1610–1614.

Tang, J., & Breaker, R. R. (1997). Rational design of allosteric ribozymes. *Chemistry and Biology*, *4*(6), 453–459.

Vincent, H. A., Robinson, C. J., Wu, M.-C., Dixon, N., & Micklefield, J. (2014). Generation of orthogonally selective bacterial riboswitches by targeted mutagenesis and in vivo screening. In A. Ogawa (Ed.), *Artificial riboswitches: Vol. 1111* (pp. 107–129). Totowa, NJ: Humana Press.

Wu, M.-C., Lowe, P. T., Robinson, C. J., Vincent, H. A., Dixon, N., Leigh, J., & Micklefield, J. (2015). Rational re-engineering of a transcriptional silencing PreQ1 riboswitch. *Journal of the American Chemical Society*, *137*(28), 9015–9021.

Zhou, L.-B., & Zeng, A.-P. (2015). Exploring lysine riboswitch for metabolic flux control and improvement of L-lysine synthesis in Corynebacterium glutamicum. *ACS Synthetic Biology*, *4*(6), 729–734.

AUTHOR INDEX

Note: Page numbers followed by "*f*" indicate figures, and "*t*" indicate tables.

A

Aach, J., 202–203
Abatemarco, J., 154
Abil, Z., 273–274
Abou-Hadeed, K., 83–84, 86–88
Achcar, F., 264–265
Adams, P.D., 249–251, 265–266, 313
Adamson, C., 15–16
Agarwal, V., 11, 83–84, 86–88
Ahmad, I., 8–9
Ahmad, I.Z., 248–249
Ahn, J.S., 42
Ahuja, M., 128–129
Aisaka, K., 95–96
Ajikumar, P.K., 180–182, 226–229, 238–239, 249–252, 262, 295, 313
Akhtar, M.K., 263
Albermann, C., 24
Alcantara, J., 149
Aldrich, C.C., 22–24
Algar, R., 295
Alkhalaf, L.M., 22–34
Allen, R.S., 151, 197–199
Allwood, J.W., 167–168, 265–266
Alonso-Gutierrez, J., 249–251, 262–263, 265–266, 286–287
Alper, H.S., 154, 182–183, 202, 226–228, 286–287
Alphey, L., 290–291, 313
Altmann, A., 78–81, 95–96
Amar, P., 32
Ammer, C., 146–148
Andersen, R.J., 24, 26*f*
Anderson, D.G., 273
Anderson, J.C., 181–182
Andorfer, M.C., 69–70, 70*t*, 72–74, 75*t*, 93–122
Annaluru, N., 313–314
Ansari, S., 248–249
Antebi, A., 209
Antoniewicz, M.R., 184–185
Antunes, M.S., 184

Ara, T., 331
Arentshorst, M., 130
Arkin, A.P., 181–182, 273
Arnaud, M.B., 128–129
Arnison, P.G., 2–3
Arnold, F.H., 1–2, 95–96, 98–99, 107, 109, 116, 119
Aron, Z.D., 87–88
Artsatbanov, V.Y., 238–239
Aryal, N., 86
Asai, K., 128–129
Asamizu, S., 22–24
Ashcroft, J., 40–42
Asokan, A., 128–129
Atanasov, A.G., 248
Auclair, K., 42
Auffray, Y., 185–186

B

Baba, M., 331
Baba, T., 331
Baccile, J.A., 43*f*
Baek, J.-M., 183–184
Baeza-Yates, R., 155
Bai, C., 32
Baidoo, E.E., 286–287, 295–296
Baidoo, E.E.K., 262–263
Baker, J.L., 320
Baldwin, G.S., 287
Baltz, R.H., 27–28, 271–272
Barbas, C.F., 273
Barber, C.J., 150–151, 199–200
Barilone, N., 22
Barken, I., 196–197
Barrett, M.P., 265–266
Barretto, R., 273
Barrick, J.E., 320, 343–345
Barrows, L.R., 42
Barton, K.W., 32
Bastian, S., 109
Bateman, A., 343–345
Batey, R.T., 321–323, 343–345

349

Battershill, C.N., 5–8
Batth, T.S., 249–251, 265–266, 286–287, 295–296, 313
Bauer, J.D., 28–29
Baulcombe, D.C., 150–151
Bazzicalupi, C., 196–197
Beaudoin, G.A.W., 144–148, 151, 153, 172–173, 196–203, 215–216, 219
Becker, K., 321
Becker, S.A., 264–265
Bellen, H.J., 290–291, 294, 313
Benkovic, S.J., 8–9
Bennett, C.S., 109
Bennett, G.N., 226–228
Bent, A.F., 11–12, 15–16
Berens, C., 321–322
Berényi, S., 144–145
Bernard, P., 291 293, 305, 307
Bernards, M., 153–154
Bernhardt, P., 94
Bernheim, R., 253–254
Bessi, I., 196–197
Bhan, N., 180–184, 187–188
Bharate, S.B., 22
Bhattarai, S., 86
Bian, X., 28–29
Bibb, L.A., 288–290, 312–313
Bibb, M.J., 2–3, 27–30, 32–33, 48–50, 272–273
Bierbaum, G., 2–3
Biessels, H.W.A., 146–148
Biggs, B.W., 226
Bikard, D., 273
Bingman, C.A., 73–74
Binkley, J., 128–129
Bird, D.A., 149
Bister, B., 66, 68–71, 71t, 95f, 96, 109
Bitto, E., 73–74
Bitzer, D., 252
Bizzini, A., 185–186
Blasiak, L.C., 73–74, 95–96, 114–115
Blazeck, J., 182–183
Bloch, P.L., 185–186
Blodgett, J.A., 279–280
Bloom, J.D., 98–99
Blount, B.A., 251–252
Boeke, J., 313–314
Boël, G., 153–154

Boese, B., 343–345
Boettcher, C., 168
Bogart, J.W., 43f
Bohlmann, J., 196–200
Bommarius, A.S., 98–99
Bonde, I., 202–203, 217
Bonde, M.T., 162–163
Bonnet, J., 313–314
Borchers, C.H., 149, 155
Bottriell, H., 22, 24
Bouchier, C., 3
Bourgault, R., 149
Bourgeois, L., 196–221
Bouwmeester, H., 249
Bower, A.G.W., 180–182
Bowers, A.A., 2–3, 43f
Bradford, M.M., 306
Brady, S.F., 22–30
Braña, A.F., 24–25, 33, 66, 69, 96–97, 122
Brasch, M.A., 287–288
Breaker, R.R., 320–322, 343–345
Breitling, R., 251, 264–266
Breüner, A., 288–290, 312–313
Brewster, R.C., 182–183
Broering, J.S., 98–99
Brøndsted, L., 288–290, 312–313
Bronine, A., 196–197, 210–211
Bross, C.D., 149, 155, 156t, 159–160, 172–173
Brown, S., 168
Brown, W.R., 290
Bruner, S.D., 83–84, 88
Buchanan, G.O., 93–94
Buck, D.E., 290–291, 298, 312–313
Budin-Verneuil, A., 185–186
Bugni, T.S., 2–3
Buhler, B., 262–263
Bulaj, G.W., 2–3, 11
Bull, D., 146–148
Bumpus, S.B., 87–88
Bunders, C.A., 93–94, 116–119
Bünner, C.M., 286–287
Burd, W., 73–74, 75t, 78–79, 97
Burge, S.W., 343–345
Burkart, M.D., 84–85, 87–88
Burns, D.C., 144–148, 151, 172–173, 199–200
Burns, R.N., 294

Buron, L.D., 161–162
Burton, E., 196–221
Butler, A., 94
Butler, L., 266, 321
Butovich, I.A., 33, 66, 69
Butterfield, D.A., 67–68
Buttner, M.J., 28–29, 32–33, 48–50, 272–273
Byrd, D.R., 287–288

C

Cai, G., 253–254
Cairns, N., 77f, 80–81, 96–97, 122
Calderone, C.T., 87–88
Calle, P.Y., 22–30
Calteau, A., 3
Cambiazo, V., 155
Cambray, G., 182–183
Campos, R., 155
Caputi, L., 43f
Carbajo, R.J., 25
Carbonell, P., 264–265
Carlsen, S., 162–163, 226, 249, 251–252
Carlson, J.C., 107
Carothers, J.M., 184, 286–287
Carpenter, E.J., 150–151, 155, 156t, 172–173
Carr, P.A., 295
Carter, O.A., 249–251, 262
Carter-Franklin, J.N., 94
Cascante, M., 263–264
Cascio, V.M., 273
Casini, A., 287
Cautha, S.C., 154
Cavalcanti, S.M.T., 93–94
Cavanagh, J., 93–94, 116–119
Ceres, P., 321–323, 343–345
Ceroni, F., 295
Cerqueira, G.C., 128–129
Chai, H.-B., 47–48
Chakrabarty, R., 150–151, 199–200
Challis, G.L., 27–28, 30–31, 272–273
Chan, A.N., 43f
Chan, R., 249–251, 265–266, 286–287, 313
Chang, F.-Y., 22–30
Chang, J.C., 290–291, 313
Chang, L., 144–173, 156t
Chang, M.C.Y., 249–251, 264
Chang, S.L., 128–129

Channarasappa, A.S., 302
Chao, Y.-P., 184–185
Chaparro-Riggers, J.F., 98–99
Chari, R., 184
Charlop-Powers, Z., 24, 29–30
Charpentier, E., 202–203
Charpin-El Hamri, G., 313–314
Charusanti, P., 30, 277, 281–282
Chater, K.F., 28–33, 48–50, 272–273
Chavez, C.L., 290–291, 313
Chávez, M.L.D., 212
Cheah, M.T., 320
Cheallaigh, A.N., 248–267, 250f
Chemler, J.A., 249–251
Chen, B., 273–274
Chen, C.-M., 261
Chen, J., 182–183
Chen, M.M., 109, 116, 119
Chen, R., 207–208
Chen, X., 144–173, 199–200
Chen, Y., 48–50, 271–272
Cheo, D.L., 287–288
Cherenson, A.R., 209
Cherouati, N., 209
Chesnut, J.D., 287–288
Chiang, C.-J., 184–185
Chiang, Y.-M., 128–129
China, H., 99, 106–107
Chisholm, J.D., 24
Chitty, J.A., 151, 197–199
Choi, J.W., 42
Choi, K.B., 145–146
Choi, S.-H., 43f, 180
Chokkathukalam, A., 265–266
Chooi, Y.-H., 42
Chow, T., 207–208
Christoffersen, M.J., 182–183
Chu, H.C., 262–263
Chu, J., 48–50
Chuang, R.-Y., 181–182, 253–254, 287
Chubukov, V., 262–263
Chueca, B., 262–263
Church, G.M., 202–203, 273
Claassens, N.J., 153–154
Clark, I.P., 266
Clastre, M., 168
Cobb, R.E., 271–283
Coelho, P.S., 95–96, 109

Cole, S.T., 22
Collens, J., 183
Collins, C.H., 107, 180–182
Collins, S.M., 180–189
Colloms, S.D., 286–314
Cong, L., 273
Connolly, M.L., 209
Connors, N.C., 184
Conrado, R.J., 153–154
Contiero, J., 185–186
Corbin, D., 82
Corey, E.J., 40
Cornelissen, S., 262–263
Cornish, V.W., 107
Corrales-Sánchez, V., 25
Corre, C., 272–273
Correa, E., 265–266
Courdavault, V., 168
Couturier, M., 291–293, 305, 307
Covello, P.S., 196–200
Cox, D., 273
Cragg, G.M., 1–2
Cram, D., 155, 156t, 159–160
Creek, D.J., 265–266
Cress, B.F., 180–184
Cromie, M.J., 320
Crook, N.C., 116, 119
Croteau, R.B., 249–251, 250f, 253, 260–262
Cserjan-Puschmann, M., 182–183
Csutorás, C., 144–145
Cuellar, M.C., 226
Cuenca-López, M.D., 25
Cummings, E.M., 226
Curran, K.A., 202
Currin, A., 264–265
Cushman, M., 43f
Cutting, S.M., 335

D

Da Silva, G.P., 185–186
Dahl, R.H., 286–287
Dairi, T., 95–96
Dalle-Donne, I., 67–68
Dang, T.-T.T., 144–145, 149, 159–160, 199–200
Dantes, A., 253–254
Dastmalchi, M., 144–173
Daub, J., 343–345

David, F., 161–165
Davidson, A.D., 128–129
Davies, J., 22, 24
Davis, E.M., 249, 250f, 253, 260–261
Davisson, V.J., 14
Day, P.J., 264–265
de Azevedo Junior, W.F., 93–94
de Crécy-Lagard, V., 345–346
De Kok, S., 181–182, 253–254
De Laurentis, W., 73–74
De Luca, V., 145–146, 168, 196–200
de Marsac, N.T., 3
De Mey, M., 182–183, 226, 251
De Paepe, B., 226
Decker, G., 145–146
DeLisa, M.P., 153–154
DeLoache, W.C., 152–154, 168, 172–173, 197–203, 207–208, 210, 214–215, 217–220
Deng, W., 48–50, 55
Deng, X., 22
Densmore, D., 287
Derewenda, Z.S., 98–99
Desai, S.K., 337–340
Desgagné-Penix, I., 144–145, 149–151, 155, 156t, 159–160, 172–173, 199–200
Dezeny, G., 48–49, 54–55
Dhaubhadel, S., 153–154
di Bernardo, M., 287
Díaz Chávez, M.L., 146–148
DiCarlo, J.E., 202–203
Diethelm, S., 94
Ding, T., 24, 26f, 28–29, 33
Ding, W., 40–42, 49, 55
Ding, X., 287–288
Dinsmore, D.R., 149, 155
Ditta, G., 82
Dittrich, H., 146–148
Dixon, N., 321–324, 326–331, 333–334, 337–340, 342–343, 346
Doi, N., 296–297
Dong, C.J., 67–68, 70–71, 73–74, 82, 109
Donia, M.S., 3–9, 11
Donninger, S.L., 199–200
Donovan, R.S., 186–187
Döring, K., 271–272
Dorrestein, P.C., 83–85, 87–88
Dossani, Z.Y., 262–263

Doudna, J.A., 202–203, 273
Drennan, C.L., 22–24, 73–74, 95–96, 114–115
Drummond, S.P., 328, 346
Du, G., 182–183
Du, J., 153–154
Du, Y.-L., 22–34, 26f
Dubouzet, E., 212
Dueber, J.E., 152–154, 168, 172–173, 181–184, 196–203, 207–208, 210, 214–215, 217–220, 286–287
Duetz, W.A., 249
Duffy, S.P., 22
Duncan, J.N., 321–322, 326–331, 346
Dunlap, W.C., 5–8
Dunlop, M.J., 262–263
Dunn, K.W., 207–208
Dunstan, M.S., 321–323, 326–333, 337–340, 343–346
Durot, M., 181–182, 253–254
Du Toit, A., 320
Dvora, H., 187–188
Dymond, J., 313–314

E

Easson, M.L., 168
Easterbrook, M., 183
Eberhardt, R.Y., 343–345
Eddy, S.R., 343–345
Edgar, S., 184–185, 226–244
Edgerton, H., 136
Eggleston, P., 290–291, 313
Ehmann, D.E., 83–84
Eichhorn, E., 66, 69, 95f, 96, 109
Eisen, J.A., 2–8, 11
Eisenreich, W., 43f
Ekins, A., 153, 168, 172–173, 196–203, 209, 212–213, 215–216, 219
El Gamal, A.A., 83–84, 86–88
Elder, J.H., 209
El-Hiti, G.A., 93–94
Ellington, A.D., 249–251
Elliott, S.J., 22–24
Ellis, D.I., 265–266
Ellis, T., 251–252, 287, 295
Endy, D., 181–182, 287, 313–314
Enghiad, B., 271–272
Englaender, J.A., 182–183

Engler, C., 181–182, 287
English, R.S., 48–50
Entwistle, R., 128–129
Eriksen, D.T., 153–154
Ermakov, S., 286–287, 295–296
Ernyei, A.J., 68, 95–96
Errington, J., 342–343
Erxleben, A., 271–272
Espah Borujeni, A., 302
Estrada, S.A., 93–94
Esvelt, K.M., 107, 273
Eustáquio, A.S., 28–29

F

Faber, K., 108
Facchini, P.J., 144–173, 156t, 196–200
Falgueyret, J.-P., 153, 168, 172–173, 196–203, 209, 212–213, 215–216, 219
Fang, J., 44–46
Farasat, I., 183
Fareed, S., 248–249
Farmer, W.R., 249–251
Farrow, S.C., 144–148, 151, 155, 156t, 159–160, 172–173, 199–200
Farruggio, A.P., 290–291, 313
Fasan, R., 109, 116, 119
Fass, R., 185–186
Fattorusso, E., 94, 96
Fedorova, N.D., 128–129
Feist, A.M., 264–265
Feling, R.H., 93–94
Fellows, B., 221
Feng, J., 184
Feng, Z., 28–29
Fenical, W., 93–94
Fernández, L., 107–108
Fernandez, P., 22
Fernández-Ábalos, J.M., 32
Fernandez-Rodriguez, J., 312–314
Fewer, D.P., 2–3
Filsinger Interrante, M., 152–153, 168, 172–173
Fink, B., 321–322
Fink, G.R., 209
Finn, R.D., 343–345
Fischbach, M.A., 1–2
Fischer, C., 286–287
Fischer, I., 78–81, 95–96

Fisher, K., 263
Fishman, A., 109
Fisinger, U., 196–197
Fist, A.J., 149, 151, 197–199
Flecks, S., 67–68, 70–71, 73–74, 82, 95–96, 109
Fogg, P.C., 313
Fonseca, B., 155
Forest, C.R., 295
Fossati, E., 153, 168, 172–173, 196–221
Fowler, K., 30–31
Fox, D.J., 27–28
Fox, R.J., 103
Franceschi, V.R., 149
Frandsen, N., 333–334
Frank, A., 116
Fraser, P.D., 297
Freeman, G., 253–254
Freestone, T.S., 182–183
Freitag, M., 22
Frese, M., 70–73, 71f, 75t, 97–98, 108, 122
Frick, S., 149, 197–199
Friedland, G.D., 286–287
Friedman, D.C., 249–251
Fritz, H., 86
Froese, J., 168
Fu, J., 28–29
Fu, L., 180–183, 185–188
Fu, P., 22
Fu, X., 24
Fu, Y., 109, 273
Furumai, T., 24
Fussenegger, M., 313–314

G

Gage, D., 97–98
Gaj, T., 273
Gajewi, M., 66–89
Galanie, S., 152–154, 168, 172–173, 181, 197–203, 215–217, 219–220
Gallivan, J.P., 337–340
Galm, U., 28–29
Galonic, D.P., 84
Gao, X., 343–345, 344f
Garcia, D.E., 286–287
García-Gonzalo, D., 262–263
Gardiner, J.M., 248–267
Garneau, P., 321–322

Garneau, S., 22–24, 70–71, 71t, 95f, 96–98
Garneau-Tsodikova, S., 83–88, 95–96
Garst, A.D., 321–323, 343–345
Gascón-Escribano, M.J., 25
Gautsch, J.W., 209
Gazda, V., 150–151, 172–173, 199–200
Gebreselassie, N.A., 184–185
Gee, P., 273
Geering, B., 313–314
Geerlings, T., 321–322, 326–331, 346
Geib, N., 84, 86–88
Geisinger, J.M., 290–291, 313
Geller, J., 196–197
Genee, H.J., 162–163
Genga, R.M., 273
Gerlach, W.L., 197–199
Gersbach, C.A., 273
Gershenson, A., 119
Gerwick, W.H., 94, 96
Gesell, A., 146–148, 212
Geu-Flores, F., 43f, 161–162
Gewain, K.M., 48–49, 54–55
Ghislieri, D., 116
Ghosh, P., 288–290, 312–313
Giard, J.-C., 185–186
Giavalisco, P., 167–168
Giaw Lim, C., 251
Gibbons, P.H., 48–49, 54–55
Gibson, D.G., 181–182, 253–254, 287
Gicquel, B., 22
Gietz, R.D., 164–166
Gilbert, L.A., 273
Gilson, C.A., 93–94
Glenn, W.S., 97–98, 114–115, 197–199
Glick, B.R., 186–187
Glieder, A., 146–148
Gocht, M., 84–85
Goh, S., 30
Gold, N.D., 154, 196–221
Goldman, P.J., 22–24
Goler, J.A., 181–182
Gomez-Escribano, J.P., 27–28, 272–273
Gomez-Juarez, M., 25
Goncharenko, A.V., 238–239
Gonzales, A.M., 152, 168, 172–173, 197–203, 207–208, 210, 214–215, 217–220
Gonzalez, D.J., 28–29
González, S., 155

Good, L., 30
Goodacre, R., 167–168
Gootenberg, J.S., 273
Goss, R.J.M., 77f, 80–81, 96–97, 122
Goto, S., 264–265
Goto, Y., 11, 15–16
Gottardi, E.M., 47–48
Gottelt, M., 272–273
Gowen, C.M., 154
Green, A.P., 116
Green, T., 153–154
Greenstein, M., 40–42
Gregory, M.A., 294
Gribble, G.W., 94
Grider, J.S., 221
Grobe, N., 196–197
Groisman, E.A., 320
Grothe, T., 146–148
Gruetzner, R., 287
Grüning, B.A., 271–272
Grüschow, S., 77f, 80–81, 96–97, 122
Gu, Q., 180–182
Guérout-Fleury, A.-M., 333–334
Gugger, M., 3
Guido, M., 183
Guimaraes, J.C., 182–183
Guleria, S., 180, 182–183
Gupta, A., 184, 202
Gust, B., 28–31
Guzowska, P.H., 70–71, 75t, 97–98

H

Habib, N., 273
Haft, D., 128–129
Hagel, J.M., 144–173, 156t, 196–197, 199–200
Hager, L.P., 94
Hahn, F., 44–46
Hajimorad, M., 181–182
Halkier, B.A., 161–163
Hall, E., 253–254
Halliwell, S., 266, 321
Halmos, B., 22
Hamill, M.J., 22–24
Hammer, K., 288–290, 312–313
Hammer, P.E., 78–82, 95–96
Han, B., 149, 155
Han, J., 149, 155

Hancox, M.I., 288–290, 312–313
Hannum, G., 264–265
Hansen, B.G., 161–163
Hansen, E.H., 286–287
Hansen, N.B., 162–163
Hardman, S.J., 266
Harris, C.M., 93–94
Harris, T.K., 343–345, 344f
Harris, T.M., 93–94
Harrison, R.G., 253
Harrison, S.J., 202–203, 217
Hartley, J.L., 287–288
Hartner, F., 146–148
Hartog, A.F., 99
Hart-Wells, E.A., 249–251
Harvey, I., 321–322
Harwood, C.R., 335
Hasan, Z., 99
Hasegawa, M., 95–96, 331
Hashimoto, J., 40–42, 46–47
Hashimoto, T., 40–42, 46–47, 181
Hatfull, G.F., 288–290, 312–313
Hathaway, B.J., 3, 5–9, 11
Hatscher, C., 66, 69–71, 71t, 77f, 78–80, 95f, 96
Hattori, M., 264–265
Hauck, A.F., 266
Haupt, C., 67–68, 70–71, 73–74, 82, 109
Hauschild, J.-P., 167–168
Hawkins, K.M., 152–153, 172–173, 197–199, 202–203, 217
Haygood, M.G., 2–9, 11
He, H., 47–48
He, W., 182–183
He, Y., 290–291, 294, 313
He, Z., 150–151, 172–173, 199–200
Heemstra, J.R., 9–11
Heide, L., 28–29
Heinzle, E., 258–259
Held, M., 321
Helinski, D.R., 82
Helm, S., 221
Herisson, J., 264–265
Hermann, I., 262–263
Hernandes, M.Z., 93–94
Herrera-Rodriguez, L.N., 93–94
Herrgard, M.J., 202–203, 217, 264–265
Herrmann, J., 73–74

Hertweck, C., 3
Hesketh, A., 30
Heyes, D.J., 266
Higo, T., 99, 106–107
Hill, A., 154
Hill, D.S., 78–82, 95–96
Hill, R., 182–183
Hillen, W., 321–322
Hillson, N., 287
Hilvert, D., 107
Hirai, M.Y., 249–251
Hirano, S., 22–24
Hirano, T., 40–42, 46–47
Hirota, H., 42
Ho, Q.H., 273
Ho, S.N., 183
Höbenreich, H., 107–108
Hochschild, A., 273
Hohaus, K., 66, 69, 78–79, 95f, 96, 109
Hohmann, H.-P., 321
Holm, D.K., 162–163
Holmes, V.F., 181–182, 253–254
Holtz, W.J., 181–182
Hölzer, M., 66, 69, 73–74, 75t, 95f, 96–97, 109
Hong Chen, M., 184–185
Hong, H., 44–46
Hong, S.H., 183–184
Honma, M., 43f
Hoover, D.M., 345–346
Hopper, A.K., 209
Hopwood, D.A., 28–29, 32–33, 48–50, 272–273
Horton, R.M., 183
Horwitz, A.A., 202–203, 217
Hosford, J., 106–108
Hoshford, J., 77–78
Hoskins, R.A., 290–291, 294, 313
Hotta, A., 273
Hotta, J., 287–288
Hotta, K., 25–27, 129
Houghten, R.A., 209
Houssen, W.E., 11–12, 15–16
Howard-Jones, A.R., 22–24
Hsiang, W.-S., 261
Hsu, P.D., 202–203, 273
Hu, P., 273–274
Hu, Q., 262–263
Hu, S., 28–29

Hu, Y., 32
Hua, Y., 273–274
Huang, F.-C., 28–29, 146–148
Huang, L., 44–46, 181
Huang, Q., 180–181, 188–189
Huang, S., 272–273
Huang, T., 84
Huang, Y., 73–74
Huang, Z., 14
Hudlicky, T., 168, 197–199
Hughes, G., 108
Hughes, J.M.X., 253
Huisman, G.W., 103
Humphreys, L., 248–267
Hung, L.H., 150–151, 155, 156t, 172–173
Hunt, H.D., 183
Hunter, P., 248
Hussain, M.S., 248–249
Hutchison, C.A.r., 42, 181–182, 253–254, 287

I

Ibrahim, M.H.A., 184–185, 187–188
Ichihara, A., 42
Igarashi, Y., 24
Ihlefeld, J., 73–74
Ikeda, H., 40–42, 46–47
Ikeuchi, M., 249–251
Ikezawa, N., 152–153, 196–197, 199, 212
Illarionov, B., 43f
Imoto, M., 22
Inglis, D.O., 128–129
Interrante, M.F., 197–202, 215–216, 219–220
Ireland, C., 2–3
Isaacs, F.J., 295
Ishida, K., 3, 273
Ishikawa, J., 25–27
Ishiuchi, K., 129
Itaya, M., 296–297
Ito, K., 196–197
Ito, M., 249–251
Ito, Y., 11, 15–16
Iwasa, K., 212

J

Jackson, B.E., 249–251
Jackson, M., 22
Jacobovitch, Y., 253–254

Jacobus, N.V., 40–42
Jadulco, R.C., 42
Jakočiūnas, T., 182–183, 202–203, 217, 273
Janata, J.W., 221
Janey, J.M., 116
Jang, J.-H., 42
Jankevics, A., 265–266
Jansen, G., 202, 204–205
Janso, J.E., 42
Janssen, R.H.A.M., 146–148
Jarvis, W.R., 116
Jaspars, M., 5–8
Jaulon, J.-L., 264–265
Jensen, J.K., 161–163
Jensen, M.K., 182–184, 202–203, 273
Jensen, N.B., 161–165
Jensen, O.R., 286–287
Jensen, P.R., 32, 93–94
Jervis, A., 248–267, 250f
Jester, B.W., 184
Jia, B., 271–272
Jia, C., 155, 156t
Jia, X., 40–42, 44–46, 54–55
Jiang, W., 273
Jiang, X., 226
Jin, Y.S., 226–228
Johnson, A.G., 146–148
Johnson, D.A., 209
Johnson, I., 208
Johnson, S.J., 97–98
Jokela, J., 2–3
Jones, D.L., 182–183
Jones, J.A., 180–189
Jones, K.L., 182, 297, 313
Jonker, H.R., 196–197
Joshi, M., 43f
Joung, J.K., 273
Jugie, G., 86
Julsing, M.K., 262–263
Jung, M.-Y., 183–184
Jung, W.E., 290–291, 313

K

Kagamiyama, H., 119
Kahne, D., 84, 86–88
Kajiwara, S., 297
Kakule, T.B., 42
Kallifidas, D., 28–29
Kallio, P., 263

Kamileen, M.O., 43f
Kaminski, F., 150–151, 172–173, 199–200
Kamocka, M.M., 207–208
Kan, Y.W., 290–291, 313
Kanagaki, S., 22
Kanchanabanca, C., 44–46
Kandzia, R., 181–182
Kang, H.-S., 24, 29–30
Kannan, R., 93–94
Karim, A.S., 202
Karow, M., 290–291, 313
Karthikeyan, C., 73–74, 75t, 97, 109, 115
Kast, P., 107
Katayama, K., 42
Katayama, T., 152–153, 168, 196–199, 214–215
Kato, N., 42
Kato, Y., 11, 15–16
Katsumata, R., 95–96
Kauffman, C.A., 93–94
Kawano, N., 152
Kawashima, S.A., 22
Kaysser, L., 94
Kearns, N.A., 273
Keasling, J.D., 10, 182–184, 202–203, 249–251, 262–266, 273, 286–287, 297, 313
Keefer, C.E., 93–94
Keith, A., 343–345, 344f
Kelemen, G.H., 32
Kell, D.B., 264–265
Kelleher, N.L., 83–85, 88, 279–280
Keller, N.P., 129, 139
Keller, S., 66, 69, 95f, 96, 109
Kelly, W.L., 40
Kennedy, J., 42
Kennedy, Z., 273
Keravala, A., 290–291, 313
Kerkman, R., 99
Kern, M., 146–148, 150–151, 172–173, 199–200
Kersten, R.D., 28–29, 83–84, 86–88
Khaldi, N., 128–129
Khaleel, T., 288–291, 307, 312–313
Khalidi, O., 181–184
Khan, M.F., 93–94, 149, 155, 156t
Kieser, T., 28–33, 48–50, 272–273
Kildegaard, K.R., 161–165
Kim, B., 153–154

Kim, D.H., 265–266
Kim, E., 180
Kim, H.J., 40, 43f
Kim, I., 183
Kim, J.H., 28–29
Kim, J.-S., 152–153, 168, 196–199, 214–215
Kim, R.-R., 43f
Kim, S.W., 182–183, 297, 313
Kimura, T., 22
King, A.J., 146–148, 151, 199–200
Kinoshita, H., 129
Kirner, S., 78–81, 95–96
Kiyota, H., 249–251
Knight, T.F., 181–182, 287
Koch, M., 42
Kock, G.R., 286–287
Koebmann, B., 32
Koehnke, J., 11–12, 15–16
Koeppl, H., 287
Koffas, M.A.G., 180–189, 249–251
Koglin, A., 73–74, 95–96, 114–115
Kohli, R.M., 83–84, 88
Kohlwein, S.D., 207–208
Koike, H., 128–129
Koju, D., 86
Kol, S., 272–273
Kolodziej, P.A., 209
Kondo, K., 297
Konermann, S., 273
Kopecka, H., 93–94
Koshino, H., 22–24
Koszelewski, D., 108
Kotera, M., 264–265
Kotzsch, A., 66, 68–71, 71t, 95f, 96, 109
Kouprina, N., 28–29
Koyanagi, T., 152, 168, 197–199, 214–215
Kramell, R., 146–149, 197–199
Kramer, C., 130
Krauss, S., 80
Krawczyk, J.M., 47–48
Kries, H., 43f
Kristensen, C., 286–287
Kristensen, M., 202–203, 217
Krivoruchko, A., 226
Kroutil, W., 108
Krumwiede, D., 167–168
Krupa, R.A., 181–182, 313
Kucukural, A., 57

Kueger, S., 167–168
Kuetchou Ngnigha, A.R., 68
Kumagai, H., 152–153, 196–197, 199
Kumaran Ajikumar, P., 226, 251
Kung, S.H., 202–203, 217
Kunzendorf, A., 66, 69–71, 71t, 77f, 78–80, 95f, 96
Kupfer, R., 22
Kuramitsu, S., 119
Kushwaha, M., 183
Kutchan, T.M., 145–149, 196–199, 212
Kuyvenhoven, A., 130
Kwon, H.-J., 44–47
König, G.M., 94, 96

L

Labthavikul, S.T., 98–99
Lachance, D.M., 180–189
Lam, C., 182–183
Lam, S.T., 80, 82
Lamichhane, J., 86
Lampel, J.S., 48–50
Lander, E.S., 202–203, 275
Lang, A., 71t, 82–83, 109, 111
Lange, B.M., 249
Lange, I., 249
Lange, O., 167–168
Lao, X., 226
Larionov, V., 28–29
Larossa, R.A., 226–228
Larsen, T., 167–168
Larson, M.H., 273
Larson, T.R., 146–148, 150–151, 172–173, 199–200
Latham, J., 73–74, 75t, 97, 109, 115
Latifi, T., 320
Laub, M.T., 312–314
Laureti, L., 272–273
Lawrence, B.M., 249
Leadbetter, J.R., 107
Leang, K., 73–74
Leao, P.N., 3, 10
Leblond, P., 272–273
Lebovich, M., 180–183, 185–188
Lee, C.Y., 43f
Lee, D., 183
Lee, E.-J., 145–146, 149–151, 155, 156t, 159–160, 172–173, 199–200

Lee, H., 153–154
Lee, J.-S., 11, 180, 261
Lee, J.W., 180
Lee, K.K.M., 42
Lee, M.-R., 261
Lee, S.H., 183–184, 203
Lee, S.-W., 183 184
Lee, S.Y., 30, 180, 277, 281–282
Lee, T.S., 181–182, 262–263, 286–287
Leguia, M., 181–182
Lei, J., 165
Leigh, J., 321, 323–324, 333–334, 337–340, 342–343, 346
Leikoski, N., 2–3
Leite, A.C.L., 93–94
Lemke, R.A.S., 167–168
Lenting, H.B., 196–197
Lenz, R., 146–148
Leonard, E., 180–182, 226–229, 238–239, 249–251, 262, 295, 313
Lequeux, G.J., 182–183
Lerner, R.A., 209
Letso, R., 153–154
Levine, R.L., 67–68
Levskaya, A., 266
Levy, C.W., 253
Lewis, J.C., 69–70, 70t, 72–74, 75t, 93–122
Leys, D., 321–323, 326–333, 337–340, 343–346
Li, B.Z., 271–272
Li, C., 273–274
Li, D.B., 84, 86–88
Li, G., 182–183, 279–280
Li, H.L., 273
Li, J., 144–173, 273
Li, J.W.-H., 42, 248
Li, L., 49, 55, 184
Li, S.-M., 25–29, 44–46
Li, X., 207–208, 217
Li, Z., 181, 184–185
Liang, J.C., 321
Liang, J.J., 343–345
Liao, D., 155, 156t
Liao, H.H., 98–99
Liao, J.C., 249–251
Ligon, J.M., 80, 82
Lim, C.G., 187–188
Lim, F.Y., 129, 139

Lim, J.-H., 183–184
Lim, K.H., 181
Lim, W.A., 203
Lin, C., 290–291, 313
Lin, C.Y., 273
Lin, G., 273
Lin, H., 107, 226–228
Lin, S., 84, 273
Lin, Y., 180–181, 188–189
Lin, Z., 3–5, 11–12, 14–16
Linder, S.J., 273
Linder, T., 248
Linhardt, R.J., 180–184
Liscombe, D.K., 146–148, 159–160, 172–173, 196–197
Liu, D., 271–272
Liu, D.R., 107
Liu, H.-W., 40, 43f, 44–47
Liu, J.J., 248–249
Liu, L., 3, 226
Liu, W., 40–60, 116, 119
Liu, Y.-N., 43f, 44–46
Liu, Z., 25–27
Lo, H.C., 128–129
Long, P.F., 5–8
Lottspeich, F., 145–146
Lotz, L.A., 273
Louerat, B., 196–197, 210–211
Louis, J.M., 8–9
Loukanina, N., 146–148, 159–160, 172–173
Lousberg, R.J.J.C., 146–148
Lowe, P.T., 321–324, 326–334, 337–340, 342–346
Lu, T.K., 312–314
Lubkowski, J., 345–346
Lucas, X., 271–272
Ludewig, H., 15–16
Luo, Y., 32, 271–274
Lussier, F.-X., 154
Lutle, A.K., 97–98
Luzhetskyy, A., 32
Lygidakis, A., 250f, 251, 253–257, 259, 261–262, 264–266
Lynch, S.A., 337–340

M
Ma, L., 25–27
Ma, P.C., 22

Ma, S.M., 42, 286–287
Macheroux, P., 146–148
Machida, M., 128–129
Mack, M., 185–186
Macleod, B.P., 146–148, 159–160, 172–173
MacNeil, D.J., 48–49, 54–55
MacNeil, T., 48–49, 54–55
Maeder, M.L., 273
Maehr, R., 273
Maertens, J., 182–183
Mahadevan, R., 154, 196–200
Maharjan, S., 86
Mahmoud, S.S., 249
Mai, Q.-A., 182–183
Maiese, W.M., 40–42
Mairhofer, J., 182–183
Makino, M., 22–24
Mali, P., 202–203, 273
Malla, S., 184
Malmirchegini, G.R., 183–184, 286–287
Manchikanti, L., 221
Mandal, M., 343–345
Mandal, R., 149, 155
Mandell, D.J., 184
Mann, G., 15–16
Manoj, K.M., 94
Mansell, D.J., 250f, 251, 253–257, 259, 261–262, 264–266
Marahiel, M.A., 84–85, 88
Maranas, C.D., 264–265
Marcano-Velázquez, J.G., 321–323, 343–345
Marillonnet, S., 181–182, 287
Marin, S., 263–264
Marisch, K., 182–183
Marraffini, L.A., 273
Marschall, M., 22
Martin, G., 86
Martin, V.J.J., 10, 152–154, 168, 172–173, 196–221, 249–251, 313
Martínez, I., 226–228
Martins, J., 3, 10
Matasci, N., 150–151, 155, 156t, 172–173
Matsuda, S.P.T., 249–251
Matsui, Y., 22
Matsumura, E., 152
Matsuzaki, C., 152
Matthews, K., 22

Matthews, P.D., 297
Maurer, M.J., 207–208, 217
Maury, J., 161–165, 184
May, G.S., 130
Mazodier, P., 275
Mazumdar, S., 183–184
McCarthy, J.E.G., 321–322, 326–331, 346
McClune, C.J., 312–314
McConkey, M.E., 260–261
McCown, P.J., 343–345
McDonald, J.H., 207–208
McEwan, A.R., 12, 15–16, 288–291, 298, 307, 312–313
McIntosh, J.A., 3–5, 9–11, 13, 15–16
Medema, M.H.M., 265–266
Medina-Stacey, D., 320–346
Meisel, L., 155
Mckhail, T., 22
Melançon, C.E., 44–47
Melander, C., 93–94, 116–119
Méndez, C., 24–25, 33, 66, 68–71, 71t, 95f, 96–97, 109, 122
Menichini, F., 146–148
Menon, B.R.K., 73–74, 75t, 97, 109, 115, 263
Menon, N., 263
Meredith, J.M., 290–291, 313
Merlo, M.E., 265–266
Merrick, C.A., 286–314
Metzgar, D., 345–346
Meyer, A., 321
Meyer, H.-P., 93–94
Meyer, V., 130
Miao, J., 32
Michener, J.K., 321
Micklefield, J., 77–78, 106–108, 320–346
Mie, T., 43f
Miersch, O., 146–148
Migaud, M., 15–16
Mikkelsen, M.D., 161–162
Milbredt, D., 66–89
Miller, J.A.C., 151, 197–199
Miller, J.H., 337–340
Millgate, A.G., 151, 197–199
Min, B.E., 183
Minami, H., 152–153, 168, 196–199, 212, 214–215
Mingardon, F., 262–263

Mino, T., 42
Minvielle, M.J., 93–94, 116–119
Mirarab, S., 150–151, 155, 156t, 172–173
Mirsky, E.A., 183, 295–296, 302
Misawa, N., 297
Mitchell, L.A., 313–314
Mitchell, W., 249–251
Mitsuda, Y., 109
Miyanaga, A., 94
Mo, M.L., 264–265
Møller, B.L., 286–287
Mondhe, M., 30
Montagnon, T., 40
Montero, J.C., 25
Montiel, D., 24, 29–30
Moon, T.S., 183–184, 286–287
Moore, B.S., 94, 180
Moore, J.C., 116
Moorthie, V.A., 42
Mootz, H.D., 88
Morawitz, F., 11
Moreira, D.R.M., 93–94
Mori, H., 331
Morishige, T., 145–146
Moriya, Y., 264–265
Morra, R., 266, 321
Morris, B., 257
Morris, D.R., 94
Morris, J.S., 144–173, 156t
Morris, N.R., 130
Mortensen, U.H., 161–165
Mosch, J., 271–272
Moser, F., 182–183
Mosurski, K.R., 202
Mou, H., 273
Moxley, J.F., 226–228
Mroczenski-Wildey, M.J., 40–42
Mucha, O., 226, 249, 251–252
Muehlbacher, M., 14
Muffler, K., 68
Muhamadali, H., 265–266
Mühlenweg, A., 47–48
Mukhopadhyay, A., 286–287
Muller, H., 313–314
Müller, N., 108
Mundorff, E.C., 116
Mutalik, V.K., 182–183
Muth, G., 275
Muto, A., 273–274
Myrand-Lapierre, M.-E., 22
Myronovskyi, M., 32

N

Na, D., 180, 183
Naesby, M., 153–154
Nagano, S., 22–24
Nair, N., 202, 217
Nair, S.K., 11
Naismith, J.H., 67–68, 70–71, 71t, 73–74, 82–83, 109, 111
Nakagawa, A., 152, 168, 197–199, 214–215
Nakano, T., 95–96
Nandi, O.I., 146–148, 159–160, 172–173
Nantermet, P.G., 248 249
Narcross, L., 152–153, 168, 172–173, 196–221
Nawrocki, E.P., 343–345
Nayak, T., 128–129, 136
Neal, T.R., 14
Neely, H., 153–154
Neidhardt, F.C., 185–186
Nelson, J.T., 2–8, 11
Neumann, S., 168
Neves, A.R., 184
Nevoigt, E., 286–287
Newill, P.L.A., 81
Newman, D.J., 1–2
Newman, J.D., 10, 249–251, 313
Nguyen, Q.T., 265–266
Nicke, T., 71t, 82–83, 109, 111
Nicolaou, K.C., 40, 248–249
Nielsen, A.A., 312–314
Nielsen, C.A., 153–154
Nielsen, J., 226
Nielsen, K.F., 167–168
Nielsen, S.V., 153–154
Nierman, W.C., 128–129
Nihira, T., 129
Niman, H.L., 209
Nimmo, D.D., 290–291, 313
Nims, E., 97–98, 114–115
Nishizaki, T., 296–297
Nishizawa, T., 22–24
Nizet, V., 28–29
Nogawa, T., 42, 59
Norholm, M.H.H., 161–163

Northcote, P., 40–42
Norville, J.E., 202–203
Nour-Eldin, H.H., 161–163
Novy, R., 257
Nowak, J., 155, 156t, 159–160
Nowroozi, F.F., 286–287, 295–296
Nudler, E., 320
Nußbaumer, B., 275
Nyman, U., 197–199

O

O'Brien, T., 22
O'Connor, S.E., 80–81, 84, 96–98, 114–115, 122, 197–199
Oakley, B.R., 128–129, 136
Oakley, C.E., 128–129, 136
O'Connor, S.E., 168
Oehl, R., 86
Ogawa, A., 324, 326–331, 346
Ogino, H., 99, 106–107
Oikawa, H., 42, 46
Okino, T., 94
Okuda, Y., 249–251
Okumura, Y., 331
Olorunniji, F.J., 286–288, 290–291, 294–298, 300, 312–313
Olsen, C.E., 161–162
Olsen, L.R., 162–163
Onaka, H., 22–24
Onoyovwe, A., 144–145, 149
Orrweaver, T.L., 164–165
Ortiz, M.E., 93–94, 116–119, 313–314
Osawa, S., 273–274
Osbourn, A., 286–288, 294–297, 300
Ose, T., 43f
Osley, M., 209
Ostrovsky, D.N., 238–239
Otey, C.R., 98–99
Oue, S., 119
Oueis, E., 15–16
Ouellet, M., 286–287
Ounaroon, A., 145–146

P

Paddon, C.J., 286–287
Pagán, R., 262–263
Page, J.E., 196–200
Paget, J., 286–314

Palomino, M., 25
Palsson, B.O., 264–265
Pampati, V., 221
Pan, H., 47–48
Pandit, S.B., 264–265
Pang, A.H., 83–84
Pang, B., 40–60
Panke, S., 321
Pannunzio, N.R., 290
Park, J.M., 180
Park, S.J., 183–184
Park, S.U., 149
Parutto, P., 264–265
Pásztor, A., 263
Patallo, E.P., 66–89, 71t, 94–96, 109, 111
Patel, A., 42
Patel, K.G., 181–182, 253–254
Patrick, B.O., 22, 24, 26f, 33
Pattus, F., 209
Pauli, H.H., 145–146
Payne, J.T., 69, 72–74, 75t, 93–122
Pease, L.R., 183
Pedersen, L.E., 202–203, 217
Peleg, Y., 253–254
Pellaux, R., 321
Pelletier, J., 321–322
Peng, C., 49, 55
Penzes, C., 146–148
Peralta-Yahya, P.P., 286–287
Pereira, B., 184–185
Perry, K., 290
Peters, N.R., 24
Peters, R.J., 249–251, 262
Petersen, P.J., 40–42
Peterson, A.C., 167–168
Petter, R., 275
Petty, M.A., 94
Petzold, C.J., 183–184, 249–251, 265–266, 286–287, 295–296
Pferschy-Wenzig, E.M., 248
Pham, V.D., 183–184
Phillips, G.N., 73–74
Phillips, R., 182–183
Phon, T.H., 249, 251–252
Pieiller, N., 12, 15–16
Pierce, E., 3–5, 9–11, 13, 15–16
Pineda-Lucena, A., 25
Pitera, D.J., 10, 249–251, 313

Platt, D.M., 202–203, 217
Poddar, S., 207–208, 217
Pogson, B.J., 197–199
Polizzi, K.M., 98–99
Polnick, S., 71t, 82–83, 109, 111
Pompon, D., 196–197, 210–211
Pontini, M., 116
Poor, C.B., 69–70, 70t, 99–106, 116, 122
Post, D.A., 48–50
Potapov, V., 207–208
Poth, D., 83–84, 86–88
Potot, S., 321
Poulter, C.D., 14
Pouwels, P.H., 130
Prasad, N., 181–182
Prather, K.L.J., 183–184
Price, W.N., 153–154
Pühler, A., 275
Pullen, J.K., 183
Punt, P.J., 130
Pyne, M.E., 196–221

Q

Qi, J., 155, 156t
Qi, L.S., 273
Qi, Z., 290–291, 313
Qian, J., 48–50
Qian, Z.G., 286–287
Qiao, K., 184–185, 230, 238–239
Qu, X., 40–42, 44–46, 54–55, 80–81, 96–98, 114–115, 122
Qu, Y., 168
Quan, J., 287
Quarmby, S.T., 167–168
Quax, T.E.F., 153–154
Quillardet, P., 3

R

Raab, A., 11
Ragan, T.J., 343–345, 344f
Raghuraman, S., 9–11
Rahman, M.A., 248–249
Ram, A.F.J., 130
Ramalingam, S., 313–314
Ramos, V., 3, 10
Ran, F.A., 273
Rao, G., 273–274
Rasko, D.A., 2–8, 11

Rasmussen, M., 122
Rattray, N.J., 265–266
Ravel, J., 3–9
Reader, J.S., 345–346
Redding-Johanson, A.M., 286–287, 295–296, 313
Redpath, P., 15–16
Reetz, M.T., 107–108
Reeves, A.R., 48–50
Reinemann, C., 321–322
Reinhart, C., 209
Reisch, C.R., 184
Reißig, H.-U., 73–74, 75t, 97
Reizman, I.M.B., 184
Renata, H., 1–2
Renirie, R., 99
Reynolds, K.A., 28–29
Richardson, S.M., 313–314
Richter, C., 196–197
Ringer, K.L., 249, 250f, 253, 260–261
Rinner, U., 197–199
Rios, X., 202–203
Rishi, H.S., 207–208
Roberge, M., 22, 24
Roberts, S.C., 226
Robertson, C.R., 11
Robins, K.T., 93–94
Robinson, C.J., 266, 320–346
Robinson, C.W., 186–187
Robinson, J.A., 83–84, 86–88
Rogosnitzky, M., 196–197
Rohr, J., 24–25, 33, 66, 69, 96–97, 122
Rolf, M., 146–148, 212
Romero, P.A., 119
Rosovitz, M.J., 3, 5–9
Rosser, S., 286–314
Roth, A., 320
Rothstein, R.J., 164–165
Rouviere, P.E., 226–228
Rowles, I., 116
Rowley, P.A., 290
Roy, A., 57
Roy, A.D., 77f, 80–81, 96–97, 122
Ruby, C.L., 48–49, 54–55
Rueffer, M., 212
Ruffner, D.E., 9
Rui, L., 109
Ruijssenaars, H.J., 99

Ruiz, M.T., 150–151
Rumpf, J., 66, 69–71, 71t, 77f, 78–80, 95f, 96
Runguphan, W., 80–81, 96–98, 114–115, 122, 197–199
Rushing, G.W., 260–261
Russ, Z.N., 152, 168, 172–173, 197–203, 207–208, 210, 214–215, 217–220
Russell, D.W., 324, 326, 331
Ruszczycky, M.W., 40, 43f
Rutherford, K., 290
Ryan, K.S., 22–34, 26f
Ryan, O.W., 207–208, 217

S

Sabatini, D.M., 275
Saeed, M., 248–249
Saini, M., 184–185
Saint-Joanis, B., 22
Saito, K., 248
Sajja, H.K., 337–340
Sakai, K., 129
Salas, A.P., 24–25, 96–97, 122
Salas, J.A., 25, 33, 66, 68–71, 71t, 95f, 96, 109
Salemink, C.A., 146–148, 196–197
Salis, H.M., 183, 295–296, 302
Salomonsen, B., 161–162
Samai, P., 273
Samanani, N., 149
Samborskyy, M., 44–46
Sambrook, J., 324, 326, 331
San, K.-Y., 226–228
Sánchez, C., 24–25, 33, 66, 69, 96–97, 122
Sanchez, J.F., 128–129, 139
Sander, J.D., 273
Sanders, T.M., 184
Santos, C.N.S., 181, 226, 295
Santoso, A.T., 22
Sardar, D., 1–18
Sasaki, Y., 287–288
Sato, F., 145–146, 152, 168, 197–199, 212, 214–215
Sato, M., 42
Sauvageot, N., 185–186
Savile, C.K., 116
Sawant, S.D., 22
Saxena, K., 196–197

Scaloni, A., 67–68
Schaart, J.G., 273
Schackwitz, W., 262–263
Schade, B., 202, 204–205
Schadt, S., 47–48
Scharl, T., 182–183
Scheuer, J., 2–3
Schiestl, R.H., 164–166
Schimmel, P., 345–346
Schmartz, P.C., 83–84, 88
Schmid, A., 262–263
Schmidt, E.W., 1–18, 42
Schmidt, J., 145–149, 197–199
Schneider, A., 68, 95–96
Schnerr, H., 66, 69–71, 71t, 77f, 78–80, 95f, 96
Schorn, M., 83–84, 86–88
Schrewe, M., 262–263
Schriemer, D.C., 149, 155, 156t
Schrittwieser, J.H., 108
Schroeder, F.C., 43f
Schubert, M.G., 202–203, 217
Schukur, L., 313–314
Schultz, A., 128–129
Schultz, J.A., 93–94
Schultz, P.G., 8–9, 116, 119
Schwandt, A., 22
Scott, D.O., 93–94
Scrutton, N.S., 248–267
Segall-Shapiro, T.H., 182–183
Seghezzi, N., 32
Seibert, P.M., 28–29
Seibold, C., 66, 69–71, 71t, 77f, 78–80, 95f, 96
Seifert, G., 68, 95–96
Seifuddin, F.T., 128–129
Sekiyama, Y., 42, 59
Sell, C., 249
Selvakumar, D., 209
Senger, C., 271–272
Seo, S.-W., 183–184
Serganov, A., 320
Serrano-Heras, G., 25
Setti, G., 287
Sewald, N., 70–73, 71f, 75t, 97–98, 108, 122
Shah, P., 128–129
Shankar, J., 266, 321
Shao, L., 40–42, 44–46, 54–55

Shao, Z., 181–182, 202, 217, 273–274
Sharp, P.M., 202
Sharp, R., 290
Sheldon, R.A., 70–71
Shen, B., 84
Shen, R., 226
Shen, W.-C., 209
Shepherd, S.A., 73–74, 75t, 77–78, 97, 106–109, 115
Sherden, N.H., 43f
Sherer, C., 22
Sherlock, G., 128–129
Sherman, D.H., 22–24
Sherratt, D.J., 307
Sherwood, P., 209
Shetty, R.P., 181–182, 287
Shi, Y., 320
Shibasaki, M., 22
Shibata, M., 196–197
Shigemizu, D., 264–265
Shiloach, J., 185–186
Shimotohno, A., 119
Shin-ya, K., 40–42, 46–47
Shirke, A.N., 180–183, 185–188
Shiro, Y., 22–24
Shirran, S., 11
Shiue, E., 183–184
Shleeva, M.O., 238–239
Siddiqui, M.S., 196–197, 219–220
Siedler, S., 184
Siegel, M.M., 40–42
Siegl, T., 32
Simeon, F., 180–182, 226–229, 238–239, 249–251, 262, 295, 313
Simionescu, R., 168
Simón, E., 153–154
Singh, M.P., 40–42
Singh, P.P., 22
Singh, S., 73–74
Sinkoe, A.L., 184–185, 187–188
Sipos, A., 144–145
Sissi, C., 196–197
Sivonen, K., 2–3
Skerker, J.M., 207–208, 217
Skidmore, C., 122
Skrzypek, M.S., 128–129
Slaby, T., 181–182, 253–254
Smith, B.M., 93–94

Smith, D.F., 185–186
Smith, H.O., 181–182, 253–254, 287
Smith, J.A.S., 86
Smith, J.M., 93–94
Smith, K., 11, 93–94
Smith, M.C., 11, 286–291, 294–300, 307, 312–313
Smolke, C.D., 152–154, 168, 172–173, 181, 196–203, 215–217, 219–220, 321
Smulders, M.J., 273
Snape, T.J., 22
Snapp, E., 203
Snyder, S.A., 40
Soetaert, W.K., 182–183
Sohng, J.K., 86
Söll, D., 153–154
Solomon, K.V., 184
Sone, T., 287–288
Song, J., 49, 55
Song, L., 27–28, 272–273
Song, X., 226
Soni, P., 107–108
Spence, M., 208
Spiller, B., 119
Sportsman, J.R., 209
Sridhar, R., 264–265
Stachelhaus, T., 83–84
Stadtman, E.R., 67–68
Stahlhut, S.G., 184
Stamminger, T., 22
Stan, G.B., 295
Stanfield, S., 82
Stanford, D.R., 209
Stanton, L.H., 181–182, 253–254
Stapon, A., 84, 86–88
Stark, W.M., 286–288, 290–291, 294–298, 300, 312–313
Starmer, J., 252
Stein-Gerlach, M., 22
Steinhauser, D., 167–168
Stenger, A.R., 184
Stentz, R., 321–322
Stephanopoulos, G., 181–182, 184–185, 226–244, 249, 251–252, 286–287, 295
Stevens, R.C., 119
Stevenson, C.E.M., 43f
Stieglitz, J.T., 182–183
Stockbridge, R.B., 320

Stoltenburg, R., 321–322
Stomp, A., 252
Storch, M., 287
Stracquadanio, G., 313–314
Stragier, P., 333–334
Straight, P.D., 87–88
Strehlitz, B., 321–322
Stremler, K.E., 14
Striedner, G., 182–183
Strotman, H., 108
Struck, A.-W., 73–74, 75t, 97, 109, 115
Strucko, T., 161–165
Su, M., 153–154
Subsoontorn, P., 313–314
Sudarsan, N., 320
Sudek, S., 2–9, 11
Suess, B., 321–322
Suga, H., 11, 15–16
Sugimoto, H., 22–24
Suh, W., 226–228
Summers, D.K., 307
Sun, H., 93–94
Sun, J., 32, 202, 217
Sun, P., 40–42, 59, 155, 156t
Sun, Z.Z., 295
Sury, G., 80
Süssmuth, R.D., 66, 68–71, 71t, 95f, 96, 109
Sutherland, A., 42
Suthers, P.F., 264–265
Suzuki, Y., 42
Swainston, N., 264–265
Szewczyk, E., 128–129, 136
Szmidt, H.L., 262–263, 313
Szostak, J.W., 164–165

T

Tabor, J.J., 266
Tabudravu, J., 12, 15–16
Taglialatela-Scafati, O., 94, 96
Taheri-Talesh, N., 136
Tait, S., 248–267, 250f
Takagi, H., 42, 59
Takahashi, S., 42, 59
Takai, Y., 331
Takano, E., 248–267, 272–273
Takeda, I., 128–129
Takemura, T., 152–153, 196–197, 199
Tam, S., 116

Tanaka, H., 196–197
Tang, J., 40–42, 44–46, 54–55, 321–322
Tang, Y., 249–251
Tang, Z., 40–60
Tange, T.Ø., 153–154
Taniguchi, S., 24
Tao, H., 107
Tao, W., 44–46
Tao, Y., 109
Tarling, T., 22, 24
Tautenhahn, R., 168
Taylor, S.V., 107
Temme, K., 28–29, 182–183
Temple, G.F., 287–288
Teodor, R.I., 146–148, 151, 199–200
Teodor, T.R., 146–148, 151
Ternei, M.A., 22–30
Teruya, K., 40–42, 46–47
Teufel, R., 94
Theuns, H.G., 146–148, 196–197
Theuns, H.L., 146–148
Thisleton, J., 151, 197–199
Thodey, K., 152–154, 168, 172–173, 181, 196–202, 215–216, 219–220, 321
Thomas, D.Y., 202, 204–205
Thomas, M.G., 84–85, 87–88
Thomas, P.M., 279–280
Thompson, C., 275
Thompson, M.L., 73–74, 75t, 97, 109, 115
Thorn, K.S., 203
Thorson, J.S., 73–74
Tian, J., 287
Tian, Z., 40–42, 44–46, 54–55
Tianero, M.D., 1–18
Till, R., 298–299
Tinberg, C.E., 184
Titus, S.A., 287–288
Tokimatsu, T., 264–265
Tokiwano, T., 46
Tokovenko, B., 32
Tong, Y., 30, 277, 281–282
Too, H.P., 226–229, 231
Toogood, H.S., 248–267, 250f
Toparlak, Ö.D., 180–188
Toyoda, A., 42, 59
Trantas, E.A., 180–181, 184
Trauger, J.W., 83–84

Trausch, J.J., 322–323, 343–345
Trenchard, I.J., 152–154, 168, 172–173, 196–202, 215–216, 219–220
Trevino, A.E., 273
Truppo, M.D., 108
Tsai, J.C., 207–208, 217
Tsai, J.H., 93–94
Tsang, S., 209
Tsao, R., 40–42
Tsodikov, O.V., 83–84
Tsuge, K., 296–297
Tsujita, T., 145–146
Tsunematsu, Y., 129
Tsunoda, S., 11, 15–16
Tuohy, T.M., 202
Turner, G.W., 128–129, 249
Tyo, K.E.J., 180–182, 226–229, 238–239, 249–252, 262, 295, 313

U

Uchiyama, N., 42
Ueno, H., 248–249
Uguru, G.C., 30, 47–48
Uhrin, P., 248
Ulber, R., 68
Ullal, A.V., 183–184, 286–287
Umemura, M., 128–129
Unger, T., 253–254
Unterlinner, B., 146–148
Unversucht, S., 67–68, 70–71, 73–74, 82, 109
Upton, M., 266, 321
Uramoto, M., 42, 59
Urban, P., 196–197, 210–211
Uto, Y., 16–17

V

Vaillancourt, F.H., 84, 95–96
van Beilen, J.B., 249
Van De Wiel, C.C., 273
van den Bogaard, M., 22
Van den Heever, J.P., 42
van den Hondel, C.A.M.J.J., 130
Van Der Donk, W.A., 18, 279–280
van der Oost, J., 153–154
van der Wielen, L.A., 226
van Dijk, J.W.A., 128–141
Van Duyne, G.D., 290
van Gent, J., 130
van Pée, K.-H., 66–89, 75t, 94–98, 95f, 103–104, 109, 113, 121
Vandamme, E.J., 182–183
Vanden Boom, T.J., 48–50
Vandenhondel, C., 130
Vansiri, A., 181–182
Varghese, A.S., 288–291, 307, 312–313
Varner, J.D., 153–154
Vasconcelos, V., 3, 10
Vassilikogiannakis, G., 40
Vederas, J.C., 248
Velasquez, J.E., 18, 279–280
Vendome, J., 11
Venkataraman, S., 181–184
Venken, K.J., 290 291, 294, 313
Venkiteswaran, S., 187–188
Venter, J.C., 181–182, 253–254, 287
Vernacchio, V.R., 180–189
Ververidis, F., 180–181, 184
Vincent, H.A., 320–346
Virolle, M.-J., 32
Vishwakarma, R.A., 22
Voigt, C.A., 28–29, 182–183, 266, 295–296, 302
Voigtländer, S., 146–148
Voinnet, O., 150–151
von Suchodoletz, H., 47–48
Vosburg, D.A., 84, 95–96
Voß, H., 70–71, 75t, 97–98
Vostroknutova, G.N., 238–239
Vouk, M., 252

W

Wachter, A., 320
Wage, T., 66, 69–71, 71t, 77f, 78–80, 95f, 96, 109
Wagner, C., 94, 96
Wagner, R., 209
Wahlsten, M., 3
Waller, J., 253
Walsh, C.T., 1–2, 22–24, 70–71, 71t, 73–74, 83–88, 95–98, 95f, 114–115
Waltenberger, B., 248
Walter, J.M., 202–203, 217
Wang, C.C.C., 128–141
Wang, H.H., 28–29, 295
Wang, J., 180

Wang, M., 46
Wang, P., 155, 156*t*
Wang, R., 165
Wang, S., 122
Wang, T., 275
Wang, W., 180–182
Wang, Y., 42, 59, 180–182, 226–229, 238–239, 249–251, 262, 271–283, 295, 313
Wang, Z.J., 1–2
Wanka, F., 130
Wardrope, C., 286–314
Waring, R.B., 130
Wasil, L.R., 288–290, 312–313
Watanabe, H., 43*f*
Watanabe, K., 43*f*, 129
Watanabe, T., 22
Wawrosch, C., 248
Weber, E., 287
Weber, T., 30, 277, 281–282
Weenink, T., 251–252
Wei, J.J., 275
Weichold, V., 66–89
Weinberg, Z., 320, 343–345
Weislo, L.J., 78–81, 95–96
Weissman, J.S., 273
Wen, F., 202, 217
Werner, S., 287
Wever, R., 99
Wever, W.J., 43*f*
Weyler, C., 258–259
White-Phillip, J.A., 44–47
Whiteway, M., 202, 204–205
Wickett, N.J., 150–151, 155, 156*t*, 172–173
Wijekoon, C.P., 150–151
Wildung, M.R., 249, 250*f*, 253
Wilkinson, D.L., 253
William, P., 71*t*, 82–83, 109, 111
Williams, D.E., 22, 24, 26*f*
Williams, D.R., 40–42
Williams, P.G., 93–94
Willies, S.C., 116
Willmitzer, L., 167–168
Wilson, I.A., 209
Wilson, I.W., 197–199
Wilson, S.A., 226
Wilson, S.E., 294
Winkler, A., 146–148

Winkler, W.C., 343–345
Winter, J.M., 94, 249–251
Winzer, T., 146–148, 150–151, 172–173, 199–200
Wipf, P., 16–17
Withers, S.T., 10, 249–251, 313
Witholt, B., 249
Witter, D.J., 42
Wohlleben, W., 275
Woithe, K., 84, 86–88
Wolfe, K.H., 128–129
Wolinski, H., 207–208
Wong, K.H., 153–154
Wong, L., 180–182, 187–188
Wong, L.S., 77–78, 106–108
Wood, T.K., 109
Woodman, M.E., 255
Woodside, A.B., 14
Worthington, R.J., 93–94, 116–119
Wu, C., 202, 204–205
Wu, G.C., 181–184, 286–287
Wu, I., 98–99
Wu, J., 48–50, 182–183, 273–274
Wu, L., 47–48
Wu, M.-C., 320–346
Wu, Q., 44–46
Wu, Z., 40–42, 44–46, 59
Wurtzel, E.T., 297
Wynands, I., 82

X

Xia, X.X., 286–287
Xiang, S., 32
Xiao, A., 122
Xiao, J., 44–46
Xiao, M., 150–151, 199–200
Xiao, W.-H., 180–182, 226–229, 238–239, 249–251, 262, 295, 313
Xie, J., 116, 119
Xiong, Y., 136
Xu, G., 295
Xu, J.H., 202, 217
Xu, K., 165
Xu, P., 180–182, 184, 187–188
Xu, T., 273–274
Xu, Y., 265–266
Xue, W., 273

Y

Yadav, V.G., 251
Yagishita, F., 42
Yahata, K., 287–288
Yamada, M.R., 207–208
Yamada, R., 99, 106–107
Yamada, Y., 145–146, 181
Yamamoto, K., 152
Yamanaka, K., 28–29, 83–84, 86–88
Yan, Y., 40–42, 180–181, 188–189
Yan, Z., 150–151, 155, 156t, 172–173
Yanagawa, H., 296–297
Yang, C., 44–46
Yang, H.J., 165
Yang, J.-S., 183
Yang, L., 312–314
Yang, M., 273–274
Yang, T., 25–27
Yang, Z., 248–249
Yano, T., 119
Yao, M., 43f
Ye, L., 290–291, 313
Ye, W., 226
Yeh, E., 22–24, 70–71, 71t, 73–74, 83–85, 88, 95–98, 95f, 114–115
Yeh, V.J., 207–208
Yeo, V., 27–28
Yin, H., 273
Yin, J., 84
Yin, P., 313–314
Yo, P., 343–345, 344f
Yonekura-Sakakibara, K., 248
Yoon, Y.J., 180
Yoshida, S., 287–288
Yoshikawa, C., 99, 106–107
Yoshimatsu, K., 152
You, L., 99
Young, D.D., 8–9
Young, L., 181–182, 253–254, 287
Young, R.A., 209
Young, T.S., 8–9
Younger, E., 288–291, 307, 312–313
Youssef, D.T., 3
Yu, F., 42, 59
Yu, J.L., 286–287, 290–291, 313
Yu, W.-L., 44–47
Yu, Y., 18
Yuan, L.Z., 226–228
Yuan, P., 290

Z

Zeder-Lutz, G., 209
Zehner, S., 66, 68–71, 71t, 95f, 96, 109
Zeng, A.-P., 321
Zeng, J., 97–98
Zenk, M.H., 196–199, 212
Zerbe, K., 83–84, 86–88
Zhan, J., 97–98, 122
Zhang, A., 273–274
Zhang, B., 47–48
Zhang, C., 24
Zhang, F., 181–182, 184, 202–203, 273, 286–287
Zhang, G., 24, 44–46
Zhang, H., 42, 44–47, 59, 184–185
Zhang, J., 184
Zhang, K., 273
Zhang, L., 30, 32, 271–272, 277, 281–282, 287–288
Zhang, Q., 18, 25–27, 44–46
Zhang, S., 122
Zhang, T.T., 165
Zhang, W., 25–27
Zhang, X., 44–46
Zhang, Y., 32, 44–46, 57, 150–151, 199–200
Zhang, Z.Y., 165
Zhao, C., 185–186
Zhao, D., 28–29
Zhao, G., 287–288
Zhao, H., 32, 98–99, 153–154, 181–183, 202, 217, 271–283
Zhao, Q., 40–42, 44–46, 49, 54–55, 59
Zhao, S., 181–184
Zhao, X., 32
Zheng, Q., 40–42
Zhong, G., 40–60
Zhong, J.J., 286–287
Zhou, H., 42
Zhou, J., 182–183
Zhou, K., 184–185, 226–244
Zhou, L.-B., 155, 156t, 321
Zhou, S., 40–42
Zhu, B., 253–254

Zhu, J., 226–228
Zhu, L., 24–25, 96–97, 122
Zhu, X., 11, 68, 73–74, 95–96
Zhu, Y., 44–46, 153, 196–203, 215–216, 219
Zhuang, Y., 48–50
Ziegler, J., 146–148, 196–197, 199
Ziemert, N., 3
Zinin, A.I., 238–239
Zollman, D., 11
Zou, R., 226–229, 231
Zulak, K.G., 149
Zumstein, G., 212
Zutz, C., 128–129

SUBJECT INDEX

Note: Page numbers followed by "*f*" indicate figures, "*t*" indicate tables, "*b*" indicate boxes, and "*s*" indicate schemes.

A
Acid chloride, pyrrolyl-S-CoA synthesis, 86
Attachment *(att)* sites
  assembled construction, 300
  attL x *attR* recombination, 293–294
  attP x *attB* recombination, 291–293, 292*f*
  central dinucleotide of, 298–299, 299*f*
  intramolecular recombination, 300
  naming convention for, 298
  primers for, 301*t*, 304*t*

B
Bacillus subtilis
  MreB morphology assays in, 342–343
  riboswitches, 333–334
    chromosomal integration verification, 335
    E. coli transformation, 335
    gDNA preparation, 334
    genomic DNA PCR, 334
    ligation, 335
    MreB morphology assays, 342–343
    mutagenesis, 335
    native gene expression constructs, 336–337
    PCR amplification, 334
    restriction digests, 334
    riboswitch mutagenesis, 334
    sequencing and glycerol stocks, 335
    transformation, 335
Benzylisoquinoline alkaloid (BIA) biosynthesis, 144–145
  candidate genes
    culture, 166–167
    DNA sequencing, 155–159
    plants and tissues selection, 155
    RNA extraction, 155–159
    selection, 159–160, 161*f*
    substrate feeding, 166–167
    transient expression constructs, 166
    yeast transformation, 166–167
  challenges, 219–221
  crop-based manufacturing, 197–199
  enzyme expression and localization, 208–209
  gene sources and selection, 199–200
  LC-MS
    analytical methods, 169–171
    analytical strategy overview, 167–168
    instrumentation and software, 168
    solvents and reagents, 168–169
    technical notes, 172
  pathway
    debottlenecking strategies, 216–219
    pathway assembly, 216
    reconstitution and optimization, 215–219
  pBOT vector, 204–205, 205*f*
  in plants, 145–148
    functional genomics, 150–152
    localization, 149
    methylene bridge formation, 146–148
    (R)-reticuline, 146–148
    (S)-reticuline, 146–148, 147*f*
    (S)-scoulerine, 146–148
  transcriptome resources, 156*t*
  USER cloning
    EasyClone system, 161–163
    PHUSER software, 162–163
    recombinant protein analysis, 165
    Ura marker excision, 165
    vectors, 163
    yeast transformation, 164–165
  in yeast (*see* Yeast, BIA biosynthesis)
BioBrick method, 287
Bioprospecting, 200–201
  of orthologous genes, 218
Biosynthetic pathway genes
  for phenotypic diversity, 296–297
  RBSs for, 295–296
Bisindoles
  analysis and purification, 33
  biosynthetic pathways, 22–24, 23*f*, 30
  chlorination, 24

371

Bisindoles (Continued)
  cladoniamide A, 26f
  gene clusters identification, 25–27
  gene disruption, 30–31
  heterologous expression
    to chassis host, 28–30
    host strains, 27–28
  isolation and combinatorial biosynthesis, 27f
  metabolite characterization, 33
  mutational biosynthesis, 30–31
  phylogenetically related biosynthetic gene clusters, 32
  product extraction, 32–33
  pyrrolinone formation, 22–24
  RebC, 25–27
  rebeccamycin biosynthesis, 24
  staurosporine enzymes, 24
  strain growth, 32–33
  Streptomyces-derived gene, 25–27
  tailoring reactions, 24
  vector construction, 29f
  vector pYD1, 30–31

C

Candidate genes, BIA biosynthesis
  culture, 166–167
  DNA sequencing, 155–159
  plants and tissues selection, 155
  RNA extraction, 155–159
  selection, 159–160, 161f
  sources and selection, 199–200
  substrate feeding, 166–167
  transient expression constructs, 166
  yeast transformation, 166–167
Carveol, 249
Catharanthus roseus, 80–81, 96–97
Cell-feeding assays, 212–214
Central dinucleotide, *att* sites, 298–299, 299f
Chromatographic separations, 105–106
Chromopyrrolic acid (CPA), 22–24
Chromosomal integration (CI), 181–182
ΦC31 integrase gene purification, 305–306
C20 isoprenoid scaffold molecule (ISMs)
  bioreactor, production by, 236–237
    feeding, 238
    HPLC analysis, 238
    initiating, 237–238

  isoprenoids purification, 238
  preparation, 237
  specifications, 237
  E. coli, production by, 228–229
    cell transformation, 234
    competent cell preparation and transformation, 235
    DNA assembly, 234
    DNA electrophoresis and recovery, 233–234
    ethanol precipitation, 234
    GCMS analysis, 236
    iodine treatment, 234
    IsoS-encoding genes, 230
    K3 basal medium, 235
    K3 master mix, 235
    1000×K3 trace elements stock solution, 235
    PCR reaction, 231–233
    primer design, 231
    seed culture, 235
    sequence of ksl_cps, 232b
    testing, 236
  oxygenation using *S. cerevisiae*
    characterization, 243
    CYP expression, 239–242
    isoprenoid-overproducing *E. coli*, 243–244
Cladoniamide A, 22, 26f
Coculture optimization, 184–185
Codeine O-demethylase (CODM), 146–148
Combinatorial construct libraries
  phenotypic diversity, 296–297
  primers for, 302–303, 303t
  RBSs degeneration, 295–296
Conjugation, 279–281
Copy number balancing (CNB), 182
CRISPR–Cas9 system, 202–203
Cross-linked enzyme aggregates (CLEAs)
  enzyme immobilization, 70–71
  gram-scale halogenation reaction, 72–73
  halogenation using, 72
  kinetic data, 70–71, 70t
  L-tryptophan, 71, 71t
  preparation, 70–72, 71f
  tryptophan to 5-chlorotryptophan conversion, 73f

Subject Index

Cultivation conditions, 219
Cyanobactin pathways
  in ascidians, 4f
  discovery of, 2–3
  in *E. coli*
    heterologous expression, 5–9, 7f
    optimization, 9–11
    tru pathway gene cluster, 6f
  modularity, 16f
  natural rules elucidation, 3–5
  peptide motifs, 16–17, 17f
  purification and characterization, 15
  RiPP pathways, 3
  in vitro, 11–16, 13f
[4+2] Cycloaddition activity, 42
  candidates coding for, 46–48, 47f
  chlE3/chlL
    assay of activities, 56–57
    expression and purification, 56
    heterologous complementation *in trans*, 55–56
    inactivation in *S. antibioticus*, 54–55
    protein natures of, 57–58
  ΔpyrE3 *S. rugosporus* mutant strain, 50
  in PYR biosynthetic pathway, 44–46, 45f
  pyrE3/pyrI4
    assays of activities, 51–54
    expression/purification from *E. coli*, 50–54
    heterologous complementation *in trans*, 55–56
    inactivation in *S. rugosporus*, 48–50
    protein natures of, 57–58
Cytochrome P450 (CYP), in *S. cerevisiae*
  characterization, 243
  coculture of, 243–244
  constructing yeast expression vector, 239–240
  expression, 239–242
  yeast transformation, 240–241
  YNB CSM-URA, 242
Cytochrome P450 reductase (CPR), 207–208

D

Dialkyldecalin synthase, 42, 46
Dickmann cyclization, 44–46
Diels–Alder reaction, 40

Dihydrosanguinarine, oxidation, 196–197
Dihydroxy ketone, enzymatic spiroketalization, 59, 60f
Dimethylallyl pyrophosphate (DMAPP), 226–228
Directed enzyme evolution, 217–218
Dithiothreitol (DTT), 12
DNA. *See also* Genomic DNA
  electrophoresis and recovery, 233–234
  extraction
    from hyphae, 132–133
    from spores, 133–134
  parts
    att sites and, 298
    binding sequence, 301–302
    combinatorial assembly of, 310–311
    PCR conditions for, 304–305
    in predefined order, 309–310
    primers for, 301–302, 301t
    purification by ethanol precipitation, 305
  synthesis, declining cost, 200–201
Dot-blot immunodetection, 209
Double-strand break (DSB)-mediated genome editing, 273
Dynamic balancing, 184
Dynamic genetic optimization, 180–181

E

Electrocompetent *E. coli*, 308
(*S*)-Enantiomer, 196–197
Endonuclease-based approaches, 287
Enzyme activity
  expression and localization, 203–209
  LC-MS analysis, 214–215
  purification, 210–212
  in vitro enzyme assays, 210–212
  in vivo cell-feeding assays, 212–214
Enzyme stability, 70–73
Erdasporine pathway (erdODP), 29–30
Escherichia coli
  chemically competent, 307–308
  CheZ motility assays in, 332–333
  C20 ISMs production, 228–229
    cell transformation, 234
    competent cell preparation and transformation, 235
    DNA assembly, 234

Escherichia coli (*Continued*)
  DNA electrophoresis and recovery, 233–234
  ethanol precipitation, 234
  GCMS analysis, 236
  iodine treatment, 234
  IsoS-encoding genes, 230
  K3 basal medium, 235
  K3 master mix, 235
  1000×K3 trace elements stock solution, 235
  PCR reaction, 231–233
  primer design, 231
  seed culture, 235
  sequence of ksl_cps, 232b
  testing, 236
  electrocompetent, 308
  gene expression regulation
    gDNA preparation, 326
    ligation, 328
    PCR amplification, 328
    primer design, 326, 327f, 329t
    restriction digests, 328
    sequencing and glycerol stocks, 328
    subcloning, 331
    transform cells, 328
  Mentha monoterpenoids, 253–254
    general assembly protocols, 254–255
    in-fusion cloning, 255–257, 256f
    vector and gene selections, 251–253
  S. cerevisiae with isoprenoid-overproduction
    bioreactor run, 244
    yeast preculture preparation, 243
Ethanol precipitation
  DNA purification by, 305
  iodine treatment and, 234

F

Fermentation optimization
  induction optimization, 186–189
  media optimization, 185–186
  temperature optimization, 186
Flavin-dependent tryptophan halogenases (FDH)
  biosynthetic pathways modification, 78–81
    thal/thdH into *P. chlororaphis*, 78–80
    in vivo modification, 80–81

  chemical substitution, 81
  CLEAs
    enzyme immobilization, 70–71
    gram-scale halogenation reaction, 72–73
    halogenation using, 72
    kinetic data, 70–71, 70t
    L-tryptophan, 71, 71t
    preparation, 70–72, 71f
    tryptophan to 5-chlorotryptophan conversion, 73f
  error-prone PCR, stability improvement by, 67–70
  5-halogenase PyrH, 68–69, 73f
  instability, 69
  mutant RebH variant 3-LR and 3-LSR, 70
  overexpression, 69
  protein engineering, 98, 109, 115
  pyrrolyl-S-PCPs, 83–84
    to carrier proteins, 87
    chemoenzymatic synthesis, 86–87
    enzymatic synthesis of, 84–86, 85f
    pyrrole-2-carboxylic acid halogenation, 87–88
  regioselectivity modification, 82–83
  substrate specificity
    7-halogenase gene *prnA*, 73–74, 77f
    6-halogenase gene *thal/thdHi*, 73–74, 77f
    halogenated arylamine, 78
    halogenation reaction, 78
    high-throughput assay, 77–78
    large scale conversion conditions, 75t
    RebH halogenation, 79f
    time course of conversion, 67–68f
    two-component system, 66
Flavin-dependent tryptophan halogenases (FDH) stability
  chromatographic separations, 105–106
  colorimetric/fluorimetric assays, 106–108, 106s
  directed evolution
    in industrial processes, 98–99
    organic solvent tolerance, 105
    thermostability, 99–105, 99s
  enzyme immobilization, 108

regioselectivity alteration
  cytochromes P450, 109
  iterative mutagenesis and screening, 110–114, 110s, 111f
    random mutations, 111–114
    targeted mutations, 114
  substrate scope
    random mutations, 115–121, 117s, 118t
    targeted mutations, 114–115
  UPLC method, 105–106
Flow cytometry, 203, 206–207
Fluorescence microscopy, 203
  advantage, 207–209
  BIAs enzyme expression and localization, 208–209
  identification of, 207
  quantitative determination of, 206–207
Functional genomics, 197–199
Fungal secondary metabolite biosynthesis pathways, 128–129

G

β-Galactosidase, riboswitches validation, 337–340, 338f
Gateway™ Cloning, 287–288
Gene homolog sourcing, 181
Genetic optimization
  coculture optimization, 184–185
  copy number balancing, 182
  dynamic balancing, 184
  gene homolog sourcing, 181
  plasmid backbone selection, 181–182
  posttranslational balancing, 183–184
  transcriptional balancing, 182–183
  translational balancing, 183
Genetic tuning, 217
Genome mining, 271–272
Genomic DNA, for PCR template, 334
  hyphae, extraction from, 132–133
  materials, 132–133
  spores, extraction from, 133–134
Genotype screening, 279–281
Geranylgeranyl pyrophosphate (GGPP), 230
Golden Gate method, 287
gp3 gene purification, 307
Green fluorescent protein (GFP), 203, 205
  and flow cytometry, 206–207

H

Halogenation, 93–94
  active halogenating species, 94
Haloperoxidases, 94–96, 95f
Heterologous expression
  bisindoles
    to chassis host, 28–30
    host strains, 27–28
  of cyanobactin pathways, 5–9, 7f
Heterologous genes
  assaying activity of, 210–211
  cytochrome P450, 204f
  evaluating activity of, 210
  functional expression, 200–201
  high-level expression of, 202
  of plant BIAs, 199
  in yeast, 200–201, 203
High-resolution liquid chromatography-tandem mass spectrometry, 167–172
Homology-based approaches, 287
Hydrocodone, BIAs, 197–199
Hypohalous acid (HOX), 67–68, 94–96

I

Immunoblotting, 203, 209
Inducer concentration, 188
Induction optimization
  inducer concentration, 188
  induction point, 187–188, 187f
  substrate delay, 188–189
Induction point, 187–188, 187f
Intramolecular recombination, and *att* sites, 300
In vitro enzyme assays
  BIA biosynthetic enzymes, 212
  LC-MS analysis of, 214–215
  microsomal preparations, 211
  purified proteins preparations, 210–211
  yeast crude cell lysates preparation, 211
In vivo cell-feeding assays
  enzyme activity, 212–214
  LC-MS analysis of, 214–215
Iodine treatment, 234
Isoprenoids, 226
  intensive research, 226–228
  MEP pathway flux, 226–228, 227f
  in microbes, 226, 227f
  purification of, 238

Isoprenoid synthases (IsoSs), construction expression vectors
  cell transformation, 234
  DNA assembly, 234
  DNA electrophoresis and recovery, 233–234
  encoding genes, 230
  iodine treatment and ethanol precipitation, 234
  PCR reaction, 231–233
  primer design, 231

J
Joint Genome Institute (JGI), 130

L
L-3,4-Dihydroxyphenylalanine (L-DOPA), 210
Limonene, and derivatives, 249
Liquid chromatography, BIA assay intermediates, 215
Liquid chromatography-tandem mass spectrometry (LC-MS) analysis
  BIA biosynthesis
    analytical methods, 169–171
    analytical strategy overview, 167–168
    instrumentation and software, 168
    solvents and reagents, 168–169
    technical notes, 172
  in vitro and in vivo assay products, 214–215
Liquid culture, of mutant strains
  glucose minimal media, 139
  materials, 138
Lissoclinum patella, 2–3
Lyophilization, 87

M
Mass spectrometry, BIA assay intermediates, 215
Meadow rue *(T. flavum)*, 149
Media optimization, 185–186
Mentha monoterpenoids
  alternative protocols, 260–261
  biosynthetic pathways of essential oil, 250f
  biotransformations, 258–261, 263f
  design–build–test cycle, 265f
  expression constructs, 261–262
  high-throughput optimization studies, 263–266
  multigene protein expression and validation, 257–258
  operon construction, *E. coli*, 253–254
    general assembly protocols, 254–255
    in-fusion cloning, 255–257, 256f
    vector and gene selections, 251–253
  reaction and detection, 259–260
  SDS-PAGE analysis, 257–258, 258f
  semisynthetic industrial-scale production, 249–251
  strain and expression optimization, 262–263
  western blot analysis, 257–258, 258f
Menthol isomers, 249
Metabolic engineering, 197–199, 202–203
Metabolic pathway optimization
  fermentation optimization
    induction optimization, 186–189
    media optimization, 185–186
    temperature optimization, 186
  genetic optimization
    coculture optimization, 184–185
    copy number balancing, 182
    dynamic balancing, 184
    gene homolog sourcing, 181
    plasmid backbone selection, 181–182
    posttranslational balancing, 183–184
    transcriptional balancing, 182–183
    translational balancing, 183
Methylerythritol pyrophosphate (MEP) pathway, 226–228, 227f
Microbes, isoprenoids in, 226, 227f
Monochlorodimedone, 94
MreB morphology assays, in *B. subtilis*, 342–343

N
Natural product biosynthesis, 248–249
Nonribosomal peptide synthetase (NRPS), 44–46
(S)-Norcoclaurine, 196–197
Noscapine synthase (NOS), 149

O
Opiates, 196–199
Orthologous genes, bioprospecting, 218

P

pBOT vector, 204–205, 205f
pCRISPomyces
    conjugation, genotype screening and clearance, 279–281
    design, of genome editing, 273–275, 274f
    evaluation, 281–282
    for multiplex gene deletion, 278–279
    for single gene disruption, 275–278
Pentacyclic scaffold, 40–42
Phenotypic diversity, biosynthetic pathway genes, 296–297
Phusion high-fidelity DNA polymerase, 304–305
PhytoMetaSyn Project, 199–200
Plants
    BIA biosynthesis in, 145–148
        functional genomics, 150–152
        localization, 149
        methylene bridge formation, 146–148
        (R)-reticuline, 146–148
        (S)-reticuline, 146–148, 147f
        (S)-scoulerine, 146–148
    secondary metabolism, 197–199, 219–220
Plasmid backbone selection, 181–182
Polyketide synthase (PKS), 44–46
Polymerase chain reaction (PCR)
    conditions for DNA parts, 304–305
    design primers for fusion, 130
    diagnostic, 138
    fusion construction
        genomic PCR, 134–135
        materials, 134
        one pot fusion reaction, 135
        two pot fusion reaction, 135–136
    genomic DNA
        hyphae, extraction from, 132–133
        materials, 132–133
        spores, extraction from, 133–134
    liquid culturing, 138–139
    recipes, 139–140
Polymerase chain reaction (PCR)-based methods, 287
Posttranslational balancing, 183–184
Primers
    for combinatorial construct libraries, 302–303, 303t
    for DNA parts, 301–302, 301t
    for RBS libraries, 302

Prochloron pat, 2–3
Proof-of-concept pathways, 180
Pseudomonas chlororaphis, 78–80
Pseudomonas minimal medium (PMM), 80
Pyrroindomycins (PYRs), 40
    biosynthetic pathway of, 41f
    candidates coding, 46–48, 47f
    [4+2] cycloaddition activity in, 44–46, 45f
    isolation from *Streptomyces rugosporus*, 40–42
PyrE3, 40–42
PyrI4, 40–42
Pyrrole-2-carboxylic acid halogenation, 87–88
Pyrrolyl-CoA thioester, 87
Pyrrolyl-S-PCPs, flavin-dependent tryptophan halogenases, 83–84
    to carrier proteins, 87
    chemoenzymatic synthesis, 86–87
    enzymatic synthesis of, 84–86, 85f
    pyrrole-2-carboxylic acid halogenation, 87–88

R

Rdc2, 97–98
RebH, 97–98
Recognition sequences (RSs), 3–5
Recombination directionality factor (RDF), 288–290, 289f
(R)-Reticuline, 196–197
    BIA metabolism, 196–197
    conversion of, 151
(S)-Reticuline, 196–197
    BIA metabolism, 153, 196–197
    conversion of, 151
    de novo production of, 152
    epimerization of, 199–200
    methylene bridge formation, 146–148
Ribosomally synthesized and post-translationally modified peptides (RiPP), 2–3
Ribosome binding sites (RBSs), 320
    for biosynthetic pathway genes, 295–296
    libraries, primers for, 302
Riboswitches
    artificial riboswitches, 322, 324
    β-galactosidase for validation, 337–340, 338f

Riboswitches (*Continued*)
  B. subtilis, gene expression regulation, 333–334
    chromosomal integration verification, 335
    E. coli transformation, 335
    gDNA preparation, 334
    genomic DNA PCR, 334
    ligation, 335
    MreB morphology assays, 342–343
    mutagenesis, 335
    native gene expression constructs, 336–337
    PCR amplification, 334
    restriction digests, 334
    riboswitch mutagenesis, 334
    sequencing and glycerol stocks, 335
    transformation, 335
  chimeric riboswitches, 343–346, 344*f*
  cloning protocols, 324
  E. coli, CheZ motility assays in, 332–333
  E. coli, gene expression regulation
    gDNA preparation, 326
    ligation, 328
    PCR amplification, 328
    primer design, 326, 327*f*, 329*t*
    restriction digests, 328
    sequencing and glycerol stocks, 328
    subcloning, 331
    transform cells, 328
  natural riboswitches, 321
  orthogonal gene expression tools, 322, 323*f*, 325*f*
  PCR synthesis, 345–346
  SELEX, 321–322
  transcriptional control, 320
  tunability and orthogonality, 321
  western blot analysis, 340–342

S

Saccharomyces cerevisiae, 197–199
  characterization
    feeding isoprenoid into yeast culture, 243
    isoprenoid-DMSO stock solution preparation, 243
    seed culture preparation, 243
  CYP expression
    construction yeast expression vector for, 239–240
    yeast transformation, 240–241
    YNB CSM-URA, 242
    with isoprenoid-overproduction *E. coli* bioreactor run, 244
    yeast preculture preparation, 243
Salutaridinol, 146–148
Secondary metabolite genes
  discovery, 128–129
  identification, 130
Secondary SIRA insertion site, 292*f*, 293–294, 303–304, 304*t*
SELEX, riboswitches, 321–322
Serine integrase recombinational assembly (SIRA), 287–288
  assembly of multiple genes, 294–295
  attL x *attR* recombination, 293–294
  attP x *attB* recombination, 291–293, 292*f*
  ΦC31 integrase gene, 305–306
  combinatorial assembly of DNA parts, 310–311
  direction for, 312–313
  DNA components, 300–305
  DNA parts in predefined order, 309–310
  E. coli
    chemically competent, 307–308
    electrocompetent, 308
  ethanol precipitation, 305
  inserting single DNA into plasmid, 294
  inserting single gene using, 308–309
  mechanism of, 288–291
  phenotypic diversity, 296–297
  purification of gp3, 307
  RBSs and, 295–296
  reaction buffer, 307
  site-specific recombination by, 289*f*
  substrate plasmids, 305
  targeted postassembly modification, 297, 311–312
Site-directed mutagenesis, 78–79, 82–83
Site-specific recombination-based method, 287–288
Site-specific recombination-based tandem assembly (SSTRA), 287–288
Small-scale experiments, 185
S-phenyl thioates, pyrroyl-S-CoA synthesis, 86–87

Spiroconjugate synthase, 40–42
Spiroketal cyclases, 59
Spirotetramate
　biosynthetic gene clusters, 46, 47f
　biosynthetic pathway of, 42
　in nature, 40–42
Spirotetronate
　biosynthetic gene clusters, 46, 47f
　natural products, 40–42
　penta-cyclic cores, 42
Static genetic optimization, 180–181
Streptomyces
　CRISPR/Cas system, 273
　DSB-mediated genome editing, 273
　gene disruption, 272–273
　growth and product extraction, 32–33
　heterologous expression of, 27–28
　natural and synthetic promoters, 32
　pCRISPomyces
　　evaluation, 281–282
　　genome editing, 273–275, 274f
　　multiplex gene deletion, 278–279
　　single gene disruption, 275–278
　S. coeruleorubidus, 80–81
　single crossover method, 272–273
　S. lividans, 294
　S. rugosporus, 40–42
　strains, 271–272
Substrate delay, 188–189
Substrate specificity, flavin-dependent
　　tryptophan halogenases
　7-halogenase gene *prnA*, 73–74, 77f
　6-halogenase gene *thal/thdHi*, 73–74, 77f
　halogenated arylamine, 78
　halogenation reaction, 78
　high-throughput assay, 77–78
　large scale conversion conditions, 75t
　RebH halogenation, 79f
Synthetic biology, 197–199, 202–203
Systems biology, 197–199

T

Temperature optimization, 186
Thebaine, 197–199
Thebaine 6-O-demethylase (T6ODM),
　　146–148
Thermostability, FDH, 99–105, 99s
1000 Plants Project, 199–200

Transcriptional balancing, 182–183
Transformation
　materials, 136
　protoplasting, 136–138
Translational balancing, 183
Trial-and-error approach, 286–287
Trichlorination, 84
L-Tryptophan FDHs, 96–97
Tryptophan 7-halogenase (PrnA), 66,
　　73–74, 95–96

U

UPLC method, 105–106, 120
Uracil-specific excision reaction (USER)
　　cloning, BIA biosynthesis
　EasyClone system, 161–163
　PHUSER software, 162–163
　recombinant protein analysis, 165
　Ura marker excision, 165
　vectors, 163
　yeast transformation, 164–165

V

Virus-induced gene silencing (VIGS)
　in California poppy, 150–151
　in opium poppy, 150–151

W

Western blot analysis, riboswitches, 340–342

Y

Yeast, BIA biosynthesis
　DNA synthesis, 200–201
　engineering, 152–155
　　codon usage, 153–154
　　industrial applications, 154–155
　　(S)-reticuline, 153
　enzyme activity
　　LC-MS analysis, 214–215
　　in vitro enzyme assays, 210–212
　　in vivo cell-feeding assays, 212–214
　enzyme expression and localization,
　　203–209
　　flow cytometry, 206–207
　　fluorescence microscopy, 207–209
　　immunoblotting, 209
　genetic considerations, 202–203
　technical protocols, 201f

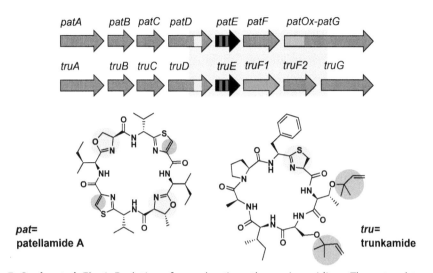

D. Sardar et al., Fig. 1 Evolution of cyanobactin pathways in ascidians. The *pat* and *tru* cyanobactin gene clusters are shown, where the precursor peptide gene *patE/truE* is in *black*. The *red bars* within the precursor peptide gene represent the variable core sequences that encode the final natural products, whereas the remaining sequence (*black*) share >80% identity. The genes flanking the precursor code for posttranslational enzymes and other functions. Outside the precursor gene, regions in *gray* are similar in sequence, whereas the colored segments represent variation in sequence. The most variability apart from the core sequence corresponds to heterocyclization (*yellow*), prenylation (*green*), and oxidation (*blue*) posttranslational chemistry (*purple box*). This variation translates clearly to structural variability in the *pat* and *tru* natural products as shaded in the same corresponding color as the genes. The *pat* pathway products carry both thiazolines and oxazoline/methyloxazolines (*yellow circles*), whereas *tru* products carry only thiazoline (*yellow circle*). In addition, *tru* products are prenylated (*green circles*) corresponding to presence of the TruF1 prenyltransferase that is absent in *pat*. Similarly, *tru* products lack oxidation of thiazolines, since the corresponding oxidase domain (*blue*) is absent in them, in contrast to *pat* products.

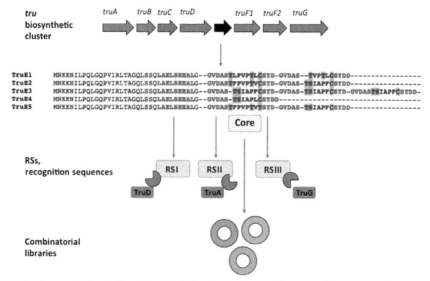

D. Sardar et al., Fig. 2 Combinatorial libraries using *tru*. The *tru* pathway gene cluster is shown, and the precursor peptide variant TruE (*black*) is magnified to its translated amino acid sequence. Observation of this sequence clearly demarcates the hypervariable regions (*blue*) that represent the core encoding the final natural products. In contrast, the rest of the precursor is highly conserved and correspond to recognition sequences (RSs, *red*) that direct specific posttranslational enzymes. This phenomenon of substrate evolution, wherein the substrates evolve to maintain a balance between variations in the core (creates diversity) and conservation of the flanking RSs (maintains modification chemistry) allows the creation of natural combinatorial libraries.

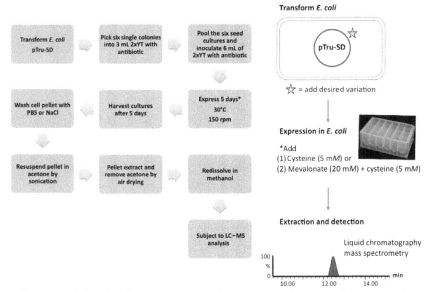

D. Sardar et al., Fig. 3 Schematic for heterologous expression in E. coli. There are three basic steps: (1) transformation of E. coli with the pTru-SD vector that carries the tru pathway. At this step, the sequence of the precursor can be manipulated to add desired motifs to the final product; (2) expression in E. coli for an optimum of 5 days. At this step, addition of cysteine or cysteine with mevalonate results in higher yields of compounds; and (3) extraction of compounds from the E. coli cell pellet and mass spectrometry-based detection.

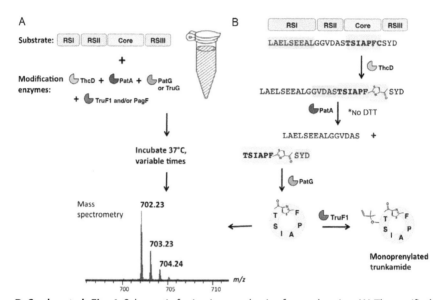

D. Sardar et al., Fig. 4 Schematic for in vitro synthesis of cyanobactins. (A) The purified enzymes and substrate are mixed in a reaction tube under optimum conditions, and the products detected by mass spectrometry. It is essential to maintain the necessary recognition sequences (RSs) in the substrate for posttranslational chemistry. (B) A detailed reaction scheme is shown with each modification step. The representative substrate carries the trunkamide core sequence, flanked by the required RSs. The heterocyclase ThcD (directed by RSI) modifies the cysteine residue in the core to thiazoline. This is followed by N-terminal proteolysis by the protease PatA (directed by RSII). It is helpful to keep the reaction medium free of reducing agents for PatA action. The subsequent protease/macrocyclase PatG (directed by RSIII) cleaves off the RSIII and joins the ends to generate the cyclic product. Further modification of prenylation is appended on the backbone by the enzyme TruF1.

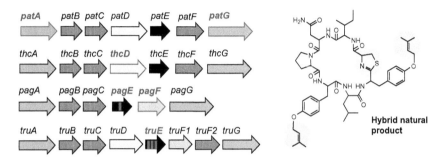

D. Sardar et al., Fig. 5 Modularity of cyanobactin pathways. Enzymes and substrates from different pathways can be mixed and matched to create hybrid natural products for combinatorial chemistry. For example, the unnatural cyanobactin shown on the *right* is derived from elements belonging to four different pathways *pat*, *thc*, *pag*, and *tru*. The specific elements involved from each gene cluster for the creation of this hybrid are highlighted in *red*. The cyclic peptide sequence is derived from *pagE* precursor sequence INPYLYP (*red bar* in *pagE*), wherein the proline is mutated to cysteine to allow introduction of a heterocycle. The recognition sequences in the precursor substrate were derived from *truE* (*red bars* in *truE*) to direct the enzymes heterocyclase (*thcD*, from the *thc* pathway), the N-terminal protease (*patA*, from the *pat* pathway), and the C-terminal protease/macrocyclase (*patG*, from the *pat* pathway). Additional modification of prenylation was introduced by *pagF* (from the *pag* pathway). This resulted in a derivative of the natural product prenylagaramide B, carrying chemistry not found in nature such as thiazoline ring (*yellow*) and double prenylation (*blue*).

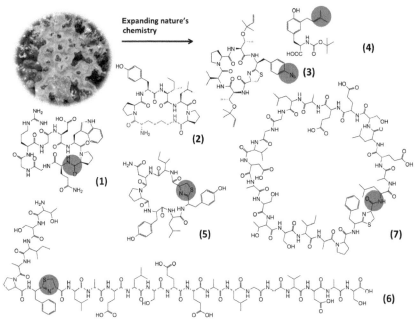

D. Sardar et al., Fig. 6 Representative examples of modified peptide motifs not found in nature created using the cyanobactin RiPP machinery both in vivo and in vitro. (1) Eptidemnamide that resembles the rattlesnake derived anticoagulant eptifibatide, (2) macrocyclic peptides with polyketide insertions, (3) macrocyclic peptides with non-proteinogenic amino acid insertions, (4) small molecules carrying isoprene units, (5) thiazoline containing cyclic peptides, (6) linear peptide with the thiazoline heterocycle at desirable positions, and (7) an unusually large macrocycle of 22 ring size.

L.M. Alkhalaf et al., Fig. 2 Analogs of cladoniamide A obtained through combinatorial biosynthesis. *Colors* link biosynthetic enzymes with the corresponding fragment of the cladoniamide structure. Major metabolites are shown, although metabolites with distinct chlorination patterns are frequently coisolated (Du, Ding, & Ryan, 2013; Du & Ryan, 2015; Du et al., 2014).

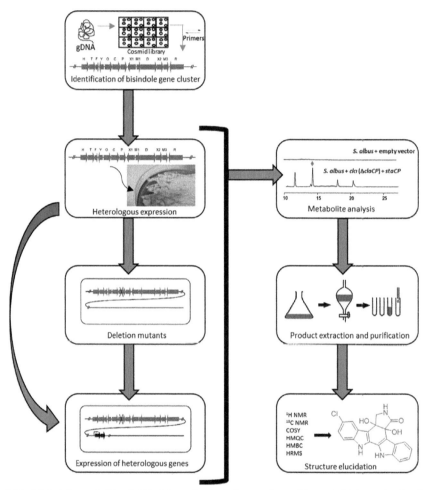

L.M. Alkhalaf et al., Fig. 3 Workflow for isolation and combinatorial biosynthesis of tryptophan-derived microbial bisindoles, from identification of a gene cluster to characterization of novel metabolites.

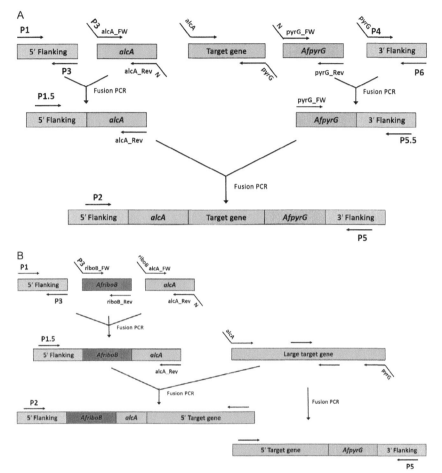

J.W.A. van Dijk and C.C.C. Wang, Fig. 1 Fusion PCR overview for genes smaller (A) and larger than 4000 base pairs (B). (A) Individual fragments are amplified from genomic DNA. Flanking fragments and the *alcA* promoter from the *A. nidulans* host, *AfpyrG* from *A. fumigatus* genomic DNA, and the target gene from its respective source. The "N" sticky ends of the alcA_Rev and pyrG_FW primers on the first line are an arbitrary number of nucleotides added to make the fusion PCRs on the middle line have nested primers on both sides. (B) The target gene is amplified in two separate pieces with 1000 base pairs overlap. In addition to the promoter, an additional marker is added to the 5′ fragment. The two fragments are mixed and added to the protoplasts for transformation.

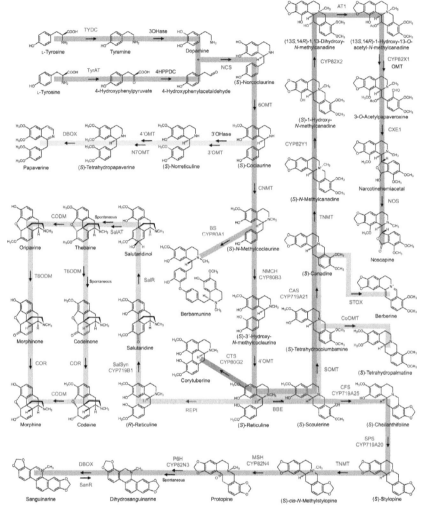

J.S. Morris et al., Fig. 1 BIA biosynthetic pathways for (S)-reticuline (*light pink*), papaverine (*yellow*), morphine (*green*), sanguinarine (*orange*), berberine (*blue*), noscapine (*purple*), berbamunine (*olive*), and corytuberine (*dark pink*). *Red* within each alkaloid highlights enzyme catalyzed. Structural changes in each compound are indicated in *red*. Cognate genes have been isolated for enzymes abbreviated in *blue*. *Abbreviations*: 3′-OHase, 3′-hydroxylase; 3′OMT, 3′-O-methyltransferase; 3OHase, 3-hydroxylase; 4HPPDC, 4-hydroxyphenylpyruvate decarboxylase; 4′OMT, 3′-hydroxy-N-methylcoclaurine 4′-O-methyltransferase; 6OMT, norcoclaurine 6-O-methyltransferase; AT1, 1,13-dihydroxy-N-methylcanadine 13-O-acetyltransferase; BBE, berberine bridge enzyme; BS, berbamunine synthase; CAS, canadine synthase; CFS, cheilanthifoline synthase;

J.S. Morris et al., Fig. 1—Cont'd *CNMT*, coclaurine *N*-methyltransferase; *CODM*, codeine *O*-demethylase; *CoOMT*, columbamine *O*-methyltransferase; *COR*, codeinone reductase; *CTS*, corytuberine synthase; *CYP82X1*, 1-hydroxy-13-*O*-acetyl-*N*-methylcanadine 8-hydroxylase; *CYP82X2*, 1-hydroxy-*N*-methylcanadine 13-hydroxylase; *CYP82Y1*, *N*-methylcanadine 1-hydroxylase; *CDBOX*, dihydrobenzophenanthridine oxidase; *CXE1*, 3-*O*-acetylpapaveroxine carboxylesterase; *MSH*, *N*-methylstylopine hydroxylase; *N7OMT*, norreticuline 7-*O*-methyltransferase; *NCS*, norcoclaurine synthase; *NMCanH*, *N*-methylcanadine 1-hydroxylase; *NMCH*, *N*-methylcoclaurine 3′-hydroxylase; *NOS*, noscapine synthase; *P6H*, protopine 6-hydroxylase; *REPI*, reticuline epimerase; *SalAT*, salutaridinol 7-*O*-acetyltransferase; *SalR*, salutaridine reductase; *SalSyn*, salutaridine synthase; *SanR*, sanguinarine reductase; *SOMT*, scoulerine 9-*O*-methyltransferase; *SPS*, stylopine synthase; *STOX*, (*S*)-tetrahydroprotoberberine oxidase; *T6ODM*, thebaine 6-*O*-demethylase; *TNMT*, tetrahydroprotoberberine *N*-methyltransferase; *TYDC*, tyrosine decarboxylase; *TyrAT*, tyrosine aminotransferase.

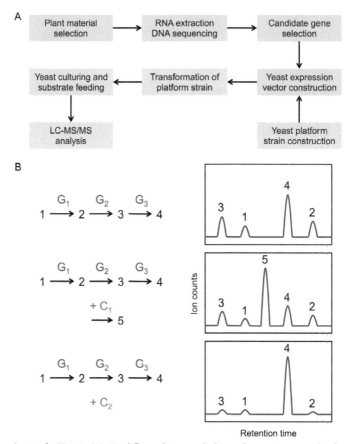

J.S. Morris et al., Fig. 2 (A) Workflow diagram linking the various methods used for acquisition and plug-and-play functional analysis of biosynthetic gene candidates in yeast platform strains with an operational BIA metabolic pathway, or parts thereof. (B) Example of a yeast platform strain able to convert a hypothetical BIA (compound 1), via two pathway intermediates (compounds 2 and 3), to an end product (compound 4). Transient expression in the yeast platform strain with a candidate gene (C_1) encoding an uncharacterized enzyme results in the accumulation of a new BIA (compound 5). Alternatively, a candidate gene (C_2) encoding an enzyme functionally similar to one already part of the platform strain increases accumulation of the end product.

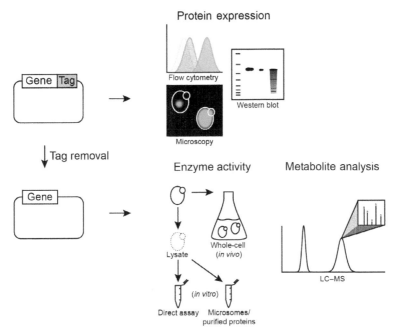

M.E. Pyne et al., Fig. 2 Overview of technical protocols outlined in this chapter. Target genes cloned as translational fusion proteins with fluorescent tags that can be used to monitor protein expression with techniques which include flow cytometry, fluorescence microscopy, and immunoblotting (Western blotting). Fluorescent tags are removed from target gene constructs for analysis of enzyme activity. Strains expressing target biosynthetic genes can be supplemented with BIA intermediates for in vivo analysis of pathway functionality. Alternatively, following cell lysis, crude cell lysates, partially purified yeast microsomes, or highly purified enzyme preparations can be analyzed for enzyme functionality in vitro. LC–MS is the preferred technique for monitoring reaction products of both in vivo and in vitro assays.

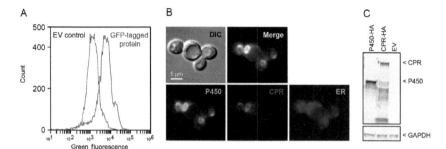

M.E. Pyne et al., Fig. 3 Representative data validating heterologous protein expression and localization in *S. cerevisiae*. (A) Detection of a heterologous C-terminal GFP-tagged cytochrome P450 reductase (CPR) by flow cytometry. Yeast cells were grown in synthetic complete medium at 30°C and 200 rpm for 4 h and analyzed for GFP fluorescence using a BD Biosciences Accuri C6 flow cytometer. Expression of the GFP-tagged CPR from a pBOT vector was observed through a shift toward higher green fluorescence (excitation 488 nm; emission 510/15 nm), compared to the empty vector (EV) control. (B) Colocalization of a heterologous plant cytochrome P450 and its cognate CPR partner to the endoplasmic reticulum (ER). The C-terminal GFP-tagged cytochrome P450 and the C-terminal mCherry-tagged CPR were coexpressed using two compatible pBOT plasmids. The ER-Tracker Blue-White DPX dye was used to visualize the ER. Superposition of fluorescence images was used to visualize cytochrome P450 and CPR colocalization. Images were visualized at 100× magnification and 700 ms exposure using a Nikon Eclipse Ti epifluorescence inverted microscope and captured using a Nikon high-resolution (4908 × 3264 pixels) color camera DsRi2 CMOS. *Red* mCherry signals were excited with LED 555 nm and filtered with the TxRed filter cube; *green* GFP signals were excited with LED 480 nm and filtered with the CFP/YFP filter cube; *blue* ER-Tracker Blue-White DPX dye signal was excited with LED 405 nm and filtered with the DAPI filter cube. Differential interference contrast (DIC) is shown in the *top left panel*. (C) Immunoblot of an HA-tagged plant cytochrome P450 (P450-HA) and its cognate HA-tagged CPR (CPR-HA) expressed independently in *S. cerevisiae* using compatible pBOT plasmids. Samples were collected from cultures grown in synthetic complete medium at 30°C and 200 rpm for 24 h. Approximately 2×10^7 cells were pelleted and suspended in 50 μL of lysis buffer (50 m*M* Tris–HCl, pH 7.4, containing 150 m*M* NaCl, 2 m*M* MgCl$_2$, and 0.1% Nonidet P-40) plus 50 μL of 2× Laemmli buffer. Samples were boiled for 5 min, resolved by SDS-PAGE (30 μL/lane), and transferred to a nitrocellulose membrane for detection of the HA-epitope using the mouse anti-HA antibody HA-C5 (Abcam). The EV control was added to exclude nonspecific binding of the anti-HA antibody. Glyceraldehyde 3-phosphate dehydrogenase (GAPDH) was used as loading control and probed using rabbit anti-GAPDH (Rockland Immunochemicals). Imaging was performed using an Odyssey imager and IRDye® secondary antibodies (LI-COR Biosciences).

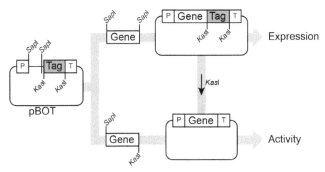

M.E. Pyne et al., Fig. 4 Custom pBOT series of S. cerevisiae expression vectors. Target genes are cloned into the pBOT series of vectors for analysis of expression and enzyme functionality. Genes can be cloned as fluorescent fusion proteins using SapI. Following validation of protein expression, fluorescent tags are removed by KasI digestion and plasmid recircularization. Alternatively, the gene of interest can be cloned directly without a fluorescent tag using SapI and KasI. Abbreviations: P, promoter; T, terminator.

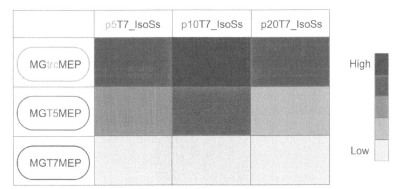

K. Zhou et al., Fig. 3 Illustration of locating optimal expression levels of MEP genes and isoprenoid synthases by modular genetic engineering of E. coli. Color indicates level of isoprenoid production. MGtrcMEP, E. coli MG1655 ($\Delta recA$, $\Delta endA$, DE3, $\Delta araA::P_{trc}$-dxs-idi-$ispDF$); MGT5MEP, E. coli MG1655 ($\Delta recA$, $\Delta endA$, DE3, $\Delta araA::P_{T5}$-dxs-idi-$ispDF$); MGT7MEP, E. coli MG1655 ($\Delta recA$, $\Delta endA$, DE3, $\Delta araA::P_{T7}$-dxs-idi-$ispDF$).

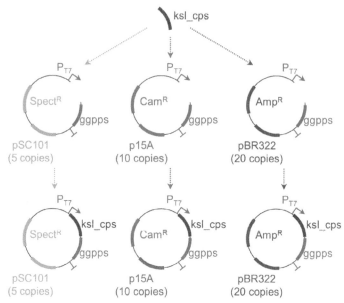

K. Zhou et al., Fig. 4 Construction of expression vectors for C20 isoprenoid synthases by using CLIVA method. *Ksl_cps*, the gene encoding miltiradiene synthase; *Ggpps*, the gene encoding geranylgeranyl pyrophosphate synthase; P_{T7}, T7 promoter; $Spect^R$, spectinomycin resistance gene; Cam^R, chloramphenicol resistance gene; Amp^R, ampicillin resistance gene; *pSC101*, pSC101 replication origin, whose copy number is *E. coli* is estimated to be 5; *p15A*, p15A replication origin, whose copy number in *E. coli* is estimated to be 10; *pBR322*, pBR322 replication origin, whose copy number in *E. coli* is estimated to be 20.

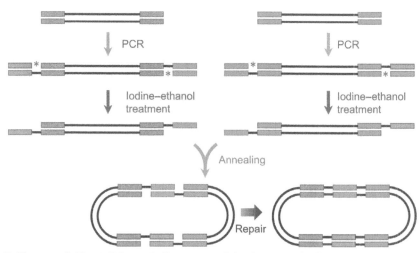

K. Zhou et al., Fig. 5 Schematic illustration of the CLIVA method.

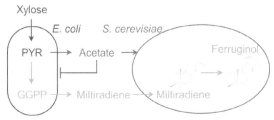

K. Zhou et al., Fig. 6 Oxygenating C20 isoprenoid scaffold molecule by using mutualistic microbial consortium. Miltiradiene is used here as an example – it is produced by *E. coli* and oxidized into ferruginol in *S. cerevisiae*. *PYR*, pyruvate; *GGPP*, geranylgeranyl pyrophosphate; *E. coli* produces acetate as a by-product that inhibits its own growth, which is used here to build a mutualistic relationship between the two species – acetate serves as carbon source of the yeast, which in return removes the inhibitor (acetate) for *E. coli*.

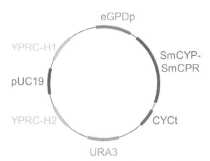

K. Zhou et al., Fig. 7 Plasmid map of pYPRC15-URA-UGPD-SmCYP_SmCPR. *pUC19*, pUC19 plasmid backbone; *YPRC-H1*, 500 bp homologous region upstream YPRC15 locus of *S. cerevisiae* genome; *YPRC-H2*, 500 bp homologous region downstream YPRC15 locus of *S. cerevisiae*; *eGPDp*, enhanced GPD promoter; *CYCt*, CYC terminator; *URA3*, URA3 marker; *SmCYP-SmCPR*, chimeric CYP-CPR gene, which is used as an example in this chapter and involved in oxygenation of miltiradiene.

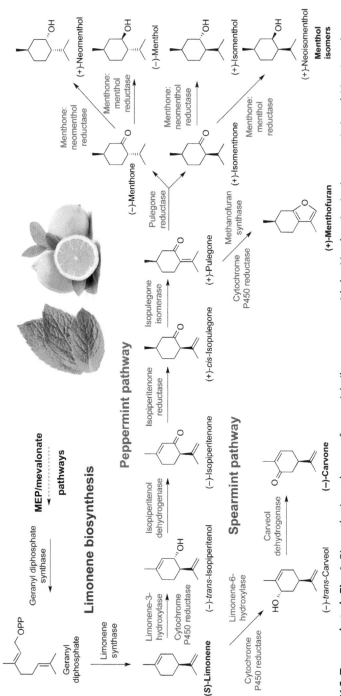

H.S. Toogood et al., Fig. 1 Biosynthetic pathways of essential oil monoterpenoids by *Mentha piperita* (peppermint) and *M. spicata* (spearmint). The enzymes involved in each pathway are color coded for limonene (*blue*), peppermint (*red*), and spearmint (*magenta*) biosynthetic pathways (Croteau, Davis, Ringer, & Wildung, 2005; Toogood et al., 2015).

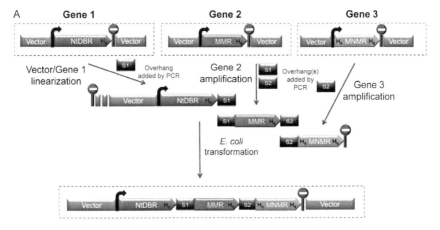
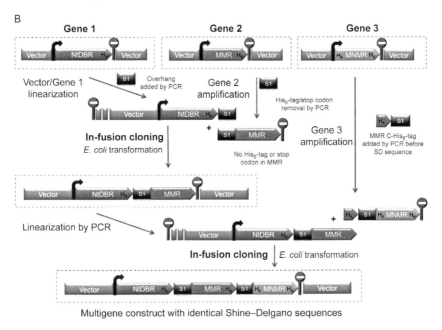

H.S. Toogood et al., Fig. 2 PCR-based In-fusion cloning to create multigene expression constructs. (A) Single In-fusion cloning step protocol for constructs without significant repeating sequences. (B) Multiple In-fusion cloning steps protocol where repeating sequences (His$_6$-tag—Shine–Dalgarno sequence) occur at the cloning termini. (C) Construction of large sequence overhangs at the 5′ end of gene MNMR by PCR. *H6*, hexa-histidine tag; *S1/S2*, Shine–Dalgarno sequences.

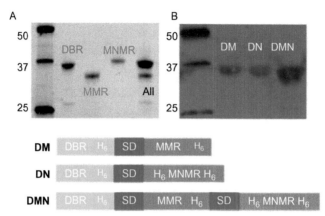

H.S. Toogood et al., Fig. 3 (A) SDS-PAGE analysis of purified recombinant enzymes. (B) Western blot analysis of cell extracts of three operons of menthol biosynthesis.

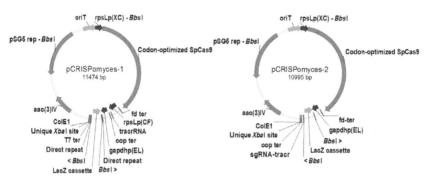

Y. Wang et al., Fig. 1 pCRISPomyces plasmids for targeted genome editing in *Streptomyces* species. *Reproduced with permission from Cobb, R. E., Wang, Y., & Zhao, H. (2014). High-efficiency multiplex genome editing of Streptomyces species using an engineered CRISPR/Cas system. ACS Synthetic Biology, 4, 723–728.*

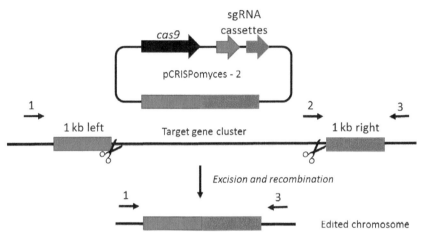

Y. Wang et al., Fig. 2 Strategy for deletion of a large gene cluster and screening for the desired genotype by PCR. Two sgRNA transcripts guide Cas9 to introduce DSBs at both ends of the cluster, and a codelivered editing template bridges the gap via homologous recombination.

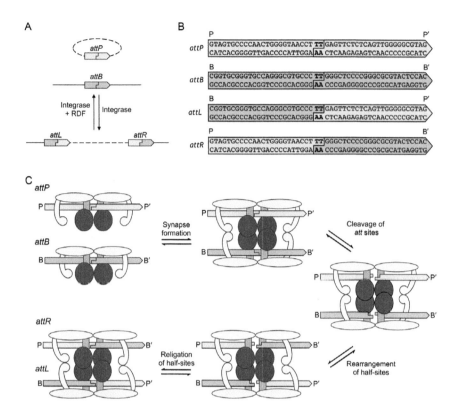

C.A. Merrick et al., Fig. 1 Site-specific recombination by serine integrases. (A) Phage integration and excision. A serine integrase recombines an *attP* site (phage) and an *attB* site (bacteria) to integrate a phage genome into a bacterial host genome generating an *attL* (*left*) and an *attR* (*right*) which flank the integrated prophage. This reaction is directional because the *attL* and *attR* sites will not recombine in the presence of integrase alone. In the presence of an accessory protein called a recombination directionality factor (RDF), the serine integrase can recombine *attL* and *attR* sites to excise the phage DNA and restore the original *attP* and *attB* sites. (B) Sequences of ΦC31 integrase *att* sites. An *attP* site consists of two half-sites, P and P', on either side of a central dinucleotide core sequence, and an *attB* site consists of B and B' half-sites, on either side of a central dinucleotide core sequence. If the central dinucleotides in *attP* and *attB* sites match, they recombine to make *attL* which consists of B and P' half-sites on either side of a central dinucleotide, and *attR* which consists of P and B' half-sites on either side of the central dinucleotide. Central dinucleotides are in *bold*, and the cleavage sites are indicated by staggered *lines*. (C) Mechanism of serine integrase recombination. The serine integrase binds *attP* and *attB* sites as a dimer, and brings the sites together forming a tetrameric protein-DNA synapse. Next, nucleophilic serine residues in the integrase N-terminal domains catalyze cleavage of all DNA strands forming covalent phosphoserine linkages with the DNA backbones and generating 3' overhangs. Next, integrase subunits rotate in relation to each other rearranging *attP* and *attB* half-sites. If the 3' overhangs of the rearranged half-sites match, they are religated to form *attL* and *attR* sites. A current theory proposes that the C-terminal domains of integrase dimers bound the *attL* and *attR* sites form intramolecular interactions that prevent formation of an *attL* × *attR* synapse. The serine integrase N-terminal domain is in *dark gray*, the connecting α-helix is in *medium gray*, and the C-terminal domain is in *pale gray*. In the color version of this figure P and P' half-sites are *green*, and B and B' half-sites are *blue*.

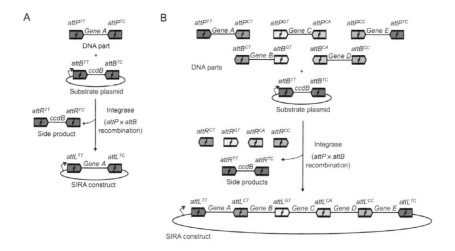
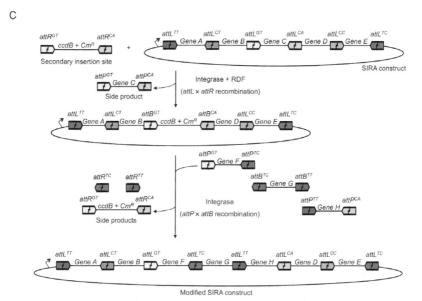

C.A. Merrick et al., Fig. 2 The serine integrase recombinational assembly (SIRA) system. (A) Inserting a single piece of DNA into a substrate plasmid by $attP \times attB$ recombination. A DNA part containing *gene A* is flanked with $attP^{TT}$ and $attP^{TC}$ in inverted orientations. The substrate plasmid contains a "SIRA insertion site" in which $attB^{TT}$ and $attB^{TC}$ are in inverted orientations on either side of *ccdB*, a gene encoding a cytotoxic protein that causes cell death. Recombination of the $attP^{TT}$ with $attB^{TT}$, and $attP^{TC}$ and $attB^{TC}$, by a serine integrase inserts the DNA part into the substrate plasmid. The DNA part replaces the *ccdB* gene enabling selection of assembled constructs after transformation of *E. coli*. (B) Assembly of five DNA parts in a substrate plasmid by $attP \times attB$ recombination. DNA parts carrying

(Continued)

C.A. Merrick et al., Fig. 2—Cont'd arbitrary genes A to E are flanked with attB sites or attP sites in inverted orientations. In the presence of integrase, attB sites will recombine with attP sites with matching central dinucleotides assembling the DNA parts in the substrate plasmid, replacing the SIRA insertion site. This means that each DNA part will be incorporated at a specific position within an assembly (here giving the gene order A-B-C-D-E), and that these positions will be separated by attL sites with different central dinucleotides in alternating orientations. (C) Targeted postassembly modification of a SIRA construct carrying genes A to E by attL × attR recombination. A "secondary SIRA insertion site" carrying a chloramphenicol resistance gene for positive selection and the ccdB gene for negative selection, and flanked by $attR^{GT}$ and $attR^{CA}$, recombines with $attL^{GT}$ and $attL^{CA}$ sites in the SIRA construct in the presence of integrase and RDF. The result of this reaction is that gene C is replaced with the secondary SIRA insertion site that, once inserted, is flanked by $attB^{GT}$ and $attB^{CA}$. The plasmid carrying the secondary SIRA insertion site can be selected through resistance to chloramphenicol. Additional DNA parts such as those carrying genes F to H can be added to the construct by attP × attB recombination, removing the secondary SIRA insertion site. Loss of the lethal ccdB gene enables selection for the new, modified SIRA construct. Different colors represent the different central dinucleotide sequences of att sites.

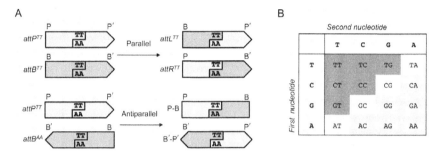

C.A. Merrick et al., Fig. 3 Central dinucleotides in *att* sites determine the polarity of recombination. (A) Alignment of *att* sites in parallel and antiparallel conformations. In this diagram, *attP* and *attB* sites have central dinucleotides that are nonpalindromic. *attP* is shown with a TT central dinucleotide, *attB* is shown with either TT or AA, the reverse complement of TT. Cleaved half-sites carry either a 3′ TT or a 3′ AA overhang. When *attP* and *attB* sites align in a parallel conformation, half-sites with complementary overhangs ligate to make *attL* and *attR*. When sites align in an antiparallel conformation, half-sites with complementary overhangs ligate to make P-B and B′-P′. $attP^{TT}$ can only recombine with $attB^{TT}$ in the parallel conformation, whereas attP^{TT} can only recombine with $attB^{AA}$ in the antiparallel conformation. In contrast, *att* sites with palindromic central dinucleotides, eg, $attP^{TA}$ and $attB^{TA}$, can recombine in both parallel and antiparallel conformations. In the color version of this figure P and P′ half-sites are *green*, and B and B′ half-sites are *blue*. (B) Nonpalindromic dinucleotides that can be used in SIRA *att* sites. The six nonpalindromic dinucleotides that are suitable for SIRA *att* sites are shaded in *dark gray*, and their reverse complement sequences are in *pale gray*. Palindromic dinucleotides that enable parallel and antiparallel recombination and are therefore not suitable for SIRA have a *white background*.

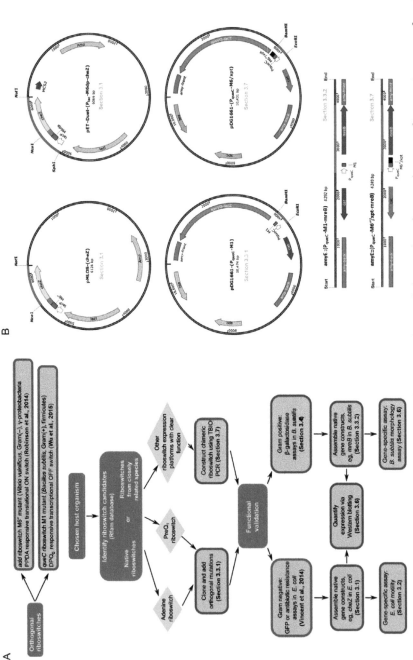

C.J. Robinson et al., Fig. 2 (A) Flowchart covering options for using orthogonal riboswitch tools to study genes of interest in your preferred host organism. Each node contains the section number under which the relevant protocol can be found. To illustrate how these protocols are applied, we describe how these tools were created, functionally validated, then were used to study *cheZ* expression in *E. coli* and *mreB* expression in *B. subtilis*. This approach is described in detail in the text. (B) Plasmid vectors used in this study. Promoters, riboswitches, genes, and homologous recombination sites are labeled. Cloned sequences are shown as *red arcs*, with relevant restriction sites indicated. Underneath each plasmid name is the section number where it is described in more detail.

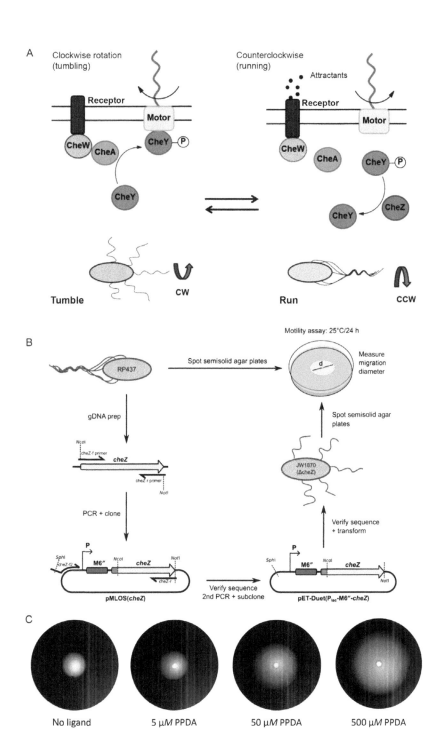

C.J. Robinson et al., Fig. 3 See legend on next page.

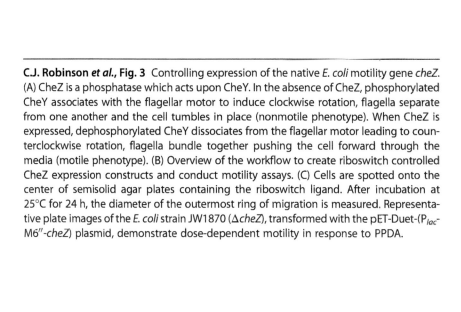

C.J. Robinson et al., Fig. 3 Controlling expression of the native E. coli motility gene cheZ. (A) CheZ is a phosphatase which acts upon CheY. In the absence of CheZ, phosphorylated CheY associates with the flagellar motor to induce clockwise rotation, flagella separate from one another and the cell tumbles in place (nonmotile phenotype). When CheZ is expressed, dephosphorylated CheY dissociates from the flagellar motor leading to counterclockwise rotation, flagella bundle together pushing the cell forward through the media (motile phenotype). (B) Overview of the workflow to create riboswitch controlled CheZ expression constructs and conduct motility assays. (C) Cells are spotted onto the center of semisolid agar plates containing the riboswitch ligand. After incubation at 25°C for 24 h, the diameter of the outermost ring of migration is measured. Representative plate images of the E. coli strain JW1870 ($\Delta cheZ$), transformed with the pET-Duet-(P_{lac}-M6''-cheZ) plasmid, demonstrate dose-dependent motility in response to PPDA.

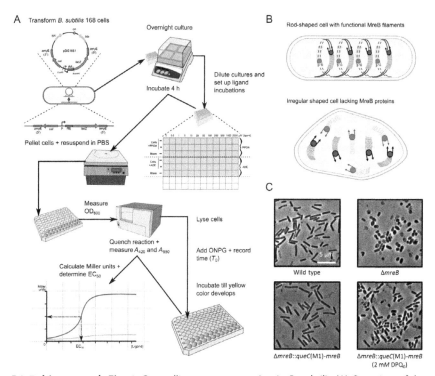

C.J. Robinson et al., Fig. 4 Controlling gene expression in *B. subtilis*. (A) Overview of the β-galactosidase assay to validate novel riboswitch function using chromosomally integrated *lacZ* expression constructs. The orthogonal *queC*(M1) and chimeric *add*(M6″)/*xpt* riboswitches were both validated by this method prior to deployment with native *B. subtilis* genes. (B) Function of the native *B. subtilis* morphology gene *mreB*. Wild-type cells have a rod-shaped morphology maintained in part by dynamic filaments of MreB (*red lines*), which are deposited in a helical orientation around the circumference of the cell, recruiting the peptidoglycan synthetic complexes (*blue circles*) to deposit the cell wall (*green crosshatch*). In the absence of MreB, the cell wall is deposited in a uncoordinated manner leading to swollen amorphous cells. (C) Phase-contrast microscopy shows the typical rod-shaped morphology of wild-type cells (*top left*) and the aberrant amorphous phenotype of Δ*mreB* cells (*top right*). Chromosomal integration of the repressible *queC*(M1)-*mreB* construct restores the rod-shape morphology to Δ*mreB* cells (*bottom left*), but the mutant phenotype reappears when the DPQ_0 ligand is added to repress *mreB* expression.

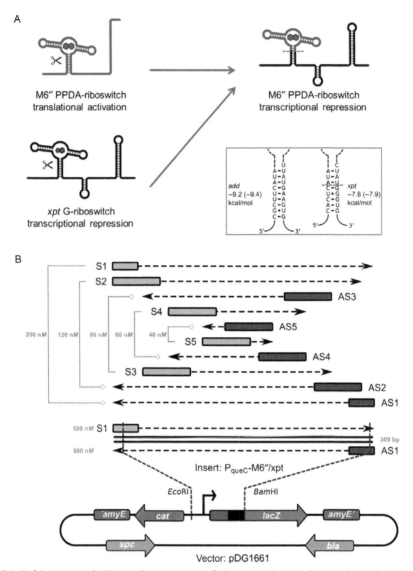

C.J. Robinson et al., Fig. 5 Construction of chimeric riboswitches as bespoke gene expression tools for diverse bacterial species. (A) A repressible chimeric riboswitch was created by fusing the orthogonal *add*(M6″) aptamer with the transcriptional OFF expression platform from the native *xpt* riboswitch of *B. subtilis*. The parent riboswitches have different P1 stem strengths (*inset*), but by preserving the switching sequence (*blue*) and P1 stem strength of the *xpt* riboswitch, a functional chimeric riboswitch was constructed (termed *add*(M6″)/*xpt*). (B) Synthesis of a constitutive promoter-driven chimeric riboswitch construct by the TBIO PCR method (Gao, Yo, Keith, Ragan, & Harris, 2003). The final PCR product was restriction cloned directly into the *B. subtilis* integrative vector pDG1661 for functional validation by the β-galactosidase assay. Following validation, the promoter–riboswitch construct was fused upstream of the native *B. subtilis* morphology gene *mreB* by overlap extension PCR.

rds Brothers Malloy
rbor MI. USA
1, 2016

Edv
Ann
July